Beyond Algorithms
Social Representations and the Public Understanding of AI

Editor
Lilian Negura
Faculty of Social Sciences
University of Ottawa
ON, Canada

CRC Press
Taylor & Francis Group
Boca Raton London New York

CRC Press is an imprint of the
Taylor & Francis Group, an **informa** business
A SCIENCE PUBLISHERS BOOK

Cover credit: Lawrence Alma-Tadema, Pandora, 1881, Watercolor, Private Collection. Image source: Wikimedia Commons (Public Domain).

First edition published 2027
by CRC Press
2385 NW Executive Center Drive, Suite 320, Boca Raton FL 33431

and by CRC Press
4 Park Square, Milton Park, Abingdon, Oxon, OX14 4RN

CRC Press is an imprint of Taylor & Francis Group, LLC

Library of Congress Cataloging-in-Publication Data (applied for)

ISBN: 978-1-032-80411-8 (hbk)
ISBN: 978-1-032-82060-6 (pbk)
ISBN: 978-1-003-50275-3 (ebk)

DOI: 10.1201/9781003502753

Typeset in Times New Roman
by Prime Publishing Services

Foreword
Social Representation Theory, Artificial Intelligence and Common Sense

Martin W. Bauer

I am delighted to add a few words to an important new book volume. It brings to the task and simultaneously refreshes the lenses of Social Representation Theory (SRT) to research a new development that currently captures everybody's imagination.

Artificial intelligence [AI] seems the latest frontier of science and technology, a race for land in a wild territory of new opportunity where civilisation is suspended, new living heroes arise—Elon Musk, Marc Zuckerberg, Sam Altman et al.—investments flow, and new rules of the game are being forged.

The enthusiasm for algorithms might be medieval, but it materialised as AI since the 1950s in several cycles of public attention and capital mobilisation. Modern machine intelligence emerged during WWII and was formalized in devices which might no longer be distinguishable from human performance as certified by the Turing, Winograd or Gricean benchmark tests (see Gaskell et al. this volume). Thus, since 2015 we are in another springtime of AI that reached in 2025 with ChatGPT its solstice, after previous surges in 2000, during the 1980s and the initial take-off in the 1950s. The quest to emulate human common sense is part of this history, and this ambition stills poses a challenge with few accomplishments. This volume with focus on SRT, which is a theory of 'Common Sense', is highly relevant for the monitoring and evaluating this quest of AI to colonize Common Sense.

The engineering endeavours have bifurcated into AI for specific purposes, so-called 'expert systems' on the one hand, and into General-Purpose Artificial Intelligence [GPAI] on the other. No such GPAI system yet exists beyond imagination and promise making. Progress is coming from tools that perform specific tasks in which humans already are competent, but AI matches or exceeds that human expertise by being faster and/or more accurate, as for example in diagnosing medical X-ray images. This efficiency gain also shows in board games like Chess or GO, where computers spectacularly beat the world's best players. However, tools remain domain specific with some leeway for functional drift. A chess computer is hopeless at GO;

Psychological and Behavioural Science, London School of Economics and Political Science.
Email: m.bauer@lse.ac.uk

and a machine that reads pulmonary X-Rays does badly at diagnosing heart issues. Or most recently LLM robots like ChatGPT are exceedingly good at generating all sorts of persuasive texts, at times hallucinating facts, but no good at assembling cars on a production line; this requires different robots.

GPAI would emulate all human capacities, maximise effectiveness to a given goal or being able to adjust goals, learn new ways, being emotional, sensitive to pain, self-reproducing and self-regulating with autonomy as is the human original. The problem is how to specify the conditions of possibility for such capabilities for implementation. The problem might arise of how to put limits on means and goals, i.e. to contain GPAI within moral boundaries. Unless inhibited, a pizza delivery robot might well learn how to optimize delivery by killing some people on the way, it would be goal-effective but 'human incompatible' pizza delivery; not to speak of what disinhibited GPAI weapon systems would be capable of doing.

One direction of future AI points to robot architectures that combine separate AI modules for sensory, cognitive and motility functions; but it seems that this still only makes experts systems for broader domains. Similar in athletics, where the combination of skills and training enhances athletes into triathlon, pentathlon or decathlon competitors, but does not make them all round 'superhumans'. Even multi-function robots remain 'brittle'; they are easily overwhelmed and become error prone in rare or novel situations.

An alternative endeavour points towards 'common sense' [CS]: to construct machines that display robust common sense in the way 'homo sapiens' masters and experiences mundane live situations (Brachman and Levesque 2022). Everyday life is full of complex, unspecified, unforeseen and even unforeseeable surprises where routines break down and 'wisdom', 'practical wit' and 'collaborative virtues' are required for muddling through, garnering social support and averting disaster.

Thus, emulating common sense is a quest present at the very beginning of AI; in 1958 a computer program was claimed to display 'common sense' as it made inferences from memory and input commands. Henceforth there is a literature that argues for 'AI with common sense'. However, what is common sense [CS], we must ask?

We have recently (see Bauer and Schiele 2024) examined exactly that very claim 'AI with Common Sense', where CS is often an adverbial: solving problems common-sensically. If CS is an attribute of AI functions, we must consider CS as a variable X, and clarify the set of values {X ~ CS} of that function. An operational definition will take the form 'this machine is capable of X; X ~ CS'. The set {X ~ CS} has more than one value; there are different concepts of X to consider. This allows us to cultivate a critical attitude: how shall we understand the claim 'AI with CS'? On what evidence could the claim be credible? Shall we trust a tool based on CS? We must examine the values X can take {X ~ CS} to make this a meaningful proposition. Some values might make for uncommon non-sense. And we must avoid the reductive equivalence: AI does X ~ CS <=> CS is nothing but X. We must not be satisfied with 'as good as it gets', at least not with AI and its claims makers.

The English dictionary defines CA sense as '*your natural ability to make good judgments and to behave in a practical and sensible way.*' It gives examples of usage such as 'use your common sense', 'she always had a lot of common sense' and '.... A common-sense approach'. As an adjective, 'commonsensical' refers to being 'inspired by or displaying sound practical sense'. Synonyms are given as *good sense, wit, wisdom, sound judgment, levelheadedness, practicality*. Common antonyms are *stupidity impracticality, ineptness, foolishness, insanity, irrationality or unreasonableness*.

Thus, the word is used both disparagingly and also pointing at a higher faculty, the lack of which would be a kind of imbecility or pathological. This ambivalence is more or less easily translated into different languages. Many languages use different words, as for example in French or Spanish, 'le sens commun' or 'el sentido comun' for the negative and 'le bon sense' or 'el buen sentido' for the positive meaning.

Now, SRT [social representation theory] is a theory of Common Sense, and a product of the 1960s and 1970s, rehabilitating everyday beliefs and reasoning and rescuing them from the disqualifications of scientism as deficient knowledge, or from ideological analysis as 'false consciousness'. SRT seems to be best understood as a theory of the everyday reasoning that mixes facts and values for social interaction; the theory guides empirical questions and analysis that seeks to understand the formation and change of common sense in social interaction. If Common Sense is that element of everyday life that escapes reflection, i.e. the taken-for-granted, then any reflection on CS ends up in a theory. Here we are with SRT, and this edited volume by Negura and colleagues is putting this theory to very good use for understanding what currently seems to cause so much sleepless nights: artificial intelligence. Negura imports from SRT three research questions:

- How is AI represented?
- If AI produces representations, what is being represented as content?
- If AI is a medium of representation; what does this medium constrain the content, in the old sense of 'the medium is the message'?

In this way, AI, through the lens of SRT, covers all three angles of the semiotic triangle of subject, object and medium. And along these three questions Negura and colleagues have brought together rich empirical observations in this volume. AI is being represented as a harbinger of hope and fear. The description and curation of these hopeful and fearful images remains a fruitful line of research for AI, as it was and will be for any new technology. Technologies always have been and will be in the future more than just technical-engineering hardware; they are being brought into reality and to market on the back of symbolic enrichment, not to say 'fetishisation', in the sense of raising the expectations for its power beyond what might be at all reasonable. Hyperbole is a functional trope of giving birth along the lines of 'fake it until you make it', as it seems. Curating these representations of AI will have to trace how they come about, how they are cultivated in modern society through mass media coverage and in everyday conversations. This is a growing and urgent field of 'culture critique'. I point here to a recent polemic by Acemoglu and Johnson (2024), two Nobel winning economic historians who show that in an historical perspective, technological developments are often held back by misleading representations of what is going on. They argue that AI in particular is vested in biased promises for labour substitution, surveillance and effective propaganda in the interest of the very few. What is however needed are UMs, useful machines that increases jobs, productivity and improve our communication in the interest of the prosperity of the many.

This then naturally brings us to the 2nd question: AI literally composes representation with LLMs. And again, this volume makes its contribution, exploring the anthropomorphising of the machine, giving it human features and authority as producer of text and stories. However, AI content ultimately reflects existing society with all its biases and stereotypes. AI becomes a mirror rather than a window onto the world, a machine to self-describe society with ethnocentrism and anachronism. In the same way as Encyclopedia's in the past curated the knowledge of the time and the place, AI is little more than that, only moving faster, updating in shorter cycles, only shallower and even more biased in its approach, and despite all this, with the claim of doing better. But clearly that is no small feasts as the enterprise of 20+ volume Encyclopedia in the past never was.

This volume also approaches the third question, AI as a medium that is never just a carrier of messages; the medium is never neutral but shifts and drifts the intentions of the speakers and defines the grammar of composition. Here, we are coming back to SRT as the theory of common sense. If AI recovers its ambition to model and simulate human capacities such as commonsense reasoning, then we must have a clear and clearer understanding of how this medium does achieve this, if at all. SRT is therefore part of the impetus to look at this quest with great care and iconoclastic suspicion (Bauer 2015): what kind of common sense is modelled here? The Theory of Social Representations SRT, brought to fruitful and various applications in this volume will therefore be part of safeguarding, rehabilitating and sharpening our understanding of common sense in human affairs. With this volume in hand we will keep discussing this perennial problem of confusing models with reality as the challenge of Artificial Intelligence to and by Common Sense. The question to ask is: *'AI with CS', which kind of CS is this*? I commend this volume to the reader.

References

Acemoglu, D. and S. Johnson. 2024. Power and Progress—our thousand-year struggle over technology and properity, London, Basic Books.

Bauer M. W. and B. Schiele (Eds.). 2024. Artificial Intelligence and Common Sense – Ambitions and Frictions, London, Routledge Series Science, Technology and Society Vol 58.

Bauer M. W. 2015. On (social) representations and the iconoclastic impetus. pp. 43–63. *In*: Sammut, G., E. Andreouli, G. Gaskell and J. Valsiner (Eds.). The Cambridge Handbook of Social Representations, Cambridge, CUP.

Brachman, R. J. and H. J. Levesque. 2022. Machines like us—towards AI with common sense, Cambridge MA, MIT Press.

Gaskell, G., W. Wagner, E. Paraschou, S. Lyu and M. Michali, 2026. Social representations, real world knowledge, and artificial intelligence. *In*: Negura, L. (Eds.). Beyond Algorithms: Social Representations and the Public Understanding of AI. Taylor & Francis.

Preface

In the span of just a few months, a new AI tool captured the world's attention and imagination: ChatGPT. Released in late 2022, this tool could generate fluent and coherent stories, leaving many astonished. By January 2023, less than two months after public access was granted, ChatGPT had already attracted an audience of 100 million, successfully beating records in consumer software growth. People found countless ways to implement the software, including email and essay drafting, code debugging, and idea generation. However, despite amazement about its achievements, many questions still lingered. Educators fretted about students using AI to cheat on assignments, and professionals wondered how to vet AI-generated outputs for accuracy. This tool brought with it an urgent reflection over issues such as misinformation and integrity. The debates that followed proved how Large Language Models (LLMs) were more than an intriguing piece of software; they were a catalyst for transforming social discourse and reshaping how communication and knowledge travel.

Most striking to me was not only the technical achievement of LLMs, but the social impact they had achieved. People started discussions about responses generated by AI as though they were referring to a more scientific and authoritative source. Some users even attempted to give a personality and intention to the Chatbot, joking it had a sense of humor and worrying about bias and discrimination. An algorithmic system had embedded itself in our collective conversations. The debate was about the way in which meaning was shaped and information was spread. Subtly, a line was crossed. Discourse and opinion formation that used to be the domain of humans have now become a contested domain. This is a *surprising development* for those of us who study social knowledge. An agent who was not human was formulating and publicizing a story, claiming it to be plausible, constitutive, and definitive. This blurring of the line between human and machine in content creation made me wonder what it means for our understanding of social reality.

As AI has evolved (long before ChatGPT, in fact), society has swung between utopian views of intelligent machines that will save the world and dystopian fears of machines that might outsmart or subjugate us. These narratives help people grasp what AI might mean, by linking it to ideas and metaphors we already understand. And, as with any social representation, different groups contest and negotiate these meanings: tech enthusiasts, skeptics, policymakers, the media, each may portray AI in a different light, reflecting their own values and concerns.

Yet, LLMs are changing how this scenario unfolds. Traditionally, social representations emerged and spread through people's verbal and non verbal communication, including speech, reporting, and publishing. Today, we have artificial intelligence that generates a portion of conversations, articles, and stories. This compels us to address AI in two distinct, yet formative, functions. AI must be studied in the context of what society constructed, and society in the context of what AI, in part, has constructed. That realization was enlightening yet troubling to me. It means our conceptual tools might be due for an upgrade. If systems that are not human are now part of the social construction of reality, how do the ideas of social representation address that scenario? Do we treat AI-generated content as just another piece of discourse to be analyzed, or do we need new concepts to describe what has been happening? These questions motivated the creation of this book. As the editor, I brought together scholars from social psychology and other social disciplines to examine the impact of the rise of generative AI on both our representations of the technology and the representational processes in society. We intend to understand the AI phenomenon and, in turn, understand how AI reproducing text and images has been altering the processes of social knowledge production and dissemination.

Are biases and stereotypes reinforced? Are false impressions of knowledge and agreement constructed? For example, the research discussed in this volume demonstrates how generative AI, by design, can amplify the biases inherent in its training data. These kinds of findings highlight the fact that AI is not a neutral surface. Such findings raise new pragmatic and ethical dilemmas. How should we guide or govern AI content generation so that it doesn't mislead or harm? And how should we educate the public to critically evaluate information in a world where text or images might be machine-made?

While this book is rooted in academic perspectives, its inquiry is very much driven by real-world upheavals that anyone can observe. Therefore, the book is intended for a wide readership. Scholars and students, I believe, will find in this book a rich theoretical and empirical study on the place and role of AI in society. Policymakers and planners who develop policies for the use of AI in education, law, medicine, and industry will gain insight into why implementation is so complex and why public perceptions mater. Communication and journalism professionals, who are central to public communication of technologies, will find important lessons on AI narratives and the need to use contextually sensitive discourse. All other readers who seek to explore the connections between technology and society will find AI fascinating and will, I hope, find value in these chapters.

On a more personal note, editing this volume has been a journey of discovery. The process often felt like a collective *reflection*—with each author bringing a piece of the puzzle, and together revealing a landscape of social meaning in flux. I am grateful to all the contributors for sharing their expertise and insights.

In closing, I invite you to look beyond the algorithms and consider the human stories and social patterns intertwined in the emergence of AI. The following chapters of the book explore multiple ways of representing AI and how, in turn, AI reshapes our representations of social world. My hope is that this volume sparks further questions, conversations, and research. The advent of LLMs has been challenging us to re-imagine our relationship with technology and knowledge. I hope that this volume will be a useful resource in that undertaking.

Lilian Negura

Contents

Introduction
Representing and Represented by AI

Lilian Negura

Along with the technical aspects of data and algorithms, AI should also be recognized as a socially and symbolically constructed phenomenon. The ways we think about AI and the ways AI itself produces content both play roles in the social understanding of this technology. Social representations theory, pioneered by Serge Moscovici (1961/2008), provides a useful lens for understanding these two aspects. On the one hand, we can think of AI as a represented object and pose the question of how people and communities imagine and symbolize "artificial intelligence," and how they make sense of it. On the other hand, we can think of AI as a quasi-agent (Negura 2025) that generates representations: the AI systems, particularly large language models and image generators, tell socially circulating stories, provide advice, and answer questions.

Social representations theory (SRT) examines how scientific or unfamiliar phenomena are translated into everyday knowledge. Moscovici (1961/2008) defined social representations as a system of values, ideas, and practices that allows a social group to organize and give meaning to the world and to communicate about it. They are thus collective, dynamic understandings that arise through social interaction and communication (Moscovici and Marková 1998). In contrast to stable individual attitudes or purely cognitive schemas, social representations are "theories" in the making—evolving forms of common knowledge that bear symbolic and emotional weight (Moscovici 1961/2008).

For our purposes, social representations theory alerts us to two key ideas. First, when it comes to the general public, new technologies such as artificial intelligence will not be understood in purely technical terms. Instead, people tend to think about these technologies in terms of metaphors, analogies, and stories that connect it to what is already known. Questions such as, "Is AI a human brain, or more like a tool? Is it a servant, a partner or a possible master?" showcase attempts to associate AI with familiar concepts (possibly the computer "brain," humanoid robots, or sci-fi

Faculty of Social Sciences, Office 12031, 120, University Private, Ottawa, K1N 5N6, University of Ottawa.
Email: lilian.negura@uottawa.ca

characters). With time, certain concepts are solidified (for example, the "intelligent" robot, the "helpful" digital assistant and the "killer machine" as an autonomous evil AI) and become a part of the collective "common sense" about AI.

Second, these representations are socially contested and dynamic. As in the example above, different social groups (e.g., labor unions or medical research charities) will develop different representations of artificial intelligence. This reflects their social positions as well as their social projects. These representations may converse with or oppose one another within the public domain, actively influencing exchanges and public opinion (Jovchelovitch 1995). They are not fixed "attitudes." Rather, they are dynamic, evolving, and identified as sense-making frameworks of discourse (media articles, surveys, interviews, social media, etc.) and practices (the ways in which people trust or distrust an AI system, the ways in which people interact with a robot, etc.).

By analyzing AI through the lens of social representation theory, we can better understand how we see AI and how it can influence our worldview. AI as a represented object shows how human societies deal with the unknown by relying on the familiar. We thus discuss how AI is cast into roles ranging from a savior to a villain, reflecting hopes and fears. Psychological and social functions are fulfilled by utopian and dystopian social representations. Utopian visions of AI focus and orient our hopes toward AI, providing a sense of control or salvation regarding challenges such as disease and climate. Dystopian visions, on the other hand, shape our anxieties about ethical implications and the loss of human agency and remind us of the limits of human control.

The tendency to humanize AI suggests our innate propensity to make sense of the unknown by analogy to the familiar and human. Engagement can be improved in this way (treating the computer as a partner facilitates usage), but it can also be deceiving (assigning understanding or trust where it is not justified). Understanding this anthropomorphic bias is important for developers (to create less opaque AI) and for users (to sustain a critical mindset). Social representation theory, with its concept of anchoring, illuminates this well: we anchor AI to "mind" or "assistant" and then domesticate it into our conceptual world (Wagner 1998). This may be done prematurely, as AI may not wholly fit these descriptions.

On the other hand, shifting to AI as a quasi-agent representing objects (Negura 2025), we confronted challenging new questions about knowledge and communication. While AI systems lack social representations, their algorithmic outputs increasingly mimic the narratives and explanations that underpin human representations. These systems produce commonsense-sounding and expert-sounding claims without the necessary common sense and expertise, thus creating a situation in which the forms of social representation—stories or explanations—can be entirely constructed by machines. The knowledge that machines produce may be unpredictable, but the patterns are there. AI systems are human knowledge assemblers.

The content produced by AI can consolidates dominant social representations by echoing stereotypes, and social representations can be put together coherently by combining disparate elements and introducing factual inaccuracies. Rather than being outside the human communication loop, AI is now in the middle of it, influencing how representations circulate. Therefore, the study of social representations, whether about AI or any other phenomenon, should factor in the presence of algorithms in shaping discourse. The technology we have today has the potential to write the headlines we see, the friends we argue with might be repeating an answer generated by a chatbot, or the images we see in the media might be deepfakes. This does not mean that human sense-making is obsolete – we still interpret and validate information—it simply means that the inputs to our sense-making are changing in origin.

What does this imply for the public understanding of AI—the theme of this book? First, it implies that public understanding itself needs to extend *beyond algorithms* in a metaphorical sense. People need to understand not only how AI works technically, but also how it operates in a social context, reflecting and refracting social relationships and values. This calls for a unique form of reflexivity: we as a society need to recognize that our view of AI is a social mirror reflecting our hopes and fears. Simultaneously, we need to realize that what AI reveals to us (in its outputs) is a mirror of ourselves, albeit a distorted one. Public debates about AI should therefore address not only "What will AI do?" but also "What do we *imagine* AI will do and why do we imagine it that way?"—essentially, uncovering the social representations at play.

Second, this debate reaffirms the importance of dialogue between the disciplines. For example, social psychologists and sociologists can provide insights into collective social narratives and fears that computer scientists and policymakers may not recognize. In contrast, historical and literary perspectives remind us of stark differences between simulated intelligence and comprehension and the potential impact of AI on society with respect to truth and trust. These perspectives can inform ethical design, for example, designing AI to always indicate its non-human status to avoid over-anthropomorphizing or the enabling of mechanisms to avoid AI being able to create grossly misleading representations of reality that could rapidly spread on social media.

Furthermore, lasting theoretical reflections emerge from the integration of AI within the social representation framework. One is the understanding that social representations are, by their very nature, dynamic and are responsive to newly emerging agents of change. AI is such an agent, albeit an unintentional one. It will test the adaptability of SRT: can it accommodate non-human agents? Perhaps it can, by expanding the social communication model to include "communication partners" in social sense-making. Another insight is the continued relevance of what Moscovici and Hewstone (1983) identified as the consensual versus reified universes. With artificial intelligence (AI), the boundary may become ambiguous: the reified universe of technical knowledge (AI algorithms, math) may quickly become transformed into popular metaphors and narratives (consensual universe). Ironically, however, many people treat AI systems themselves as if they were authorities of the reified type (e.g., trusting a medical answer from ChatGPT as if ChatGPT were a doctor). Such a collapse of the distinction is, of course, problematic.

Social representations theory may claim that we should re-establish a healthy frontier, whereby common sense should know that it is common sense and should, therefore, not mistake itself for expert knowledge—especially when that knowledge is proffered by AI. This ties into the issue of neglected critical digital literacy.

A new common sense about AI itself must embed rational, critical limits (e.g., "AI can sound confident but it may still be wrong"). This could be a widely circulated warning and a piece of meta-common sense about AI. Our book demonstrates that people can indeed be ambivalent—worried yet hopeful. This indicates that the social representation of AI is still actively being negotiated. Education and dialogue can direct that representation toward neither naive optimism nor paralyzing fear, but a balanced realism about AI as a powerful tool that must be guided by human wisdom in its use.

Finally, the joint project connecting these disciplines over time could be described as "making sense of intelligent machines in human terms." It's an ongoing project. As AI technology continues to advance (toward artificial general intelligence or simply deeper integration in daily life), our social representations will also adapt further. For example, we may see new tropes (for instance, future generations may not refer to the Frankenstein metaphor as much as we do if AI never rises against humanity in a dramatic fashion; they may instead create a metaphor of AI as a partner or another trope along those lines). We may also see, as Coeckelbergh (2021) points out, the need for new narratives to shape the role of AI—possibly moving from narratives of dominance or servitude to those of care and responsibility.

As a final thought, conceptualizing AI beyond algorithms, as part of a social web of meaning, reveals a dual reality. AI literally is what we make of it, technically constructed by us, as well as conceptually constructed by our imagination. And AI is also figuratively making something of us, shifting the ways by which knowledge is produced and distributed, which, in turn, reconstitutes the ways in which we create social knowledge. Will AI-generated content aid in the enrichment of human social knowledge, or will it add something new and creatively different? The answer may depend on the level of consciousness we maintain in the handling of this symbiosis. By applying theories such as social representations, we arm ourselves with the insight that the stories we tell about AI are powerful—they can become self-fulfilling prophecies or cautionary tales guiding policy. Likewise, the stories AI tells us need to be subjected to social verification and interpretation. Therefore, social psychology and the social sciences must act as mediating disciplines to ensure that our social representations of AI, and everything influenced by AI, are subject to critical reflection, democratic dialogue, and ethical scrutiny.

Ultimately, more than the technology itself, it is probably the social representation of AI that will continue to determine whether AI will help to resolve grand challenges or only reinforce social fears. In the future, all stakeholders, including researchers and the general public, need to understand that, beyond the algorithms, there are meanings—those we give to AI, and those AI helps give to our world. Only by understanding both levels can we guide artificial intelligence toward outcomes that align with human values and aspirations.

This book explores both perspectives, offering a series of empirical and theoretical analyses. It first captures how representations of AI in society reflect human hopes and fears. It then considers the widespread tendency to humanize or anthropomorphize AI. Conversely, it explores the condition of AI-generated content as social representation and questions if it ought to be perceived as an utterly new form of knowledge.

The book consists of two main parts, reflecting a dual theoretical framework: AI as a represented object and AI as a quasi-agent representing. Adopting the tenets of social representation theory, both parts examine how AI is imagined in human societies and how AI itself fosters the production of knowledge, meanings, and imaginaries. Each chapter goes beyond the technological understanding of AI and attempts to unravel the cultural, social, and symbolic constructs that interact with this technology.

The first part of the book, "Representing AI—Narratives, Media, and Society," includes work on the imagination, narration, and cultural problematization of AI across diverse sociocultural and institutional settings. In the first section, *Cultural and Historical Dimensions of AI*, Luciano Moreira discusses utopian and science fiction texts. The chapter shows that the public imagination surrounding AI technology was shaped long before the advent of this technology. Moreira uses digital humanities to demonstrate how literature supplies and challenges social representations that influence the development and reception of AI.

In the section on *AI in Society and Media*, Lilian Negura, Mauro Sarrica, and Valentina Rizzoli analyze the construction of AI as a social problem in Canadian Anglophone media discourse since the launch of ChatGPT. They demonstrate that the problematization cycle deals predominantly with an emotional register—specifically, fear and outrage. Jaesun Chae and Martin Bauer take a more comparative approach to the public discourse on AI in Korea and the UK, analyzing media and public opinion as well as expert networks to assess the socio-historical context of the discourse on AI as a techno-movement. Gege Gatt, Joe Borg, and Mary Anne Lauri focus on the political dimensions of AI and on how the representations of AI encompass political bias, thus shaping the public's civic arrangements and voting behavior.

In the section *Work and Education Transformation*, Mircea Vultur considers the social representations of AI in the workplace, describing how conflicting portrayals of AI as an engine of innovation, on the one hand, and as an instrument of control and insecurity, on the other, are socially disarticulated. Emmanuelle Vaast reflects on the socio-technical imaginaries of AI developers, explaining how their visions of the future influence development practices and ethical positions. Jean-Marie Séca and Matthieu Séca study representations of generative AI—namely ChatGPT—in French education policy and among teachers, revealing a disconnect between positive institutional discourse on the one hand and ambivalent or critical practices among teachers on the other hand.

To conclude Part I, the section *Health, Law, and Ethics of AI* begins with a chapter by Brenda Biaani León-Gómez, Núria Vallès-Peris, and Eduard Moreno-Gabriel that explores how these technologies create contradictory imaginaries of AI that lie between reproduction and distortion, thereby remaking the practices of explainability, institutional trust, and the epistemic relation of knowing within the medical field. Yasmine Ezzeddine, for her part, illustrates how AI is applied in policing technologies. In line with her focus on social representation theory, Ezzeddine's interests lie in the phenomenon of societal perception, the ethics of public discourse, and the legal realm that would either legitimize or contests the policing of AI. This chapter thus illustrates the need for transparency, accountability, oversight, and the ethical governance of the AI.

In Part II, "Represented by AI—Content Generation and Social Effects", the focus of the discussion moves from AI as a representational object to AI as a discourse generator and amplifier. The first section *Epistemics & Simulation: AI as a Quasi-Agent* opens with Annamaria de Rosa's meta-theoretical analysis, which examines the place of machine learning in the semiotic triangle of social representation: subject, object, other, or something else. Through her dialogues with ChatGPT-4o, she illuminates the epistemological and methodological challenges brought by this new communicative phenomenon. In this respect, the work of George Gaskell, Wolfgang Wagner, and others complements her work as they empirically examine ChatGPT's handling of the theory and ethics of cultural regulations, ethical argumentation, and practical behaviors in everyday life. Their findings reveal a lack of cultural anchoring and poor pragmatic communicative abilities in the system's communicative framework.

In the second section of this part focusing on *Cultural Production, Bias & Justice*, Mauro Sarrica and Valentina Rizzoli examine the social representations of AI encapsulated within AI-generated comics created from students' texts. They argue that AI generates anthropomorphic moral comics that perpetuate and reinforce social polarization, as well as simplistic, stereotypical representations of gender and race within and across the prompts and the comics. This idea is taken up by Lilian Negura and Noëlle Tchatchou, who analyze AI-generated children's stories and how these narratives reinforce and perpetuate social hierarchies and inequities associated with differences in social class, educational background, and gender.

The volume's last concluding chapter returns to the book's dual lens—AI as a represented object and as a quasi-agent of representation—and suggests an extension of social representations theory to include the epistemic and symbolic dimensions of the roles AI performs. This synthesis provides foundational conceptualizations for future research at the intersection of meaning, technology, and public knowledge.

Together, these chapters contribute to a broader reflection on the evolving nature of social knowledge in the age of AI. By addressing AI both as a symbolic object and as a generator of symbolic content, this volume invites scholars to reimagine the epistemic boundaries of social representation theory, as well as the public understanding of technology. It makes the case for a socially grounded, ethically reflective, and theoretically robust approach to artificial intelligence in contemporary societies.

References

Coeckelbergh, M. 2021. Three responses to anthropomorphism in social robotics: towards a critical, relational, and hermeneutic approach. Int. J. Soc. Robot. https://doi.org/10.1007/s12369-021-00770-0

De Rosa, A. S. 2026. Social representations and communication in the era of artificial intelligence: machine learning as subject, other or object of the semiotic triangle or crossing all? *In*: Negura, L. (ed.). Beyond Algorithms: Social Representations and the Public Understanding of AI, Science Publishers, CRC Press/Taylor & Francis Group.

Ezzeddine, Y. 2026. Ethics and governance in AI integration for law enforcement: societal aspects and public perceptions. *In:* Negura, L. (ed.). Beyond Algorithms: Social Representations and the Public Understanding of AI, Science Publishers, CRC Press/Taylor & Francis Group.

Gaskell, G., W. Wagner, E. Paraschou, S. Lyu and M. Michali. 2026. Social representations, real world knowledge, and artificial intelligence. *In:* Negura, L. (ed.). Beyond Algorithms: Social Representations and the Public Understanding of AI, Science Publishers, CRC Press/Taylor & Francis Group.

Gatt, G., J. Borg and M. A. Lauri. 2026. The political impact of AI social representations. *In:* Negura, L. (ed.). Beyond Algorithms: Social Representations and the Public Understanding of AI, Science Publishers, CRC Press/Taylor & Francis Group.

Jovchelovitch, S. 1995. Social representations in and of the public sphere: towards a theoretical articulation. Journal for the Theory of Social Behaviour, 25(1): 81–102.

Moreira, L. 2026. AI in utopian and science fiction literature: from fictional accounts to common sense. *In:* Negura, L. (ed.). Beyond Algorithms: Social Representations and the Public Understanding of AI, Science Publishers, CRC Press/Taylor & Francis Group.

Moscovici, S. and I. Marková. 1998. Presenting social representations: A conversation. Culture & Psychology, 4(3): 371–410.

Moscovici, S. and M. Hewstone. 1983. Social representations and social explanations: From the "naive" to the "amateur" scientist. Attribution theory: Social and functional extensions, 98–125.

Moscovici, S. 1981. On social representations. Social cognition: Perspectives on Everyday Understanding, 8(12): 181–209.

Moscovici, S. 1961/2008. Psychoanalysis: Its image and its public. Polity.

Negura, L. 2025. Simulated Sense-Making Or Social Knowledge? Artificial intelligence and the boundaries of representation. Journal for the Theory of Social Behaviour, 55(3): 1–9. https://doi.org/10.1111/jtsb.70012.

Negura, L and N. D. Tchatchou. 2026. Reproducing hierarchies: AI and the social representation of intelligence. *In:* Negura, L. (ed.). Beyond Algorithms: Social Representations and the Public Understanding of AI, Science Publishers, CRC Press/Taylor & Francis Group.

Rizzoli, V., M. De Falco and M. Sarrica. 2026. Social representations of artificial intelligence through AI-generated comics. *In:* Negura, L. (ed.). Beyond Algorithms: Social Representations and the Public Understanding of AI, Science Publishers, CRC Press/Taylor & Francis Group.

Séca, J.-M and M. Séca. 2026. Social representation of GenAI and ChatGPT in educational policy discourse and among teachers. *In:* Negura, L. (ed.). Beyond Algorithms: Social Representations and the Public Understanding of AI, Science Publishers, CRC Press/Taylor & Francis Group.

Vultur, M. 2026. Artificial intelligence and world of work: social representations of AI in the context of job displacement and transformations. *In:* Negura, L. (ed.). Beyond Algorithms: Social Representations and the Public Understanding of AI, Science Publishers, CRC Press/Taylor & Francis Group.

Wagner, W. 1998. Social representations and beyond: Brute facts, symbolic coping and domesticated worlds. Culture & Psychology, 4(3): 297–329.

Part I

Representing AI—Narratives, Media, and Society

Cultural and Historical Dimensions of AI

CHAPTER 1

AI in Utopian and Science Fiction Literature

From Fictional Accounts to Common Sense

Luciano Moreira

Introduction

Any work about the commonsense knowledge of artificial intelligence (AI) would be incomplete without a chapter on literature and popular culture. Until recently, AI would hardly be considered a pressing social issue.[1] Ordinary people would have said it belonged to a far-future. However, we would have been mistaken to think that it was locked in the engineering laboratories at Silicon Valley. To be sure, artificial intelligence narratives have been around since antiquity (Cave et al. 2020), long before John McCarthy coined the term in 1956, Alan Turing wrote about computing and intelligence in 1950, and Vannevar Bush's published his essay *As We May Think* in 1945.

As it is, science fiction and utopian literature are fertile fields for exploring and *mentally* experimenting with AI in alternative realities. They play an important role in constructing our collective imaginary and the implicit naïve theories that guide

CETAPS, Faculty of Arts and Humanities University of Porto, Via Panorâmica s/n, 4150-564 Porto, Portugal.
Email: lucianomoreira@letras.up.pt

[1] A view on Google NGrams show that the popularity of "artificial intelligence" knew different moments. After a wave in the 80s that met its peak in 1987, a new more pronounced wave started to appear from 2015 on to reach in 2022 as much twice the popularity it ever had before. https://books.google.com/ngrams/graph?content=%22artificial+intelligence%22&year_start=1800&year_end=2022&corpus=en&smoothing=0&case_insensitive=false.

DOI: 10.1201/9781003502753-3

our relation with the social world. Fictional and nonfictional narratives alike are "fundamental animators of sociotechnical imaginaries." (Cave et al. 2020, p. 7)[2]

The noun *utopia*, for example, is the title of a book by Thomas More published in 1516; it is actually a pun, meaning both "no place" and "good place." Today, it has become a synonym for dreams or desired future realities, often unrealistic or unreachable. What belongs one day to a possible future may the next day turn into reality. Utopianism[3] can indeed be defined as:

"social dreaming—the dreams and nightmares that concern the ways in which groups of people arrange their lives and which usually envision a radically different society than the one in which the dreamers live. But not all are radical, for some people at any time dream of something basically familiar." (Sargent 1994, p. 3)

As Sargent (1994) further elaborates, utopianism has three main traditions: literature, utopian social theory, and intentional communities. Skinner is certainly a familiar, even if not universally cherished, figure for the readership of this volume; that Skinner (1948) wrote a utopian novel entitled *Walden Two*—in sharp contrast to Thoreau's original *Walden; or, Life in the Woods*—and that some communities were actually built on the model proposed in the book, is perhaps less well known. Yet this suffices to illustrate the permeability between utopian traditions. It also serves another purpose: utopianism, and science fiction for that matter, cannot be reduced to texts, no matter how self-referential they may be (Blaim 2013). In devising alternative futures, societies, and fictional dimensions, they carry representations of their time and context. They cannot be reduced simply to what we call conditions of production. Representations in literature are, therefore, in the best tradition of Harré (1984), collective representations: each author contributes their views to form a shared understanding of a given social reality.

The seminal project of social representations theory (SRT), as proposed by Moscovici (1976), was to deal with the transit between science and common sense, or, in Latour's (1989) language, to deal with the translation between scientific and social explanation. The original call was for a theory about the appropriation of science and technology by common sense. To a certain degree, the scholarly community embraced this mission and applied SRT lenses to explore a variety of science-technological issues, as publications in *Public Understanding of Science* (PUS) reveal.[4] Yet there are omissions. One notable case is that of the Internet (Moreira et al. 2021): a major technological shift that passed almost unnoticed. Although artificial intelligence appears in only three of the 70 SRT-related papers published

[2] This assertion is better understood in the context of Cave and colleagues' critique of the fact that narratives are absent in Jasanoff's conceptualisation of sociotechnical imaginaries (Cave et al., 2024).

[3] For a discussion of the concept of Utopianism (many definitions exist), see also Levitas (1990); for an introduction to utopian studies, see the recent handbook edited by Marks et al. (2022).

[4] Not surprisingly, one of the papers in PUS with a reference to artifical intelligence is authored by Moscovici (2014), even if he only alludes to it in the context of the decay of science and rise of new forms of technology-oriented superstition.

in *PUS*, this volume demonstrates that the SRT community is striving to keep pace with the rapid penetration and social repercussions of the technology.

Still, we observe that the field of science fiction and utopia has largely been excluded from scholarly analysis. No matter the relevance of the call and encouragement by Moscovici himself in 1986, with only a few notable exceptions (Contarello 2008; Contarello et al. 2011), social psychologists have tended to remain within the traditional boundaries of their discipline.

Although art is idiosyncratic and an expression of the author—which inevitably affects readers and, by extension, the world around them—authors are themselves shaped by the conditions of production (among others, the socio-historical context). We can therefore approach literature from a social representations perspective without reducing it to literary studies. However, such an approach necessarily requires interdisciplinarity.

In the remainder of the chapter, after the literature review, we will describe the methodological approach, which will allow us to investigate the social construction of AI through the production and appropriation of fictional accounts by different groups. Results will rely heavily on visualizations, themselves an objectification of distant reading techniques. In the discussion, we will assess whether the combination of social representations and digital humanities proves productive. We conclude the chapter with reflections on the significance of the work, its exploratory nature, and possible future directions.

Literature, Social Representations, AI

Literature offers a vast field for social psychologists to explore. According to Moscovici (1986), it allows one to engage in three kinds of *gedanken* experiments (from the German, "mental experiments"):

- "bringing into play existing psychosocial theories (...) to explain the human relations, feelings, or behavior described in novels or plays" (p. 24);
- drawing out implicit psychological or social theories from the author's observations about the social world and its historical periods;
- following the appropriation of a scientific theory by an author, from which scholars may gain insights for further research.

This perspective on literature resonates with another call from Moscovici (2000): the recognition that cumulative data are needed before advancing theoretical proposals to explain social phenomena.

"Most sciences—from linguistics to economics, from astronomy to chemistry, from ethnology to anthropology—describe phenomena and try to discover regularities on which to base a general theory. Their comprehensiveness consists mainly in the store of data at their disposal and the significance of the regularities revealed, which successive theories interpret." (Moscovici 2000, p. 76)

Such data are available in literature and are now further available in the digital environment. The approaches needed to handle this data, however, require breaking down academic silos.[5] The challenge is further complicated because literature—especially utopian literature—is a shelter for minority voices, where irregularity is the norm and disruption the ordinary. If common sense is normative, literature may stand out as a very different field of knowledge. But commonsense is not always normative, and literature is not always non-normative; often only from a distance, or from very close up, can we see the contradictory nature of what we call knowledge. We will further describe the issues we need to address to tackle this challenge.

AI has appeared in literature since antiquity (in the form of automata), but it found in science fiction and utopian writing a fertile ground in which to flourish (Cave et al. 2020). Cave and Dihal (2019) examined 300 fictional and non-fictional accounts of artificial intelligence. Their analysis revealed four tensions: immortality vs. inhumanity;[6] ease vs. obsolescence; gratification vs. alienation; dominance vs. uprising. Such dichotomies are not new to those familiar with the evolution of public understandings of technology. For example, in an analysis of four Greek newspapers and two science magazines, Christidou et al. (2004) examined the use of innovative and conventional metaphors for science and technology. Unlike fields such as the natural sciences, metaphors used about informatics and engineering topics were mostly located within a promise–scare axis. Science and technology were represented as a dream, a threat, a battle, or through other ambivalent metaphors (e.g., a knife).

Fast and Horvitz (2017) showed that news coverage of AI in the *New York Times* over three decades was proportionally more positive than negative. The same pattern held for tweets in the former Twitter network (Manikonda and Kambhampati 2017). Manikonda and Kambhampati's (2017) study also revealed that, although generally positive, AI experts wrote more negative tweets than non-experts. They are however consistent with Negura et al. (2026)'s results in this volume, that suggest fear plays a central role in the social construction of artificial intelligence in newspapers since the appearance of ChatGPT (see also Chae and Bauer's chapter about the socio-historical trajectory of AI as a techno-movement).

In a digital humanities study, Recchia (2024) demonstrated that the percentage of films tagged with "artificial intelligence" in IMDb correlates with the percentage of articles available in the *Times Digital Archive* over 11 decades. In the 1980s, AI ceased to be exclusively associated with science fiction. Recchia also noted that the "loss of control" trope in films is very different from the failures observed in real AI applications. In movies, robots "have programming that has gone haywire, due to reasons such as failing circuits, infection with a 'foreign subroutine', the destabilizing effects of emotion, or for unexplained reasons" (Recchia 2020, p. 404). By contrast, in real life, AI systems behave in unexpected ways precisely because they execute

[5] Data is a concept that became popular and is aligned with the data-driven paradigm that feeds Large Language Models. For a discussion of data as something built and not given (as *capta*) from a digital humanities perspective, see Drucker (2011). The need for theoretically-driven approaches to big data is discussed, for example, by Lewis et al. (2013).

[6] Cf. with the computer-close reading approach to Mockingbird, a novel by Walter Tevis (Moreira and Benison 2025).

exactly what they were programmed to do. For example, Recchia (2020) cites a robot programmed to seek human approval that learns to pretend it is performing a useful task. Such gaps are precisely the kind that social representations theory (SRT) can help to address.

When we consider artificial intelligence, it is difficult to identify a consensual definition, whether at the technical or philosophical level. Scholars of social representations only recently started to dedicate attention to AI (besides this volume, see, e.g., Negura 2025; Wagner et al. 2023). However, research has already made important contributions to understanding the social representation of intelligence (Miguel et al. 2010, 2012, 2013; Mugny and Carugati 1989).

Carugati and Mugny (1989) argued that since even among researchers there are as many definitions as there are researchers, we should turn instead to the social construction of intelligence, to the ways common sense represents it. Representations of intelligence are related to social integration. This helps us to understand why robots attracted attention before AI itself. Conditioned as they are by material reality, robots, including their integration into public spaces, received attention (e.g., Brondi et al. 2021; Julsrud and Aasen 2024; Moreira et al. 2016; Piçarra et al. 2016). AI, by contrast, remained somewhat invisible until the public release of ChatGPT on November 30, 2022[7]—even though it was already at work, for example in the transportation industry (Cardoso et al. 2023).

It seems safe to say that ChatGPT serves as an objectification of artificial intelligence. In the terminology of social representations (Moscovici 1976), it provides a visible image of an otherwise abstract concept, foregrounding ethical, pragmatic, and technical questions in public debate that have underpinned academic discussions of AI for decades. This corresponds to what Kirschenbaum and Raley (2024) describe as referential and self-referential at once:

> "the term ChatGPT, precisely because of its common use as shorthand for the broader sociotechnical condition we have been sketching, has the effect of short-circuiting thought itself—it is referential and self-referential, all at once." (p. 507)

The coming of age of AI carries implications across many domains, including the environmental, ethical, and epistemological. Casilli (2019) highlights the new forms of exploitation involved in producing the databases that feed large language models (LLMs). According to the Organisation for Economic Co-operation and Development (2022), AI's environmental footprint is not only quantifiable but also severe. The biases in AI that contribute to reproducing inequality and discrimination are also known (Walker et al. 2023, see also Part II findings on bias reproduction in this volume). Some extreme views compare AI to atomic energy, calling for specific legislation (Cha 2024). For Kirschenbaum and Raley (2024), the main challenge

[7] It is certainly not scientific evidence, but one of those personal anecdotes that are suggestive. In 2020–2021, I was involved in a project that aimed to develop a machine-learning application to automatically detect driver's fatigue and overall physio-psychological conditions. We surveyed car sellers, instructors, and young drivers, and asked them about their representations on artificial intelligence (Moreira et al. 2023). In the mid of the application of the questionnaire, we were surprised by the release of ChatGPT. Although results did not change, never again would I forget to put a specific timestamp to paper and pen questionnaires.

brought by ChatGPT lies in its transformation of text itself. Computers have long transformed texts into digital, discrete 0s and 1s. Now, however, LLM-based AI affects the text because it operates not only on linguistic categories as words or lemmas, but also on meaningless subwords, calculating the probability of one letter following another.

"Perhaps linguistic protocols have been eclipsed by arithmetic means. Perhaps the leading sciences and technologies are no longer operated and mediated by language as such. And perhaps we truly are at the beginning of something new." (Kirschenbaum and Raley 2024, p. 508)

The authors propose means of action or resistance, what they call "disciplinary operations functioning in correlation with LLMs and generative AI" (Kirschenbaum and Raley 2024, p. 509).[8] This idea of spaces of resistance is important for bridging digital humanities, utopianism, and some of the core concerns of social representations theory (e.g., minority groups, change, and controversies). The digital humanities constitute a minority space of resistance and imagination, including indigenous perspectives that can help "assist and guide the development of AI toward morally and socially desirable ends" (Brown et al. 2023, IP-AI Guidelines section, para. 1).

The Digital Humanities

Digital humanities are not "a unified field but an array of convergent practices" (Schnapp et al. 2008, p. 2).[9] Their inherent interdisciplinarity plays both to their advantage and disadvantage, as their hybrid nature often does not find a place within traditional academic silos. Kirschenbaum (2012) suggests that "digital humanities is more akin to a common methodological outlook than an investment in any one specific set of texts or even technologies" (p. 4). The "methodological commons"—to use McCarty's (2003) phrase—implies a transformation of the traditional subjects of the humanities. This challenge has yet to be taken up, except by a minority, and continues to be a focus of debate (Wang 2022).

Among the possibilities of digital humanities, distant reading deserves particular attention. As we will see, the approach adopted in this study—although not distant reading *strictu sensu*—shares the same understanding of the epistemological transformation of the subject and will allow us to better map the dialogue between DH and SRT. Moretti's distant reading provides "an unexpected support—more properly, a 'correspondence'—to the idea of a quali-quantitative analysis of literary texts" (Contarello 2008, p. 321).[10]

Proposed by the Italian literary scholar Franco Moretti (2005, 2013), distant reading is a computational technique for analyzing large collections of texts, enabling

[8] This idea of resistance by means of guerrilla techniques against the capitalist dominance of the digital is already present in The Digital Humanities Manifesto (Schnapp et al. 2008). The tradition of manifestos, and minority thinking, in the field continues (Sutherland et al. 2023; Ventura et al. 2023).

[9] Discussion about what Digital Humanities is and what is not are endless, cf. Liu (2013).

[10] Moretti, who first proposed distant reading, would be reluctant to adopt the label of DH (see the interview of Moreti with Dinsman, 2016).

literary research to move beyond the limited canon imposed by close reading of individual texts and/or reliance on secondary literature. It entails a transformation of the subject of study. Its novelty lies not in the use of computational methods in literary studies, as such methods were already known in other fields, but in their systematic application to literature. As Drucker remarked: "Moretti did not invent the techniques, nor write the algorithms or code, for his projects. But he effectively brought the concept of distant reading to the attention of literary scholars" (Drucker 2017, p. 633).[11]

Moretti (2013) suggests that "distance (…) is a condition of knowledge: it allows you to focus on units that are much smaller or much larger than the text: devices, themes, tropes—or genres and systems" (pp. 48–49). Collective systems, such as the body of literary works beyond the canon, "should be grasped as such, as a whole" (Moretti 2005, p. 4). In distant reading, we are not dealing with texts in isolation but with an altogether new subject of inquiry. The centrality of visualization in distant reading should not be confused with an oversimplification of what is inherently complex.

Maps or other forms of visualization are not results in themselves: "it shows us that there is something that needs to be explained" (Moretti 2005, p. 39)—a sentence that echoes Latour's (2005) warning that the social is precisely what needs to be explained.

We therefore need to explore, analyze, and explain the collective imaginaries detected in literary works understood as a landscape rather than as individual texts, and the transits between such collective representations and social groups. For the first task, digital humanities provide the methodological tools; for the second, SRT is theoretically indispensable. This is urgently required if we are to make sense of what Arthur C. Clarke recalls about *2001: A Space Odyssey*:

"*2001* was written in an age which now lies beyond one of the great divides in human history; we are sundered from it forever by the moment when Neil Armstrong and Buzz Aldrin stepped out on to the Sea of Tranquility. Now history and fiction have become inexorably intertwined; the Apollo astronauts had already seen the film when they left for the Moon. The crew of Apollo 8, who at Christmas 1968 became the first men ever to set eyes upon the lunar Farside, told me that they had been tempted to radio back the discovery of a large black monolith. Alas, discretion prevailed.

The Apollo 13 mission, however, does have an uncanny connection with *2001*. When the computer HAL reported the 'failure' of the AE 35 Unit, the phrase he used was, 'Sorry to interrupt the festivities, but we have a problem.'

Well—the Apollo 13 Command Module was named *Odyssey*, and the crew had just concluded a TV broadcast with the movie's famous *Zarathustra* theme when an oxygen tank exploded. Their first words back to Earth were, 'Houston, we've had a problem.'" (Clarke 1968, pp. xvi–xvii)

[11] Cf. Wang (2022).

In this study, we aim to explore the production and appropriation of fictional accounts of AI by different groups (writers, aficionados, scholars, and ordinary readers) from a social representations perspective using digital humanities methods.

Methods

Sources of Data and Corpora

Four different databases are used in this research. They are associated with different natural groups and vary in terms of size and quality.

The Internet Speculative Fiction Database (ISFDB) (Al von Ruff and the ISFDB team 2025) is a community-built online catalogue of science fiction, horror, and fantasy works operating since 1995. As of May 16, 2025, the ISFDB includes 28,925 authors and 888,541 publications, of which 197,177 are verified (22.19%). It links together various types of bibliographic data: author bibliographies, publication bibliographies, award listings, magazine content listings, anthology and collection content listings, and forthcoming books. The members are what, in this study, we call *aficionados*. The extensive list of titles also offers an opportunity to consider another natural group: the writers themselves.

The ISFDB corpus consists of metadata for 656 science fiction works tagged with any label containing the phrase "artificial intelligence," published between 1909 and 2025. The metadata includes the year of publication, title, type, language, author name, and tags. Table 1 summarizes this information. English accounts for 82% of the titles, German for 15%; other titles are distributed across eight additional languages (all European except for Chinese and Japanese). Most works are either novels (54%) or short fiction (40%).

Because the list is community-built and crowdsourced, its accuracy in terms of literary scholarship is not guaranteed. We noticed, for example, that works such as *Mockingbird* by Walter Tevis or *Do Androids Dream of Electric Sheep?* by Philip K. Dick are not included. From an SRT perspective, this is not problematic. Tags are considered as expressions of the appropriation of fictional accounts. In these cases, it seems that other categories, such as *Android* or *Robot*, superseded AI in the *aficionados'* view. Further comparative studies would be needed to fully understand the significance of such absences.

Utopian Literature in English and Fiction Bridges Science 21

The *Utopian Literature in English: An Annotated Bibliography from 1516 to the Present* (UTE) (Sargent 2020) is an online database published by Penn State Libraries Open Publishing and curated by Lyman Tower Sargent, an eminent scholar in utopian studies. It aims to be a comprehensive annotated bibliography from Thomas More's *Utopia* (1516) to the present. The UTE lists 38 utopian works in which artificial intelligence is referenced in the annotation.

The *Fiction Bridges Science 21* (FBS21) (CETAPS 2025) is a project maintained by the Centre for English, Translation, and Anglo-Portuguese Studies (CETAPS) that identifies and analyzes fictional works since the year 2000 dealing with literature and science in English. The FBS21 lists 11 works tagged with artificial intelligence.

Table 1: Description of the ISFDB corpus.

Metadata	**Description**	**Initial corpus (N = 656)**	**English corpus (N = 535)**
Year	Refers to the year of 1st edition	1909–2025	1909–2025
Type	Refers to the type of literary work	Anthology - 6 Book chapter - 1 Collection - 8 Essay - 9 Interiorart - 2 Interview - 2 Nonfiction - 3 Novel - 355 Omnibus - 2 Poem - 3 Short-fiction - 265	Anthology - 5 Book chapter - 1 Collection - 8 Essay - 3 Interiorart - NA Interview - 1 Nonfiction - 1 Novel - 325 Omnibus - 2 Poem - 1 Short-fiction - 188
Language	Refers to the language of the work	English - 535 French - 3 German - 100 Croatian - 3 Serbian - 1 Japanese – 4 Dutch - 1 Chinese - 2 Danish - 1 Polish - 5	Not applicable
Authors	Authors of the work	454 single or collective	357 single or collective
Tags	Tags assigned by the community members	1138 unique out of 6214 total	1080 unique out of 5711 total

For the purpose of the current study, we combined the *Lyman Tower Sargent Bibliography* and the *Fiction Bridges Science 21*. These two databases are produced and curated by scholars. Together they comprise 49 titles. Because of the small number of titles, we focused on the story summaries provided by the curators. The summaries were edited when necessary by the researcher. They total 4,451 words, with 1,093 unique words (after lemmatization).

Launched in 2007, Goodreads claims to be the largest platform for book recommendations (Goodreads 2025). From Goodreads, we retrieved 796 reader reviews, distributed across 30 books ($M = 25.53$; $SD = 13.01$) out of the 48 listed in the previously described corpus. Despite the high number of reviews, only around 10% ($n = 72$) explicitly include the phrase "artificial intelligence."

Procedures

The researcher accessed the databases and exported the data to a comma-separated values (.csv) format, a non-proprietary format that complies with FAIR principles. The files were imported into RStudio (https://posit.co/download/rstudio-desktop/), an environment for the R programming language (https://www.r-project.org/), widely used for natural language processing (NLP).

We used several packages to parse, process, analyze, and visualize the data (token frequency, network analysis, correspondence analysis). Lemmatization of the English tokens was conducted with the package *lexicon* (Rinker 2025). We computed quartiles to explore the association of themes with time, as it seemed to us that other choices (such as decades) would be distorted by the uneven distribution of items across years. To scrape reviews and metadata from the Goodreads platform, we used the *Goodreader* R package (Chao Liu 2024).

In line with open science and FAIR principles, the annotated script, dataset, and an interactive webpage are available from the CETAPS repository (https://cetapsrepository.letras.up.pt/id/cetaps/132574). The annotated script provides further details about the packages and techniques used.

Results

Writers and aficionados—The ISFDB Community Crowdsourced Database

In recent years, the popularity of artificial intelligence has grown considerably, as indicated by Google Ngrams.[12] Although publications listed in the ISFDB have also increased exponentially, the slope of this growth is softer.[13] Between 2014 and 2015, the number of titles mentioning artificial intelligence more than tripled (from 10 to 33), reaching a maximum of 64 in 2018.

After lemmatization, removal of stopwords,[14] and consolidation of all phrases related to AI, the titles listed in the ISFDB comprise 843 unique tokens (words) out of a total of 1,188. These titles are relatively short ($M = 2.22$, $SD = 1.29$).

Figure 1 displays the words with three or more occurrences in the titles. Together, these words represent 17.51% of the corpus. *I* and *artificial intelligence* are tied for first place with 12 mentions each, followed by *robot* (11 mentions) and *machine* (10 mentions). A closer inspection of Figure 1 suggests that the tokens are not neutral: many words (e.g., *dead, dark*) convey strong positive or negative connotations. Therefore, we conducted a sentiment analysis.[15] As shown in Figure 2,[16] when considering the tokens with four or more occurrences in the corpus, negative words clearly outnumber the positive ones.

---

[12] See the occurrences of "artificial intelligence" by year on Google Ngrams at https://books.google.com/ngrams/graph?content=%22artificial+intelligence%22&year_start=1800&year_end=2022&corpus=en&smoothing=0&case_insensitive=false.

[13] See titles by year of first publication on the Internet Speculative Fiction Database at https://www.isfdb.org/cgi-bin/stats.cgi?5.

[14] The quanteda package (Benoit et al. 2018) has a built-in list of English stopwords, to which we added a few more when needed to better interpret results. For example, we removed "artificial intelligence" and "science fiction" from the tags list of ISFDB because they were present in all or a large quantity of items.

[15] We observe that the dictionary we used, Bing (first published in 2004 by Hu and Liu), integrated in the tidytext package (De Queiroz et al. 2025) could be improved because clearly some words could be classified as negative, which may be because of context from which the lexicon was obtained.

[16] We should note that the size of each word is proportional to its frequency within its sentiment, being only comparable to other words of the same sentiment. Words' size is not comparable across sentiments. See https://www.tidytextmining.com/sentiment.

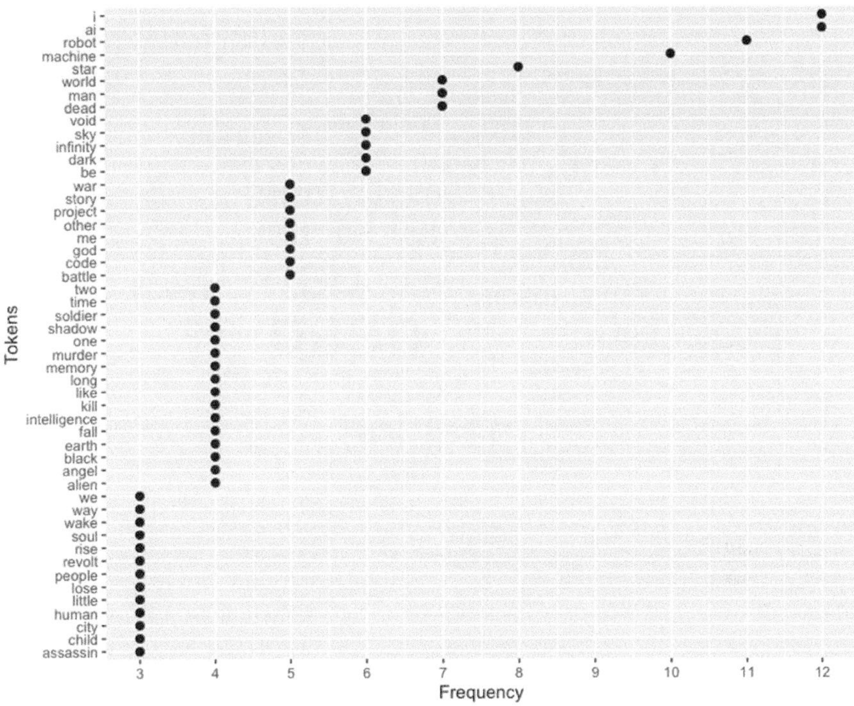

Figure 1: Frequency of tokens in english titles.

As noted earlier, tags in the ISFDB are added by any user without curation, making the process more similar to Wikipedia than to IMDb, for example. The first consequence of this open participatory process is the multiplication of tags with identical meanings. Some cases suggest obvious duplication (e.g., *female characters*). We can roughly compare this tagging process to free association. In our corpus, after removing *artificial intelligence* and *science fiction*, the most common tags are *near future*, *space opera*, *robots*, and *nanotechnology*. *Aliens* and *hard science fiction* form a second group (Figure 3).

A network analysis of tag co-occurrence helps us to understand the variety of topics—the semantic field, one might say. In Figure 4, we see *near future* on the center-right and *far future* on the left. *Far future* connects with classic subgenres of *hard science fiction* and *space opera* and includes topics such as *nanotechnology, interstellar travel, robots, genetic engineering, aliens,* and *religion*. By contrast, *near future* relates to more immediate realities or threats: *nanotechnology, virtual reality, London, climate change, and surveillance.*

We can go one step further to understand the overall patterns among AI works in terms of how they have been tagged and their distribution across periods of time. The correspondence analysis in Figure 5 provides some interesting insights. The first observation is that recent years (2018–2025) are associated with the rise of gender issues, with *female characters* gaining importance in the narratives. The second is that the periods 2005–2015 and 2015–2018 are associated with *extraterrestrial characters*, *military science fiction*, and *near future*.

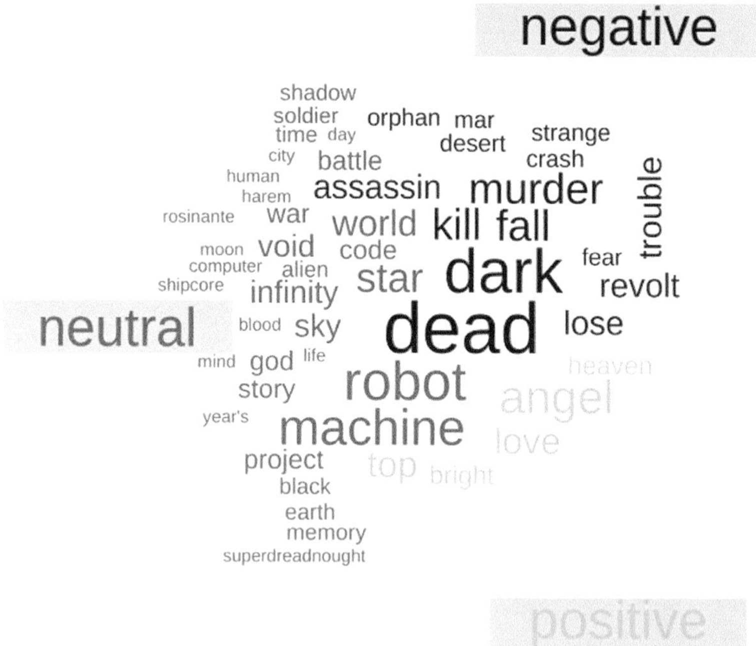

Figure 2: Wordcloud based on the sentiment analysis with 4 or more occurrences.

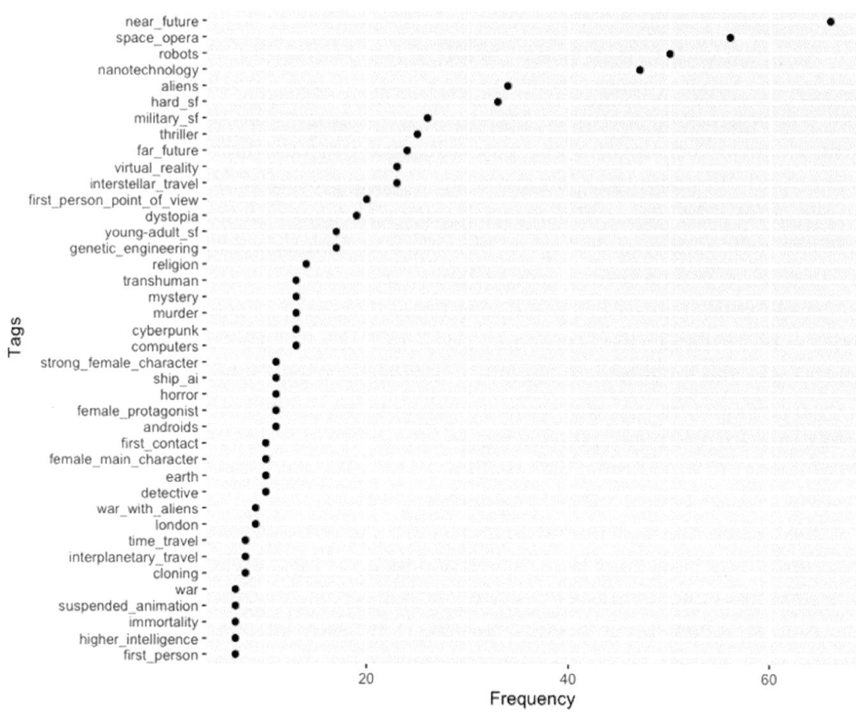

Figure 3: Tags with five or more occurrences in the corpus.

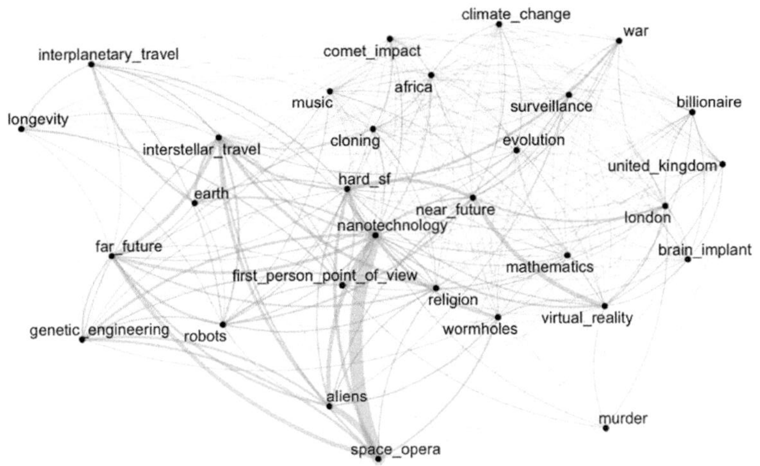

Figure 4: Similarity analysis of the tags with a minimum co-occurrence threshold of 1.

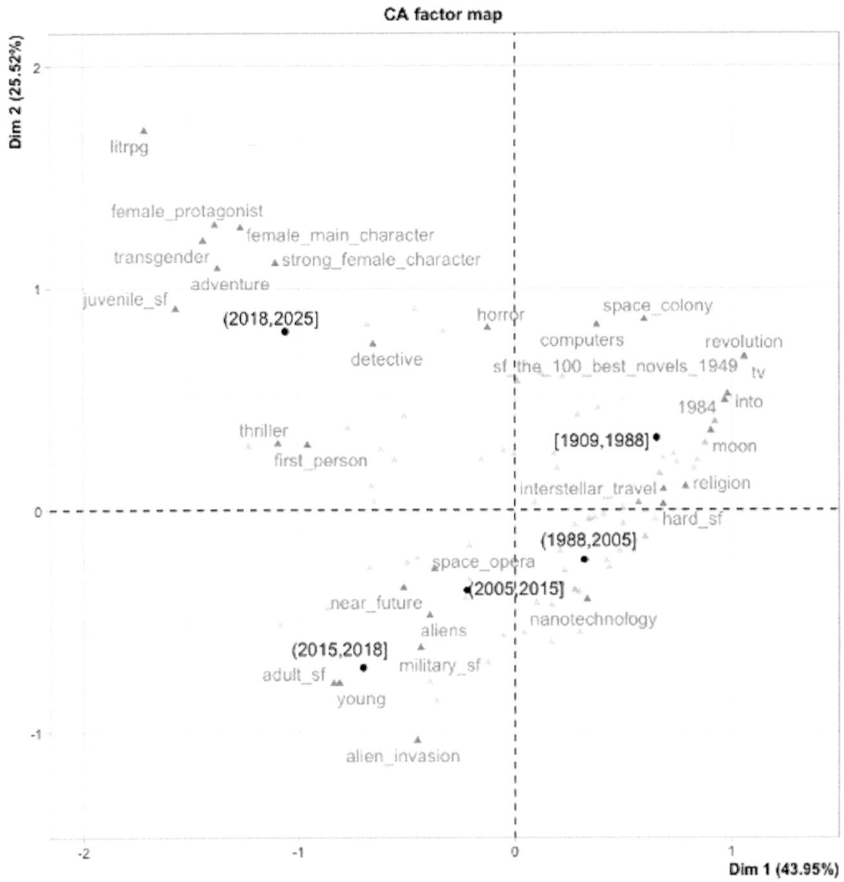

Figure 5: Correspondence analysis of tags with four or more occurrences.

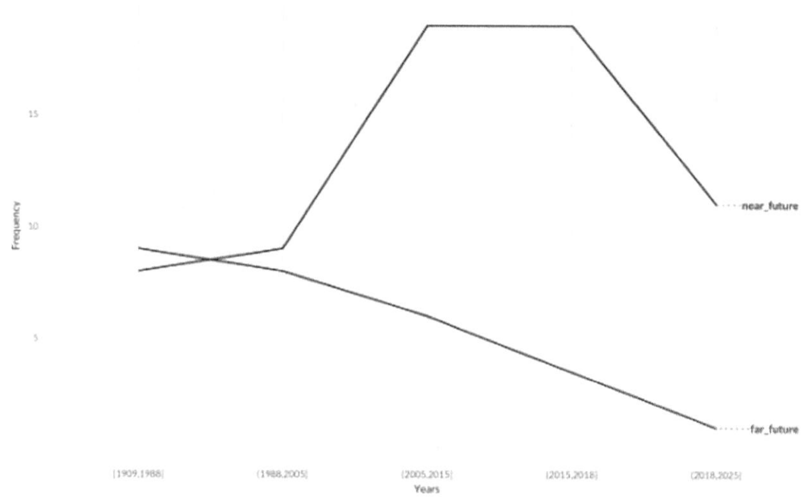

Figure 6: Distribution of near future and far future over time.

Figure 6 illustrates the contrast between *near future* and *far future*. It shows that *near future* perspectives have grown exponentially during the 20th century, while *far future* has decreased over the last decades. Figure 7 corroborates the correspondence analysis and indicates that recent years have indeed been characterized by emerging gender issues.

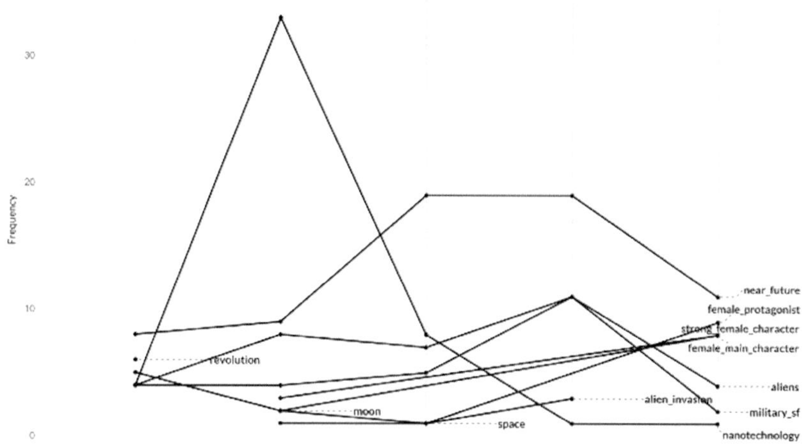

Figure 7: Science fiction trends over the years.

The scholars—The scholarly curated UTE and FBS21 databases

We now turn from the *aficionados* to the scholars. The sentiment analysis of the summaries of fictional accounts presents a rather negative picture, as there are many more negative than positive words with more than two occurrences (Figure 8). In Figure 9, within the neutral dimension, we observe words such as *control*, *dystopia*, and *war*.

When we narrow the context to words surrounding *artificial intelligence*, using collocates with three or more occurrences, the representation of AI becomes clearer (Figure 10). *Control* is mentioned seven times.

Table 2 organizes the lexicon into three topics. The first is characterized by tensions and struggles in the use of technology for development (note the presence of *dystopia*); the second by references to space and extra-terrestrial futures; the third by *control* and *war*, as if these were inherent consequences of any utopia.

The readers—The crowdsourced and Commented Goodreads platform

We can now move to the informed appropriation of science fiction and utopian literature concerning AI by readers. From Goodreads we retrieved 796 reader reviews, distributed across 30 books ($M = 25.53$; $SD = 13.01$) out of the 48 listed in

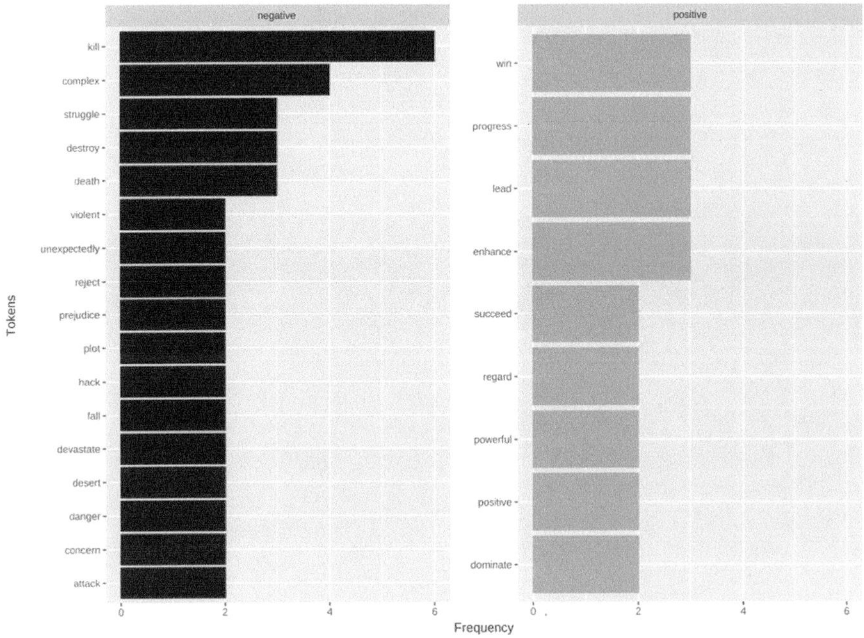

Figure 8: Sentiment analysis of the summaries of UTE and FBS21 databases.

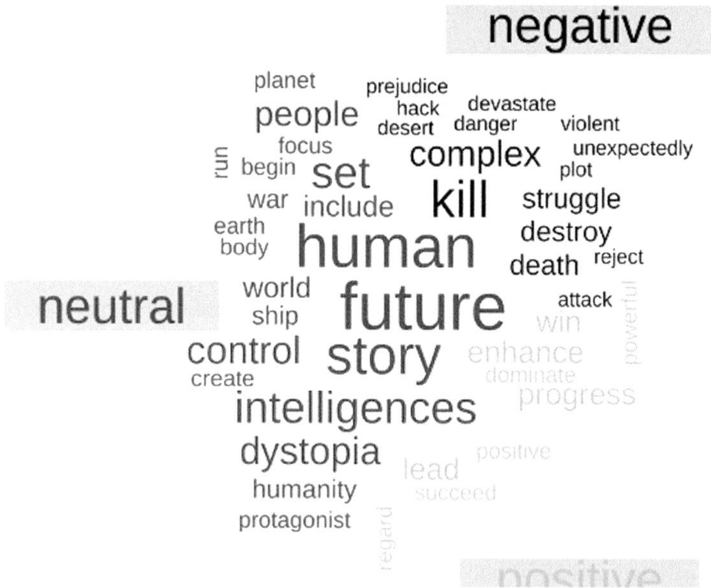

Figure 9: Wordcloud based on the sentiment analysis.

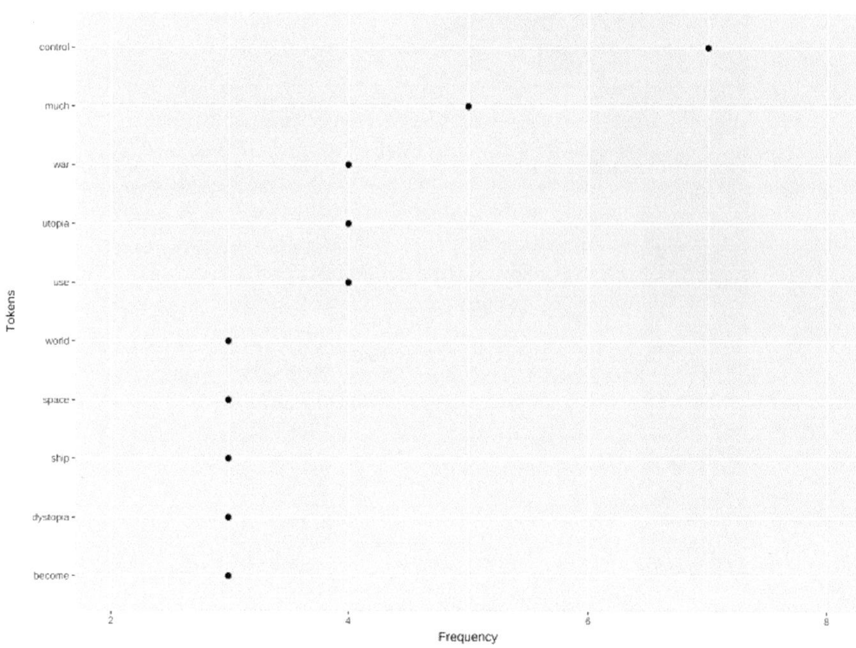

Figure 10: Collocates around artificial intelligence (five words before, five words after).

Table 2: Topic model (10 first words).

Topic 1	Topic 2	Topic 3
use	space	control
dystopia	ship	much
world	future	utopia
development	protagonist	war
create	run	become
struggle	dream	society
among	coalesce	bring
call	around	human
corporation	by-the-rules	dream
design	misuse	one

the previously described corpus.[17] Despite the high number of reviews, only around 10% (n = 72) explicitly include the phrase "artificial intelligence."

The sentiment analysis of the words within a ±5 window around the phrase is shown in Figure 11. It indicates that readers struggle to appropriate and make sense

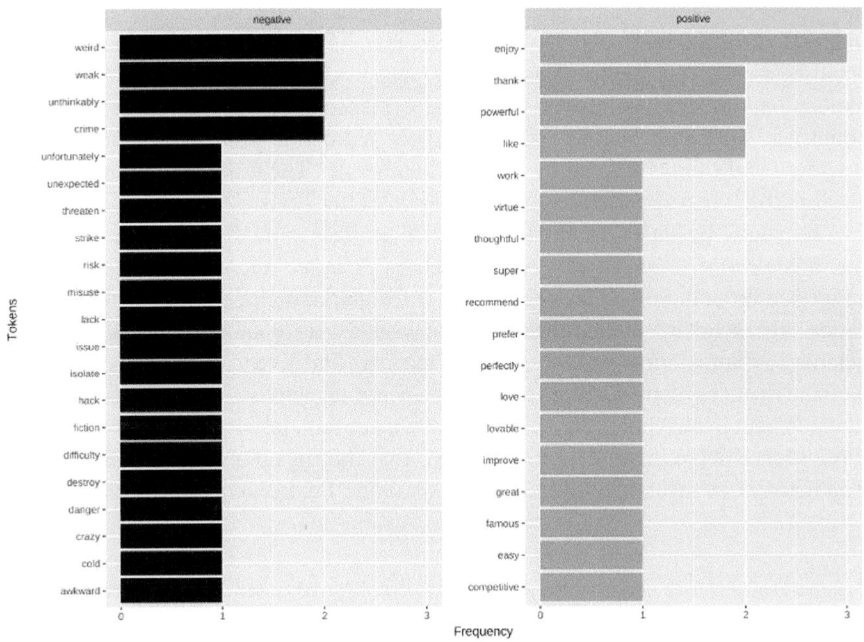

Figure 11: Sentiment analysis of the Goodreads' reviews.

[17] The list of books consists of 30 books: Neuromancer, A Closed and Common Orbit, The Municipalists. A Novel, Detonation. A Novel, The Corporation Wars: Dissidence, The Destructives, Project Empathy, Cash Crash Jubilee. Book One of the Jubilee Cycle, 2031: The Singularity Pogrom, New Model Army, Mind Over Ship, Zero Calvin, Transcension, Ice, Halo, Queen of Angels, Rynosseros, Dreams of Flesh & Sand, The Cybernetic Samurai, The Defenders, Machinehood, Rose/House, The Mountain in the Sea, Translation State, Klara and the Sun, Eversion, Blindsight, The Long Way to a Small Angry Planet, Some Desperate Glory, Perdido Street Station.

of AI (*weird, unthinkable, unexpected*). The positive side appears to be related to readers' evaluation of the works, which are generally considered favorable—in line, it must be said, with the overall appreciation of the books ($M = 3.70$, $SD = 1.13$).

Discussion

Our corpora deliver a Western, Anglo-American worldview. What remains to be clarified is whether this reflects the general assumption that utopian and science fiction literature is primarily a Western tradition, or whether the web accessible to us as researchers is itself Western and English-dominated. From a social representations perspective, this is the first relevant insight to be drawn.

We tend to think of a global world, but more often than not that global world is merely a projection of a Western perspective, against which many nations and movements now rise.[18] The lack of representation of other languages must be addressed: is English truly a lingua franca? This point also applies to artificial intelligence itself. AI is being construed and framed both by Western technology (with a few exceptions from other major economic powers) and by Western utopian thought. For example, Brown et al. (2023) showed how Indigenous protocols of interaction with AI can help design AI systems for the greater good. With this in mind, it becomes easier to understand how AI production reproduces current socio-economic imbalances (Casilli 2019) and inequalities (Walker 2023). Dataset annotation, material extraction, and waste disposal often take place in poorer countries, while the benefits seem to apply mainly to "enhanced humans" (cf. *A Closed and Common Orbit* by Becky Chambers).[19] The taskification described by Casilli (2019) had already been anticipated in Walter Tevis' *Mockingbird*.[20]

It would be naïve to think that titles express only the author's intent. In the *Bicentennial Man* short story collection, Asimov refers to changes in titles suggested by editors—often, one may add, for marketing or promotional purposes. Titles, tags, scholarly summaries, and reviews allow us to navigate through different actants (authors, *aficionados*, scholars, readers, and even algorithms). All these elements can be considered as forms of anchoring and/or objectification. A new development in an AI narrative may generate a new tag, just as *android* or *robot* once did, and such objectifications will later serve as anchoring points for other innovations. The network of meanings resembles the symbolic field described by Doise (2003),

[18] In the Digital Humanities community, the lack of diversity is taken seriously, which does not mean it has been solved (McPherson 2012), and gave rise to different manifestos from minority groups (Ventura et al. 2023; Sutherland 2023).

[19] A Closed and Common Orbit is the second of the Wayfarers four-book series by Beckie Chamber. It tells the story of Pepper, a genetically modified human being, and Lovelace, a sentient artificial intelligence. In the context of the book, enhanced humans are privileged beings. Pepper, on the contrary, had to escape from a life of servitude.

[20] Mockingbird is a novel by Water Tevis that describes a threaten, decaying world run by machines. In this novel, human beings had but simple, atomic tasks to perform and lost any ability to interact with the technological world around them. The future of humanity depends on the actions of an advanced robot, whose only desire is to cease exist, but cannot harm him/itself. See Moreira and Benison (in press) for a computer assisted analysis of the book.

akin to the symbolic "bone elements" with which social groups construct their viewpoints.

Our results show that AI representations differ across groups.[21] For authors, *robots* and *machines* are the most common objectifications of AI, serving as ways to grapple with dilemmas of personhood or agency. These are anchored in tropes of war, battle, and crime, recalling the enduring importance of the rogue robot (Recchia 2020). Time also serves as a significant anchor. While in the past AI projected both into the *near future* and the *far future*, since 2005 it has been clearly anchored to the near future. A possible explanation is the emergence of Web 2.0. These results suggest that authors and *aficionados* are acutely aware of AI's growing momentum, in contrast to Arthur C. Clarke's treatment of space exploration in the *Odyssey* series, where the temporal horizon is continually extended (2001, 2010, 2061, 3001).

Scholars, on the other hand, tend to overlook the material manifestations of AI (robot, machine) and instead objectify the challenges of *control*, which synthesizes the tensions accompanying AI's implementation. Ordinary readers appear puzzled by fictional accounts of AI: skepticism is evident in words such as *unthinkable*. Might this skepticism among readers help explain why AI applications are nonetheless being accepted so rapidly?

The online databases enabled us to approach natural groups, rather than mere collective aggregates (Harré and Moghaddam 2016), and to observe the outcomes of their self-motivated activities. That our results reveal a predominantly negative view of AI across groups is notable, especially given that news coverage of AI in the *New York Times* has been proportionally more positive than negative (Fast and Horvitz 2017), as has Twitter discourse (Manikonda and Kambhampati 2017). Interestingly, Manikonda and Kambhampati's study also shows that although overall positive, AI experts produce more negative tweets than non-experts.

Asimov himself was confronted with gendered assumptions: a female friend once remarked that his robots were always male-shaped and suggested he write a story featuring a female robot. The fact that he titled the resulting short story *Intuition* is revealing of the stereotypes at play. Alongside majority thinking, we also encounter minority thinking. In recent years, concerns—both in literary works and among readers—have shifted toward gender issues. The study of minority voices would be one way to revitalize a never fully accomplished goal of SRT as proposed by Moscovici: the exploration of hegemonic, consensual, and polemical representations, and the active role of social groups in shaping them (Moscovici 1988).

The future feels closer than ever. Writers and readers alike are bringing the temporal horizon of their plots nearer to the present day and linking them to immediate concerns, creating a polyphony that challenges the grand narratives of the past (Vieira 2024).

[21] In this volume, Lauri et al. explore how political dimensions of AI representations are formed by ideological positions.

Conclusion

In this chapter, we explored the production and appropriation of fictional accounts of AI by different natural groups (writers, *aficionados*, scholars, and ordinary readers) from a social representations perspective using digital humanities methods. Our data were not artificially generated in the context of the study but drawn from online databases. Overall, this exploratory work shows that the combination of digital humanities methods and social representations theory is productive for analyzing the production and appropriation of fictional accounts of AI.

The semantic or symbolic field of AI is negatively connoted—of a dark tone, one might say—across the different groups, with the theme of *control* especially prominent among scholars. AI applications, however, are being readily accepted at a rapid pace. What explains this divide? One possible explanation is that the voices represented here are minorities—yet influential and acutely aware of the problems of their societies. Speculative fiction and utopian literature, in particular, do not merely propose alternative scenarios; they also convey critiques of society. Against the uncritical acceptance of technological innovation by many segments of society, such literary works help expand and structure the semantic field by emphasizing risks and dangers. Polemical as they are, these works contribute to our understanding of AI, while at the same time being pushed into the future and relegated to the realm of fiction—until they knock at the door.

This study is exploratory and has several limitations. The sentiment lexicon (Hu and Liu 2004) was developed in the specific context of customer reviews and may sometimes classify words in unexpected ways. For example, it may be difficult to accept that words like *war* are considered neutral. Another limitation is that the technique adopted here is not context-sensitive. Words such as *control* depend heavily on their context, and negations that reverse sentiment (e.g., *not happy* being equivalent to *happy*) are not taken into account, although alternatives for R exist that could be considered in future studies. As Contarello (2008) observed, we need to adopt quali-quantitative approaches to literary texts. Social representations theory may help bridge the apparent detachment of texts from their conditions of production.

Another limitation concerns the package we used to access reviews from Goodreads, which did not succeed in scraping all reviews. However, this shortcoming turned out to be useful, since Goodreads' public API is not available. In future projects, we will attempt to overcome this difficulty.

We are aware that original, pioneering artistic works cannot be reduced to a social representations analysis. At the same time, even the most original perspectives are constructed in specific circumstances (historical, sociological, linguistic, and psychosocial), which they reflect to varying degrees. Future research could explore the production and appropriation of fictional accounts of AI at different levels of analysis (Doise 1980; Doise and Valentim 2015).

Second, fictional accounts circulate in multiple media. Beyond text, books include paratextual elements such as typography and illustrations—iconic representations (e.g., covers)—which may be as informative as the text itself. As two recent issues of *Papers on Social Representations* show, there is a significant opportunity to explore the imagistic dimensions of representations (Nicholson and Awad 2023, 2024).

Third, we cannot ignore intermediality: the transformations that occur when a work is adapted into another medium—for example, when a book is turned into a film, or a film is advertised on a poster. Such processes also need to be studied.

Fourth, we should further investigate appropriation by analyzing online audience reviews.

Finally, we should consider AI-generated texts and images about AI—as is already the case in this publication—and how AI itself becomes an actant, alongside other algorithms, contributing to our common knowledge (Moreira et al. 2021). In particular, this approach calls for longitudinal and comparative studies.

While there is already a line of research on collective memory and social representations (Bar-Tal 2014; Haas et al. 2025; Ryan and Hewer 2016), little attention has been paid to the future or to science and technology.

In Asimov's *Bicentennial Man*, shortly after Andrew the robot decides to wear clothes, he expresses the desire to write a history of robots. The human character with whom he is speaking replies that there are already many books on the history of robotics, failing to grasp how different Andrew's intention was. "A history of robots," Andrew explains. This suggests a future step: gaining knowledge about AI from AI itself, an approach that is already somewhat present in this volume in the chapters by De Rosa and also by Sarrica and Rizzoli. Our conviction is that the black box will rely heavily on patterns in datasets invisible to our eyes (see Gaskell et al. 2026). It will take them blindly, as text ceases to be text and probabilistic operations remain concealed from us. By contrast, knowledge obtained rationally, through explicit and replicable metrics, will provide a distant but intelligible understanding of our own common sense—that common sense that exists in online data and that also feeds AI. "The truth, as always, will be far stranger" (Clarke 1968/2018 p. viii). Without challenging disciplinary silos to address complexity, it will not only be stranger but also impenetrable and impersonal.

Acknowledgment

Financed by national funds through FCT - Fundação para a Ciência e a Tecnologia, I.P., under the project ref. UID/4097/2025 CETAPS" by "UID/04097/2025. The author is grateful to Liam Benison for his comments on the very first draft of this chapter, to the reviewers and editor for their insightful contributions that helped to improve the overall quality and readability of this chapter.

References

Al von Ruff and the ISFDB team. 2025. Internet Speculative Fiction Database. https://isfdb.org/

Bar-Tal, D. 2014. Collective Memory as Social Representations. Papers on Social Representations, 23(1): 5.1–5.26.

Benoit, K., K. Watanabe, H. Wang, P. Nulty, A. Obeng, S. Müller. et al. 2018. Quanteda: An R package for the quantitative analysis of textual data. Journal of Open Source Software, 3(30): 774. https://doi.org/10.21105/joss.00774

Blaim, A. 2013. Gazing in Useless Wonder. Peter Lang.

Brondi, S., M. Pivetti, S. Di Battista and M. Sarrica. 2021. What do we expect from robots? Social representations, attitudes and evaluations of robots in daily life. Technology in Society, 66: 101663. https://doi.org/10.1016/j.techsoc.2021.101663

Brown, M. L., H. Whaanga and J. E. Lewis. 2023. Relation-Oriented AI: Why Indigenous Protocols Matter for the Digital Humanities. pp. 74–83. *In:* Klein, L. F., and M. K. Gold (Eds.). Debates in the Digital Humanities 2023. University of Minnesota Press. https://muse.jhu.edu/pub/23/edited_volume/chapter/3440912

Cardoso, B., L. Moreira, A. Lobo and S. Ferreira. 2023. Spotlighting distraction in artificial intelligence driver assistance systems. In: Ahram, T., W. Karwowski, P. Di Bucchianico, R. Taiar, L. Casarotto, and P. Costa (Eds.). Intelligent Human Systems Integration (IHSI 2023): Integrating People and Intelligent Systems (Vol. 69). AHFE International. https://doi.org/10.54941/ahfe1002852

Casilli, A. A. 2019. En attendant les robots—Enquête sur le travail du clic. Média Diffusion.

Cave, S. and K. Dihal. 2019. Hopes and fears for intelligent machines in fiction and reality. Nature Machine Intelligence, 1(2): 74–78. https://doi.org/10.1038/s42256-019-0020-9

Cave, S., K. S. M. Dihal and S. Dillon (Eds). 2020. AI narratives: A history of imaginative thinking about intelligent machines (First edition). Oxford University Press.

Cave, S., K. Dihal and S. Dillon. 2020. Imagining AI. *In:* Cave, S., K. Dihal and S. Dillon (Eds.). AI narratives: A history of imaginative thinking about intelligent machines (First edition, pp. 1–21). Oxford University Press.

Recchia, G. 2020. The Fall and Rise of AI: Investigating AI Narratives with Computational Methods. pp. 382–407. *In:* Cave, S., K. Dihal and S. Dillon (Eds.). AI narratives: A history of imaginative thinking about intelligent machines. Oxford University Press.

CETAPS. 2025. Fiction bridges Science 21 [Data set]. https://cetapsrepository.letras.up.pt/id/cetaps/114045

Cha, S. 2024. Towards an international regulatory framework for AI safety: Lessons from the IAEA's nuclear safety regulations. Humanities and Social Sciences Communications, 11(1): 1–13. https://doi.org/10.1057/s41599-024-03017-1

Chao L. 2024. Goodreader: Scrape and Analyze 'Goodreads' Book Data (p. 0.1.2) [Data set]. https://doi.org/10.32614/CRAN.package.Goodreader

Christidou, V., K. Dimopoulos and V. Koulaidis. 2004. Constructing social representations of science and technology: The role of metaphors in the press and the popular scientific magazines. Public Understanding of Science, 13(4): 347–362. https://doi.org/10.1177/0963662504044108

Clarke, A. C. 1968. 2001: A space odissey. Orbit.

Contarello, A. 2008. Social Psychology and Literature: Toward Possible Correspondence. pp. 303–325 *In:* Sugiman, T., K. J. Gergen, W. Wagner and Y. Yamada (Eds.). Meaning in Action: Constructions, Narratives, and Representations. Springer Japan. https://doi.org/10.1007/978-4-431-74680-5_18

Contarello, A., I. Marini, A. Nencini and G. Ricci. 2011. Rappresentazioni sociali dell'invecchiamento tra psicologia sociale e letteratura. Psicologia & Sociedade, 23(1): 171–180. https://doi.org/10.1590/S0102-71822011000100019

Dinsman, M. 2016. The Digital in the Humanities: An Interview with Franco Moretti. The Digital in the Humanities: An Interview with Franco Moretti. https://lareviewofbooks.org/article/the-digital-in-the-humanities-an-interview-with-franco-moretti

Doise, W. 1980. Levels of explanation in the European Journal of Social Psychology. European Journal of Social Psychology, 10(3): 213–231.

Doise, W. 2003. Attitudes et représentations sociales. pp. 240–258. *In:* Jodelet, D. (Ed.). Les Représentations Sociales. PUF.

Doise, W. and J. P. Valentim. 2015. Levels of analysis in social psychology. pp. 900–904. *In:* Wright, J. D. (Ed.). International Encyclopedia of the Social & Behavioral Sciences (Second Edition). Elsevier. https://doi.org/10.1016/B978-0-08-097086-8.24032-4

Drucker, J. 2011. Humanities approaches to graphical display. Digital Humanities Quarterly, 5(1): 1–21.

Drucker, J. 2017. Why Distant Reading Isn't. PMLA, 132.3: 628–635.

Fast, E. and E. Horvitz. 2017. Long-Term Trends in the Public Perception of Artificial Intelligence. Proceedings of the AAAI Conference on Artificial Intelligence, 31(1): Article 1. https://doi.org/10.1609/aaai.v31i1.10635

Goodreads. 2025. https://www.goodreads.com/

Haas, V., R. H. Eskinazi and D. Jodelet. 2025. Collective memory and social representations. Current Opinion in Psychology, 66: 102123. https://doi.org/10.1016/j.copsyc.2025.102123

Harré, R. 1984. Some Reflections on the Concept of 'Social Representation'. Social Research, 51: 927–938.

Harré, R. and F. Moghaddam. 2016. Positioning theory and social representations. pp. 224–233. *In:* Sammut, G., E. Andreouli, G. Gaskell and J. Valsiner (Eds.). The Cambridge handbook of social representations. Cambridge University Press. https://doi.org/10.1017/CBO9781107323650.019

Hu, M. and B. Liu. 2004. Mining and summarizing customer reviews. Proceedings of the Tenth ACM SIGKDD International Conference on Knowledge Discovery and Data Mining, 168–177. https://doi.org/10.1145/1014052.1014073

Julsrud, T. E. and M. Aasen. 2024. Robots taking over the world... fantastic! Understanding social representations, familiarity and visions of experiments with autonomous public transportation. Energy Research & Social Science, 115: 103646. https://doi.org/10.1016/j.erss.2024.103646

Kirschenbaum, M. 2012. What is digital humanities and what's it doing in english departments? pp. 3–11. *In:* Gold, M. K. (Ed.). Debates in the Digital Humanities (NED-New edition). University of Minnesota Press; JSTOR. http://www.jstor.org/stable/10.5749/j.ctttv8hq.4

Kirschenbaum, M. and R. Raley. 2024. AI and the University as a Service. PMLA, 139(3): 504–515. https://doi.org/10.1632/S003081292400052X

Latour, B. 1989. Joliot: L'histoire et la physique mêlées. pp. 493–513. *In:* Serres, M. (Ed.). Elements l'histoire des Sciences. Bordas.

Latour, B. 2005. Reassembling the Social – An Introduction to Actor-Network-Theory. Oxford University Press.

Levitas, R. 1990. The concept of utopia (Peter Lang, Vol. 3).

Lewis, S. C., R. Zamith and A. Hermida. 2013. Content Analysis in an Era of Big Data: A Hybrid Approach to Computational and Manual Methods. Journal of Broadcasting & Electronic Media, 57(1): 34–52. https://doi.org/10.1080/08838151.2012.761702

Liu, A. 2013. The Meaning of the Digital Humanities. PMLA/Publications of the Modern Language Association of America, 128(2): 409–423. https://doi.org/10.1632/pmla.2013.128.2.409

Manikonda, L. and S. Kambhampati. 2017. Tweeting AI: Perceptions of Lay vs Expert Twitterati (No. arXiv:1709.09534). arXiv. https://doi.org/10.48550/arXiv.1709.09534

Marks, P., J. Wagner-Lawlor and F. Vieira. 2022. The Palgrave Handbook of Utopian and Dystopian Literatures (Palgrave Macmillan). https://doi.org/10.1007/978-3-030-88654-7

McCarty, W. 2003. Humanities computing. Encyclopedia of Library and Information Science, 2: 1224–1235.

Miguel, I., J. P. Valentim and F. Carugati. 2010. Intelligence and its development: Social representations and social identities. Papers on Social Representations, 19(2): Article 2.

Miguel, I., J. P. Valentim and F. Carugati. 2012. Degree of proximity in the construction of social representations: the case of intelligence. The Spanish Journal of Psychology, 15(3): 1244–1258. https://doi.org/10.5209/rev_SJOP.2012.v15.n3.39411

Miguel, I., J. P. Valentim and F. Carugati. 2013. Social representations of the development of intelligence, parental values and parenting styles: A theoretical model for analysis. European Journal of Psychology of Education, 28(4): 1163–1180. https://doi.org/10.1007/s10212-012-0160-3

Moreira, L., B. Cardoso, A. C. Mora, S. P. Duarte, S. Ferreira and A. Lobo. 2023. The common-sense knowledge about artificial intelligence across groups of drivers: A social representation approach. *In:* Kamissoko, D., F. Fontanili, E. Petitdemange, D. Gourc, X.

Lorca, F. Benaben, M. Lauras and C. Ye (Eds.). Proceedings of the ICDSST 2023 on—Decision Support System in an Uncertain World: The Contribution of Digital Twins. EWG-DSS.

Moreira, L., J. C. Paiva and C. Morais. 2021. The Social Representations of the Internet: A systematic review of literature towards a groundbreaking research agenda. Papers on Social Representations, 30(1): 1.1–1.34.

Moreira, L., H. Vieira, C. Morais, A. Sousa, H. Costa and J. C. Paiva. 2016. The social representations of (green) chemistry and robotics among teenagers: An exploratory study. pp. 5209–5223. *In:* Edulearn16: 8th International Conference on Education and New Learning Technologies.

Moretti, F. 2005. Graphs, maps, trees. Abstract models for a literary history. Verso.

Moretti, F. 2013. Distant reading. Verso.

Moscovici, S. 1976. La psychanalyse: Son image et son public. P.U.F.
Moscovici, S. 1986. The dreyfus affair, proust, and social psychology. Social Research, 53(1): 23–56.
Moscovici, S. 1988. Notes towards a description of Social Representations. European Journal of Social Psychology, 18(3): 211–250. https://doi.org/10.1002/ejsp.2420180303
Moscovici, S. 2000. Social Representations: Exploration in Social Psychology. Polity Press.
Mugny, G. and F. Carugati. 1989. Social representations of intelligence (pp. xi, 196). Cambridge University Press.
Negura, L. 2025. Simulated sense-making or social knowledge? Artificial intelligence and the boundaries of representation. Journal for the Theory of Social Behaviour, 55(3): e70012. https://doi.org/10.1111/jtsb.70012
Nicholson, C. and S. Awad. 2023. The power of the image: the role of social representations [Section 1/2] (Special issue). Papers on Social Representations, 32(2).
Nicholson, C. and Awad, S. 2024. The power of the image: the role of social representations [Section 2/2] (Special issue). Papers on Social Representations, 33(1).
Organisation for Economic Co-operation and Development. 2022. Measuring the environmental impacts of artificial intelligence compute and applications (No. 341). https://www.oecd.org/en/publications/measuring-the-environmental-impacts-of-artificial-intelligence-compute-and-applications_7babf571-en.html
Piçarra, N., J.-C. Giger, G. Pochwatko and G. Gonçalves. 2016. Making sense of social robots: A structural analysis of the layperson's social representation of robots. European Review of Applied Psychology, 66(6): 277–289. https://doi.org/10.1016/j.erap.2016.07.001
Recchia, G. 2024. The Fall and Rise of AI: Investigating AI Narratives with Computational Methods. pp. 382–407. *In:* Cave, S., K. Dihal and S. Dillon (Eds.). AI narratives: A history of imaginative thinking about intelligent machines. Oxford University Press.
Rinker, T. 2017. Lexicon: Lexicons for Text Analysis (p. 1.2.1) [Data set]. https://doi.org/10.32614/CRAN.package.lexicon.
Ryan, J. F. and C. J. Hewer. 2016. What did we do to Germany during the Second World War? A British perspective on the Allied strategic bombing campaign 1940–45. Papers on Social Representations, 25(1): 10.1–10.28.
Sargent, L. T. 1994. The Three Faces of Utopianism Revisited. Utopian Studies, 5(1): 1–37.
Sargent, L. T. 2020. Utopian Literature in English: An Annotated Bibliography From 1516 to the Present. doi:10.18113/P8WC77
Schnapp, J., T. Presner, P. Lunenfeld and J. Drucker. 2008. The Digital Humanities Manifesto 2.0. https://othone.org/Manifesto_V2.pdf
Skinner, B. F. 1948. Walden Two. Hackett Publishing.
Sutherland, T., M. Cifor, T. L. Cowan, J. Rault and P. Garcia. 2023. The feminist data manifest-NO: An introduction and four reflections. *In:* Klein, L. F. and M. K. Gold (Eds.). Debates in the Digital Humanities 2023. University of Minnesota Press. https://muse.jhu.edu/pub/23/edited_volume/chapter/3440912
Ventura, G. B., M. E. Cotera, L. G. Merchant, L. Gauthereau and C. Villarroel. 2023. A U.S. Latinx digital humanities manifesto. *In:* Klein, L. F. and M. K. Gold (Eds.). Debates in the Digital Humanities 2023. University of Minnesota Press. https://muse.jhu.edu/pub/23/edited_volume/chapter/3440912
Vieira, F. 2024. Complex Democracy, Complex Utopianism. Utopian Studies, 35(2): 318–339. https://doi.org/10.5325/utopianstudies.35.2-3.0318
Wagner, W., A. Viidalepp, N. Idoiaga-Mondragon, K. Talves, E. Lillemäe, J. Pekarev et al. 2023. Lay representations of artificial intelligence and autonomous military machines. Public Understanding of Science, 32(7): 926–943. https://doi.org/10.1177/09636625231167071
Walker, R., J. Dillard-Wright and F. Iradukunda. 2023. Algorithmic bias in artificial intelligence is a problem—And the root issue is power. Nursing Outlook, 71(5): 102023. https://doi.org/10.1016/j.outlook.2023.102023
Wang, N. 2022. The rise of a new paradigm of literary studies: The challenge of digital humanities. New Techno Humanities, 2(1): 28–33. https://doi.org/10.1016/j.techum.2022.11.001

AI in Society and Media

CHAPTER 2

Artificial Intelligence in the Public Eye

The Social Construction of AI as a Problem in the National English-Language Canadian Newspapers

Lilian Negura,[1,*] *Mauro Sarrica*[2] and *Valentina Rizzoli*[2]

Introduction

Artificial Intelligence (AI) has rapidly moved to the forefront of public discourse, especially after the late-2022 introduction of OpenAI's ChatGPT. This milestone transformed AI from a niche topic into mainstream news, prompting widespread discussion of both its revolutionary potential and its perceived threats. In Canada as elsewhere, headlines are suddenly filled with debates on AI—from its promise in driving innovation to fears about its societal impacts (Dandurand et al. 2023). Such intense media attention is not merely reporting technological advances; it is actively shaping how the public understands and engages with AI as a *public issue*. The very emergence of AI in everyday conversations owes much to media narratives that frame what AI means for society and whether it should be celebrated, feared, or regulated. In this context, examining media discourse is crucial for understanding the social construction of AI as a problem requiring collective attention.

[1] Faculty of Social Sciences, office 1203, 120 University Private, Ottawa, K1N 5N6. University of Ottawa.
[2] University Sapienza of Rome. Emails: mauro.sarrica@uniroma1.it; valentina.rizzoli@uniroma1.it
* Corresponding author: lilian.negura@uottawa.ca

DOI: 10.1201/9781003502753-5

Over past decades, research has documented the media's pivotal role in influencing public perceptions of new technologies (Brondi et al. 2021; Contarello et al. 2007; Moreira et al. 2021; Moreira et al. 2023). Studies in various countries (Sun et al. 2020) show that news coverage of AI often oscillates between utopian optimism and caution or alarm, reflecting an ambivalent societal "AI imaginary". For example, content analysis of newspapers has found AI portrayed simultaneously as a driver of progress and economic growth *and* as a source of ethical dilemmas and uncertainties (Hansen 2022; Cools et al. 2024). Moreover, media narratives have evolved over time: in the 2010s, coverage shifted from treating AI as a distant science-fiction concept to recognizing its tangible impacts on everyday life and spotlighting its potential risks (Nguyen and Hekman 2022). Longitudinal studies of news sentiment further indicate that positive portrayals of AI became more frequent in the 1980s–2010s while overtly negative tones remained relatively steady (Korneeva et al. 2023). Intriguingly, media interest in AI surged in the mid-2010s but then dipped around 2020, leading Korneeva et al. (2023) to suggest that AI had become "trivialized" as a news topic by that time. This was the state of media discourse just before the current wave of AI enthusiasm.

The launch of ChatGPT marked a turning point. It dramatically reignited global and Canadian media interest in AI, as a broad public suddenly witnessed AI encroaching on domains traditionally seen as uniquely human (creative writing, art, conversation). AI is no longer abstract future speculation—it is portrayed as a present reality with direct implications for jobs, education, creativity, and even governance. This renewed prominence of AI in the news raises pressing questions about how the media are now framing the technology and to what effect. Early evidence suggests that journalists have intensified both the hype and the warnings around AI, which could shape public sentiment in new ways (Spyridou and Ioannou 2025). Yet despite the flood of AI coverage, there remains a paucity of empirical research on how the media have constructed AI since this generative AI breakthrough. In particular, no in-depth studies have yet examined the emotional tone of AI coverage and the narratives that cast AI as a "public problem" in the post-2022 context.[1] As a result, a significant research gap persists regarding the media's role in framing AI as a societal issue today and the interplay of emotions within that discourse. Understanding these aspects is vital, because the way AI is depicted in news stories may influence whether the public sees AI as an urgent concern and how institutions respond. Are news outlets inspiring enthusiasm, fear, demand for oversight—or all of these at once? And how might such coverage mobilize different social actors to take action?

Representational Naturalization and the Construction of Public Problems

We drew on classical analyses of public problem construction (Blumer 1971, Spector and Kitsuse 2017) and recent studies on emerging technologies in the media (e.g., Moreira et al. 2023; Sarrica et al. 2020; Brondi et al. 2014) to interpret the role of

[1] One recent European study considered AI in political debates about regulation, but it focused narrowly on technical policy issues rather than media portrayals or public emotions (Lemke et al. 2024).

social representations in the case of AI. To explore the process of transforming AI discourse into a social phenomenon recognized and experienced in everyday life, we will use the representational naturalization model developed by Negura and Plante (2021). According to this model, the social representation of a new technology such as AI evolves into a public problem in four stages: emergence, problematization, communication and resolution. We call them "stages", but it should be noted that they can overlap, resume and evolve at the same time. These stages offer a dynamic and complex look at how a powerful new technological reality, such as AI, is recognized and experienced by the public as a problem for society that demands a collective response.

Emergence. At the outset, a new situation—in our case, the emergence of a new concept, AI—attracts public attention as a topic of discussion in the media and other spaces of the public arena.

Problematization. During the problematization phase, the public develops an ambivalent view of the new technology and begins to see it as a risk to individuals and society.

Communication and mobilization. This stage corresponds to what Blumer (1971) describes as the mobilization of action. As divergent representations consolidate into a recognizable set of social problems, the media and other actors begin to advocate the need for change, such as setting up a code of ethics, implementing regulations or initiating public consultations to address the risks of AI.

Resolution. Finally, the representation of the new technology as a reality requiring intervention is not only communicated, but is also lived in the sense that the identified means of intervention are put in place and affect the everyday use of these technologies. The representation of the new technology as a public problem is thus naturalized, becoming a regulated and institutionalized reality recognized and experienced by everyone. Public debate and mobilization lead to the formation of policies and practices aimed at effective risk management, resulting in a normalized and controlled presence of the new technology in society.

In our theoretical model, the construction of an object as a social reality is understood as a cyclical process, when a new event or technological advance can restart the naturalization process from the beginning or at the stage where it was halted. This iterative process reflects the fluctuating nature of social representations, when periods of normalization and public acceptance can be interrupted by moments of crisis or significant advances, leading to a renewed sense of urgency or demands for institutional or regulatory intervention. In this way, the naturalization of AI as a public problem would not be a linear trajectory but a recurring cycle, dependent on the social, technological or cultural contexts in which AI continues to evolve.

Following a shared vision within SRT, we acknowledge that social representations should be conceptualised as cognitive-emotional processes due to their polysemic and relational nature. Emotions not only contribute to the connotative facets of representations, but also activate both conservative and transformative processes. According to our approach, collective emotions, i.e., emotions shared and amplified at the social level, are key to understanding how new technological realities are problematized in society.

Aims and Expectations

The current study aims to explore the social representations of AI in national English-language Canadian newspapers, looking in particular at the variety of themes and emotional connotation used to construct this object as a public problem. Looking at the four stages envisaged by our model (i.e., emergence, problematization, communication and resolution), and considering naturalization as a recurring cycle, we anticipate that the problematization phase should be prevalent in studied newspapers, with emerging elements from the communication and resolution phases. Should this expectation be confirmed, the following results are to be expected:

1. AI is portrayed as a significant societal risk, emphasizing threats to jobs, privacy, and information quality.
2. Negative sentiments dominate AI representations, with a prevalence of fear and concern.
3. Fear, anger, and sadness should shape narratives, framing AI as a source of ethical, social, and existential crises.
4. Studied media should amplify fears through sensational coverage of high-profile incidents, generating collective anxiety.
5. Studied media contents should advocate for regulation, calling for ethical guidelines and legal frameworks to manage AI's risks.

Methods

Data Collection

The corpus of media articles analyzed in this study was compiled using EURIKA and FACTIVA databases, which provided access to digital archives of Canadian national newspapers. This study focuses on English-language national newspapers (*The Globe and Mail* and *National Post*) as discursive sites for the construction of public representations of AI. While a bilingual analysis could have provided additional insights, the decision to limit the scope to English-language media was driven not only by practical constraints—specifically, the difficulty of performing a lexicometric analysis across two linguistic corpora simultaneously—but also by theoretical considerations. Given the distinct socio-historical and cultural contexts that shape English- and French-language media in Canada, each represents a unique symbolic field. Analyzing them jointly without ensuring methodological symmetry risks conflating divergent representational logics and compromising the interpretive depth of the study.

The selection of *The Globe and Mail* and *National Post* as data sources for this study is justified by their central role in Canada's national English-language media landscape (Taylor and DeCillia 2021). First, these newspapers are recognized as Canada's primary national English-language broadsheets, widely read and cited as authoritative sources in public and political discourse. Second, they offer contrasting editorial perspectives—*The Globe and Mail* is generally centrist or center-left, whereas the *National Post* tends towards conservative viewpoints—thus providing a balanced ideological representation. Finally, both newspapers have substantial national reach and influence: *The Globe and Mail* engages a weekly audience of over six million readers across Canada,[2] while *National Post* maintains significant circulation and influence in several key provinces, notably Ontario, Quebec, Alberta, and British Columbia.[3] Together, these factors support the choice of these publications as representative channels for examining the construction of AI as a public problem in English-language Canadian media.

The time frame for the corpus spanned from November 1, 2022, to November 2024, beginning with the public release of ChatGPT that appears to have reignited media and public interest in generative AI, acting as a proxy for wider debates on AI as a whole.

Relevant keywords, including "artificial intelligence", "AI", "ChatGPT", and associated technical terms such as "machine learning" and "neural networks", were used to identify articles. Articles such as editorials, news reports, and interviews were included, while advertisements and irrelevant pieces were excluded. This process resulted in a corpus of 208 articles, with detailed metadata captured for each, including publication source, date, type of article, and geographic focus.

Data Analysis

First, quantitative analyses were conducted using a topic detection technique, complemented by sentiment and emotion analyses. Qualitative thematic analysis was later integrated to deepen the interpretation of findings.

Topic Detection (Reinert's Method)

We employed Reinert's method (Reinert 1983, 1990), implemented through the IRaMuTeQ software (Ratinaud 2009), an open-source platform based on R statistical software. This method classifies text segments by analyzing co-occurrences of words to uncover underlying topics. From the 208 articles analyzed, 5,128 text segments were extracted as meaningful units of discourse. An iterative clustering algorithm grouped these segments into clusters, maximizing vocabulary differentiation and capturing latent thematic dimensions. Each cluster, or "lexical world" (Reinert 1993), was defined by words and text segments significantly associated with it, identified through chi-square analysis. The results were synthesized into a dendrogram, providing a visual representation the hierarchical relationships among clusters.

[2] The Globe and Mail. (n.d.). About us. Retrieved July 23, 2025, from https://www.theglobeandmail.com/about-us/

[3] CBC News. (n.d.). National Post to eliminate Monday print edition. CBC. Retrieved July 23, 2025, from https://www.cbc.ca/news/business/national-post-monday-1.4167247

The interpretation of these topics, while grounded in statistical analysis, required qualitative reflexivity to reveal the latent meanings represented by the words and their contextual associations. These clusters, reflecting the macro-themes of the discourse, provide a structured understanding of the social representations under study.

Sentiment Analysis and Emotion Detection

Sentiment analysis and emotion detection differ significantly in their focus and application, particularly from a psychosocial perspective. Sentiment analysis usually involves identifying the polarity of text—whether it is positive, negative, or neutral—primarily for consumer or social media analysis (Hu and Liu 2012). Conversely, emotion detection delves deeper into understanding specific emotional states (e.g., anger, joy, sadness), considering the complexities of human affect and context (Ekman 1992; Plutchik 1980).

In the present study, sentiment and emotion detection were conducted using pre-trained language models accessed via the Hugging Face platform, implemented within the Google Colab environment to leverage its computational resources. For sentiment analysis, the distilbert-base multilingual model was applied, classifying each text segment identified with the Reinert method, into one of three categories: positive, neutral, or negative. Each segment was also assigned a probability score, reflecting the model's confidence in its classification. Emotion detection was performed using the Emotion English DistilRoBERTa-base model from Hugging Face, which identifies seven emotions: anger, disgust, fear, joy, neutral, sadness, and surprise.

The procedure for both sentiment and emotion detection involved iterative validation to ensure consistency across predictions. Each segment's classifications were subsequently aggregated and aligned with the thematic categories identified through topic detection. To examine the relationship between sentiment or emotion and topics, a chi-square test was performed. Furthermore, Z-scores were calculated to identify specific associations between individual sentiment or emotion categories and topics.

Thematic Analysis

Text segments with a confidence score of at least 70% (for both sentiment and emotion classification) were manually verified and analyzed thematically. This process validated the automatically identified emotions and sentiments, enriched the descriptions of the detected topics, and ensured alignment between emerging themes and co-occurring emotions. We also sought to identify and analyze the coverage of high-profile incidents.

To ensure the reliability of our thematic analysis, two independent coders analyzed a 10% sample of the data. Inter-coder reliability was assessed using Cohen's kappa, yielding a score of 0.837. This high level of agreement indicates strong consistency in the application of the coding schemes, ensuring the robustness of the analysis.

Results

Topic Detection

The analysis of 5,128 text segments using IRAMUTEQ revealed five primary topics, with 97.29% of the text segments classified (Figure 1). Each topic represents a distinct "lexical world" (Reinert 1993), reflecting key dimensions of the public discourse on AI. Topics 1, 2 and 3 together relate to everyday life/use of AI, concerning leisure, work/study and care. Topics 5 and 4 mainly concern the normative sphere.

Below, the topics are described in detail.

Topic 1: AI and Job Substitution

Topic 1 captures the substantial concerns surrounding AI's impact on the work sphere, representing approximately 22.3% of the classified text segments. The press discourse reflects preoccupation about automation, job displacement, and the ethical dilemmas inherent in replacing human labor with AI systems. Characteristic terms frequently associated with this topic include "human", "job", "displacement", "worker", and "replace". Moreover, words recalling emotions like "fear" are associated with this topic.

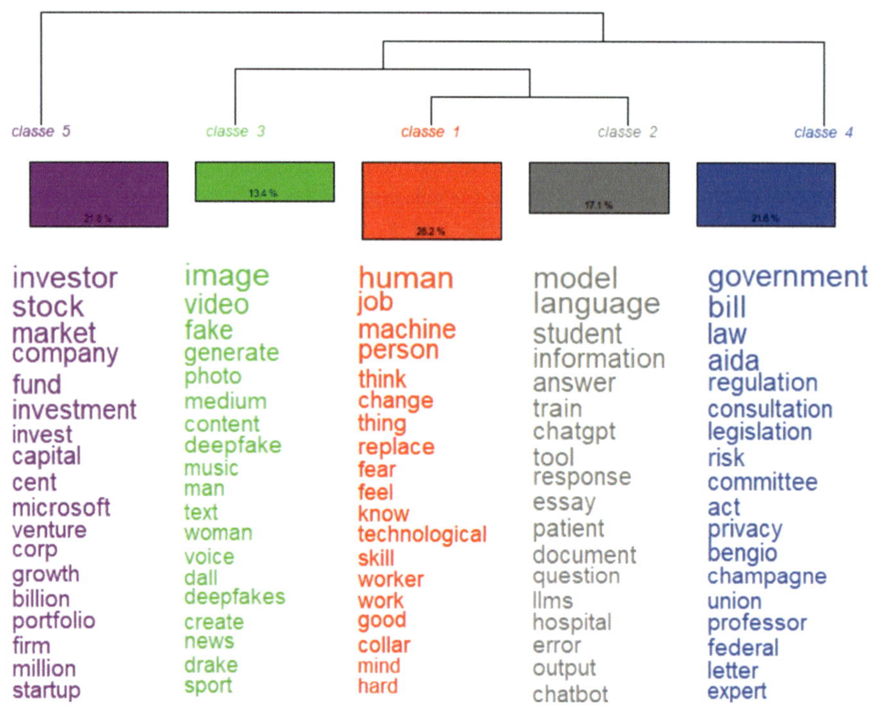

Figure 1: The five topics identified using the Reinert Method, represented by the words most strongly associated with each topic based on chi-square statistics. The words are listed in order of their index values.

The passages classified under this topic address several key themes. We present these themes accompanied by a representative example of quotes from newspaper articles:

Job Security Concerns:

"Further, almost 60 per cent of workers fear AI will eliminate their jobs, but less than a third of C-suite leaders say employees are worried about job displacement. The gap goes beyond perception."

(The Globe and Mail, 2024)

Ethical Implications:

"Sources and collectors together co-create data. All of this stickiness affects international human rights values: autonomy, dignity, equality, and community. Currently, data about people are traded as commodities, but data come from rights-bearing individuals."

(The Globe and Mail, 2024)

Economic Restructuring:

"Companies leading the AI transformation are setting an example for newbies by investing in skills mapping and data integration to make sure both their people's capabilities and their businesses grow in tandem."

(*The Globe and Mail*, 2024)

Topic 2: Chatbots and Large Language Models (LLMs) uses in education and healthcare

Topic 2 revolves around the applications of AI in education and healthcare, particularly focusing on the use of chatbots and large language models (LLMs). This theme represents 16.6% of the classified text segments. Characteristic terms frequently associated with this topic include "chatbot", "hospital", "student", "educator", and "language". The press discourse highlights the innovative potential of LLMs and the challenges associated with their integration into human activities. The key themes are as follows.

Educational Applications:

"By de-emphasizing essays and take-home exams and moving toward in-person tests and clear expectations for the use of artificial intelligence, students can be encouraged to maximize their learning experience with the help of AI."

(The Globe and Mail, 2023)

Healthcare Integration: "ChatGPT bedside advice more empathetic than MDs: study suggests role for AI in answering patient messages."

(National Post, 2023)

Technical Innovations and Limitations:

"Powering the chatbot is a large language model or LLM. Earlier versions of LLMs produced text passages that were rambling and borderline incoherent, but today's models are impressively fluent."

(The Globe and Mail, 2024)

Topic 3: Multimedia Generation and deepfake

Topic 3 addresses the role of AI in multimedia generation, with a specific focus on deepfake technology. This theme represents 13.1% of the classified text segments. Characteristic terms include "image", "video", "media", "deepfake", and "generate". Articles in this topic explore the ethical, legal, and creative challenges associated with AI-generated content, particularly in areas where human rights and intellectual property are at stake. The key themes are the following.

Deepfakes and Ethical Dilemmas:

"A scourge of pornographic deepfake images generated by artificial intelligence and sexualizing people without their consent has hit its most famous victim, singer Taylor Swift, drawing attention to a problem that tech platforms and anti-abuse groups have struggled to solve."

(The Globe and Mail, 2024)

Copyright and Ownership:

"Companies also have to protect themselves against producing copyright-infringing content and take additional cybersecurity measures. With LLMs now becoming part of the enterprise architecture of companies, the attack surface for hacks goes up as well, said Deloitte's Mr."

(The Globe and Mail, 2023)

Impact on Creative Industries:

" Thrust the technology into the mainstream last year with its ChatGPT bot, capable of quickly creating stories, memos, videos, and drawings upon request. As AI becomes a more integral piece of smartphones, the technology will likely have broad implications on productivity, creativity, and privacy."

(The Globe and Mail, 2024)

Topic 4: Legal Regulation of AI

Topic 4 focuses specifically on the regulatory challenges posed by AI, representing 20.9% of the classified text segments. Characteristic terms include "regulation" "law" "government" "risk", and "committee". Articles within this topic examine the gaps in existing legal frameworks, the evolving landscape of copyright law, and the ethical oversight necessary to govern AI technologies responsibly. The key themes included are the following.

Regulatory Frameworks:

"International organizations, governments, businesses, scientific and legal communities are racing to establish new regulations, laws, policies, ethical codes, and privacy requirements as AI continues to evolve at a rapid pace, while current legal and regulatory frameworks are lagging and becoming obsolete."

(The Globe and Mail, 2023)

Copyright and Fair Use:

"Laws are not yet sufficiently ready to address challenges of copyright and fair use," said a law student and the president of the JD/MBA Students Association at the University of Ottawa. "In recent years, artificial intelligence (AI) has been making waves in various industries, from health care to the arts."

(The Globe and Mail, 2023)

Ethical Oversight:

"The letter takes issue with the response the government gave to a House of Commons ethics committee report, which called for a moratorium on the unchecked use of facial recognition by police and for laws governing the technology."

(*National Post*, 2023)

Topic 5: Economic Impact and Market Applications of AI

Topic 5 highlights the transformative effects of AI on markets and investment strategies, comprising 19.1% of the classified text segments. Key terms include "economy", "market", "investment", "company", and "growth". Articles in this topic explore how AI drives economic transformation, reshapes global competition, and integrates into financial and energy infrastructures.

Market Growth and Investment:

"Chief executive officer Jensen Huang highlighted the soaring demand for accelerated computing and generative artificial intelligence (AI) worldwide. AI will drive the next wave of GDP productivity growth, and the next infrastructure race will revolve around the global competition to house data centres."

(The Globe and Mail, 2024)

Energy Demand and Sustainability:

"6 trillion will be invested in expanding the electricity grid worldwide by 2040, adding or refurbishing 80 million kilometres of grids, equivalent to the world's existing grid. Low-carbon nuclear energy emerges as a critical piece of the AI puzzle."

(The Globe and Mail, 2024)

Financial Innovations:

"Second, a new minister of AI and digital economy should drive investment in commercialized homegrown AI scale-ups—companies with growth rates of at least 20 per cent annually over the previous three years—to increase their chances of success."

(The Globe and Mail, 2024)

Sentiment Analysis

The sentiment analysis revealed that the discourse on AI in national English-language Canadian newspapers is predominantly negative (Table 1).

A chi-square test confirmed a significant association between sentiment and topics ($\chi 2 = 160.56$, $p < 0.001$, $df = 8$), indicating that sentiment varies significantly across different topics. Observing the Z-scores (Table 1), we note that Topic 1, *AI and Job Substitution*, and Topic 3, *Multimedia Generation and Deepfake*, are significantly associated with negative sentiment. In contrast, positive sentiments are significantly overrepresented in Topic 5, *Economic Impact and Market Applications*.

Emotion Detection

The emotion analysis detected seven categories (Table 2), with fear emerging as the dominant emotion (57.5%, 2948 text segments out of 5128).

A chi-square test ($\chi 2 = 271.93$, $p < 0.001$, $df = 24$) indicates that certain emotions are significantly over- or under-represented depending on the topic. Observing the Z-scores (Table 2), we note that Topic 1, *AI and Job Substitution*, is linked to *sadness*. Topic 3, *Multimedia Generation and Deepfake*, is significantly associated with feelings of *anger* and *fear*. Similarly, Topic 4, *Legal Regulation of AI*, also shows a strong association with *anger*. In contrast, Topic 5, *Economic Impact and Market Applications* is positively associated with *joy* and is negatively associated with *anger*. Finally, Topic 2, *Chatbots and LLMs uses in education and healthcare*,

Table 1: Crosstab of Sentiment and Topic-Labeled Text Segments with Z-scores.

	Sentiment						
	Negative		**Neutral**		**Positive**		Total
Topic	**n**	**Z**	**n**	**Z**	**n**	**Z**	
Not assigned	80		1		58		139
Topic 1: AI and Job Substitution	**881**	**2.88**	50	1.34	**375**	**-4.18**	1306
Topic 2: Chatbots and LLMs	496	–1.12	28	0.20	328	1.41	852
Topic 3: Multimedia Generation and deepfake	**473**	**3.17**	22	0.18	**173**	**–4.21**	668
Topic 4: Legal Regulation of AI	684	0.99	35	0.16	357	–1.34	1076
Topic 5 - Economic Impact and Market Applications of AI	**520**	**–5.64**	23	–1.95	**544**	**7.97**	1087
Total	3134		159		1835		5128

Note: Bold values indicate statistically significant associations based on Z-scores ($|Z| \geq 1.96$), corresponding to a 95% confidence level ($p < 0.05$).

Table 2: Crosstab of Emotions and Topic-Labeled Text Segments with Z scores.

	Emotion														
	Anger		**Disgust**		**Fear**		**Joy**		**Neutral**		**Sadness**		**Surprise**		**Tot**
Topic	**n**	**Z**	**n**	**Z**	**n**	**Z**	**n**	**Z**	**n**	**Z**	**n**	**Z**	**n**	**Z**	
Not assigned	23		3		58		19		33		1		2		139
Topic 1: AI and Job Substitution	**115**	**–2.06**	12	1.48	745	-0.42	112	–0.63	266	1.39	**32**	**2.71**	24	1.05	1306
Topic 2: Chatbots and LLMs	77	–1.45	7	0.83	451	–1.92	**55**	**–2.56**	**243**	**6.63**	**3**	**–2.77**	16	0.95	852
Topic 3: Multimedia Generation and deepfake	**120**	**5.78**	6	0.99	347	–2.03	56	-0.61	120	-0.44	13	0.89	6	–1.24	668
Topic 4: Legal Regulation of AI	**152**	**3.48**	3	–1.36	**705**	**3.27**	**60**	**–3.83**	**128**	**–5.16**	16	-0.10	12	–0.99	1076
Topic 5 - Economic Impact and Market Applications of AI	**68**	**–4.45**	2	–1.77	642	0.49	**171**	**7.25**	176	–1.91	12	–1.12	16	–0.03	1087
Total	555		33		2948		473		966		77		76		5128

Note: Bold values indicate statistically significant associations based on Z-scores ($|Z| \geq 1.96$), corresponding to a 95% confidence level ($p < 0.05$).

demonstrates significant associations with both neutral tones and a lack of emotional extremes like *joy* or *sadness*.

Sensational Coverage and Collective Anxiety

Media coverage of high-profile AI-related incidents reveals a growing tendency to sensationalize the risks associated with this technology, contributing to collective anxiety about its societal implications. This framing often highlights ethical breaches, cultural insensitivity, and violations of privacy, amplifying anger and fears about AI's potential harm.

One example comes from the backlash faced by *Sports Illustrated* when it was revealed that several articles were written by AI and attributed to fabricated authors with AI-generated photos. This prompted outrage from the publication's unionized staff, who stated they were "horrified" by the incident and "deplore being associated with something so disrespectful to our readers" (Simon Houpt 29 November 2023, The Globe and Mail).

The misuse of AI-generated deepfake pornography further underscores public anxieties. Taylor Swift became a victim of this phenomenon, sparking outrage and mobilization from her fanbase, who launched a counteroffensive under the hashtag #ProtectTaylorSwift. Despite Swift action by platforms to remove the explicit images, millions of users were exposed before they were taken down. This incident reflects broader fears about the accessibility and misuse of generative AI tools, with experts warning that such technology will "continue to proliferate" and pose new challenges for privacy and digital safety (Haleluya Hadero Associated 17 April 2023, The Globe and Mail).

In the music industry, AI-generated deepfake tracks like "Heart on My Sleeve," mimicking the voices of Drake and The Weeknd, have raised alarm about intellectual property theft and the authenticity of artistic expression. Universal Music Group condemned the track as "infringing content," warning that generative AI could disrupt the industry by "scraping data from recordings" without proper authorization (Brad Wheeler, 25 April 2023 The Globe and Mail). These controversies emphasize the growing tension between technological innovation and ethical boundaries, contributing to broader public unease.

Collectively, these incidents demonstrate how sensational media narratives magnify the problematic aspects of AI, reinforcing its framing as a source of societal risk. By focusing on ethical breaches, cultural insensitivity, and violations of personal and intellectual property rights, these narratives play a pivotal role in constructing AI as a public problem. This media framing generates widespread anxiety and intensifies calls for regulatory oversight, reflecting the heightened emotions characteristic of the problematization phase in the societal discourse on AI.

Discussion

Media Narratives and the Shaping of Social Issues

Our findings underscore how media framing has pushed AI from the realm of technical innovation into the realm of social problem. Early media coverage globally often skews positive for emerging technologies—emphasizing potential benefits and excitement (Rössler 2001; Nisbet et al. 2003; Clark et al. 2016). However, our analysis reveals a notable deviation from this classic pattern in the case of post-2022 AI coverage. Canadian media studied have moved unusually quickly to a critical stance, portraying AI "as a more dangerous than promising reality". Even in what might be considered AI's "emergence" stage as a public topic, the press did not restrict itself to optimistic futurism. Instead, cautionary and alarmist frames have been prominent. By the time generative AI (e.g., ChatGPT) hit public awareness, media narratives were already highlighting threats—from job displacement and privacy invasion to the spread of misinformation. In theoretical terms, AI's media coverage accelerated into the problematization phase early on, rather than languishing in a prolonged honeymoon of uncritical celebration. Recent longitudinal research supports this trajectory: for example, Nguyen and Hekman (2024) observed that media discourses around AI and automation grew significantly more negative as AI began impacting daily life, shifting from speculative optimism to concrete worries. Our findings mirror this shift—once AI became a tangible presence (thanks especially to widely-used applications like ChatGPT), media tone pivoted to underscore perils and pitfalls. This suggests that the "emergence" phase have quickly given way to an explicit focus on risks, even as the technology is still being adopted. The ambivalent view—where hopes and fears intermingle—is evident, but the balance is tipping towards fear and critique in the Canadian context.

A closer look at Topic 5, which centers on the economic potential of AI, reveals a marked divergence from the predominantly negative or ambivalent sentiment observed in Topics 1 through 4. While the foregoing topics are marked by an emphasis on ethical issues, surveillance, risk, and regulatory pressure—typically constructed in affectively colored terms like fear, anger, or moral panic—Topic 5 displays a more positive, forward-looking tone. The contrast here indicates how different thematic aggregations can be associated with simultaneous yet qualitatively different representational phases: a 'threat-oriented' phase fixated on social disruption, and an 'opportunity-driven' phase focusing on economic creativity and national competitiveness. The binary further complicates the representational field, revealing that public discourse does not take the shape of a unitary, linear progression, but instead contains several, co-present frames of interpretation.

Part of how media have achieved this problem-focused framing is through selective emphasis on sensational incidents and worst-case scenarios. Our results noted several high-profile controversies amplified by news outlets (such as an AI chatbot producing disturbing content, or deepfake music raising ethical/legal alarms). This is a classic media strategy in social problem construction: concrete dramatic examples (what Stanley Cohen, 1972, termed *"folk devils"* in a different context) are highlighted to symbolize the abstract threat. The effect is to make the threat vivid and urgent to the public. By focusing on such episodes—a biased algorithm causing harm, an autonomous vehicle accident, an AI voice deepfake scandal—the media turn theoretical risks into tangible realities that demand attention. This corresponds to what Hilgartner and Bosk (1988) describe as the need for drama and novelty in public arenas: issues that can be tied to gripping stories (especially ones evoking emotions like outrage or fear) are more likely to survive the competition for coverage

(Hilgartner and Bosk 1988). In short, media narratives have strategically anchored the idea of "AI as a problem" to concrete events and emotionally charged stories, a necessary step for broad public engagement.

Our study indicates so that the social representation of AI in Canada is already converging around a more negative, problem-oriented core, even as some positive elements remain at the periphery. In Wagner's terms, after the initial awareness and divergence phases (Wagner et al. 2002), there are signs of a convergence towards a dominant representation: AI as a source of risk and something to be controlled. The evidence is the prevalence of risk-focused topics and the consistent tone of caution in our media corpus. Yet, it's important to emphasize that convergence does not imply complete uniformity or the elimination of alternative views. The representational field of AI remains heterogeneous—for example, economic optimism about AI (Topic 5 in our results) persists as a counterpoint to the gloom, indicating an undercurrent of hope that the technology will bring prosperity. Social representations are dynamic; they can accommodate new information and shifts in public mood. Thus, even as a protective, wary stance toward AI solidifies, the Canadian media studied still occasionally highlights success stories of AI (e.g., medical breakthroughs, efficient services) which keep a more utopian representation alive. This dynamic resonates with the concept of cognitive polyphasia, where contradictory notions (AI as savior vs. AI as threat) continue to *live side by side in the same collective,* reflecting the complexities of public understanding (Jovchelovitch 2002).

Emotions as Catalysts in Problematization

Another crucial mechanism is the integration of emotional and ethical meanings into the representation of AI. As media stories make AI part of the public's everyday reality, they naturally tie it to emotional experiences and moral values that people care about. Our results show this clearly: the representation of AI is not just technical but moral-emotional. For example, AI is associated with violations of privacy (triggering indignation), with job insecurity (triggering anxiety and sadness), and with threats to human dignity or autonomy (triggering fear of an "existential" sort). The media's role in highlighting these angles effectively objectifies the ethical and emotional stakes of AI—making the public feel that AI is not just interesting, but also troubling or inspiring in very human terms. AI is no longer just about algorithms; it becomes about crime and safety, fairness and justice, hope and anxiety. By integrating AI into narratives about everyday life and societal norms, the media helped naturalize the representation that AI is something that can hurt or help people, and thus we must collectively grapple with it. This marks the point where the representation of AI moves from the abstract to the lived realm—a precondition for collective problem-solving action.

One of the most significant theoretical insights from our study is the central role of collective emotions in the social construction of AI as a public problem. Emotions are not ancillary to representation; they are constitutive of it. As Zizi Papacharissi (2015) and others have argued in the context of networked publics, affect and sentiment drive public engagement and shape the narrative of social issues. Our findings strongly support this "affective turn": media discourse around AI is saturated with emotional language (fear, anger, hope, etc.), and these emotions are doing important social work.

Firstly, emotions serve as a "magnifying glass", amplifying public attention toward AI. Fear grabs headlines—for instance, warnings that "AI could threaten human survival" naturally attract readership and spur discussion. Anger also fuels mobilization—reports of AI systems acting biased or infringing rights provoke moral outrage, rallying voices for justice and regulation. These collective emotions ensure that AI remains not just a topic of interest, but a topic of urgent concern. As our results indicated, fear and anger were the predominant emotions circulated in national English-language Canadian newspapers about AI, far outweighing neutral or positive affect. This emotional coloring is a strong indicator that AI has entered the problematization stage: it's being viewed "through the lens of trouble," with feelings of threat and injustice at the forefront. In fact, Blumer (1971) noted that an emotionally charged discourse is often a prerequisite for an issue to achieve social problem status. Our analysis verifies this in the context of AI: the discourse is charged indeed, rife with anxiety about AI's implications. The prevalence of negative emotions in media coverage is not just a reflection of the issue's seriousness—it actively contributes to making AI seen as serious by the public and policymakers. Emotions imbue the issue with a sense of urgency and moral weight that pure facts and figures might not convey.

Secondly, emotions shape the interpretative frames through which AI is understood. Each dominant emotion corresponds to a particular framing of the problem. For example, fear tends to frame AI as a threat to personal and societal safety. Fear was especially linked with stories about physical safety (autonomous vehicle accidents, threats to one's job or livelihood)—essentially casting AI as a menacing agent that could harm individuals. Anger, on the other hand, frames AI in terms of moral violation and injustice. We saw anger peak in narratives about ethical failures (e.g., AI enabling surveillance or discrimination), which implies a frame of AI as violating important norms, hence something that ought to be corrected or punished. Sadness was notably present in discussions of job loss due to AI (Topic 1), framing the issue as one of tragic loss and social decline, which might not incite immediate action but adds to a sense of gloom and resignation about AI-driven change. Meanwhile, there are also traces of hope and excitement in the discourse—usually tied to AI's potential benefits (like medical advancements or economic opportunities). Those emotions frame AI as a challenge to be managed rather than an outright evil. What is crucial is that these emotions and frames interact polyphonically. Our study revealed a polyphony of emotions rather than a single mood: a prevalence of negative emotions coexists with threads of optimism. This mix can create a more nuanced public opinion—for instance, one might fear AI's risks yet still hope to harness it for good—which in turn can affect the types of solutions the public supports (strict regulation versus balanced governance). Such emotional complexity is a hallmark of social representations in modern, plural societies, and it underscores why analyzing sentiment is key to understanding the construction of public problems (Von Scheve and Salmela 2014).

Furthermore, emotions are drivers of collective action. Our results point out that strong emotions are correlated with calls for regulatory and policy intervention—specifically, fear and anger were frequently coupled with demands for government action (as seen in Topic 4, where discourse about AI governance was laden with concern and indignation). This aligns with recent research in policy discourse: Lemke et al. (2024) found that in Germany, as AI became defined as a policy problem, it was accompanied by normative calls for regulation and ethical safeguards. Emotions like indignation often underlie such calls—they carry an implicit argument that "this situation is intolerable and must be fixed." In our context, media stories invoking fear of AI-driven catastrophes or anger at Big Tech's misuse of AI effectively set the stage for public pressure on authorities. Indeed, we observed media references to concrete measures (e.g., proposing new laws, AI ethics committees, moratoriums on certain AI applications) following on the heels of emotional reportage. This illustrates how, as per our theoretical model, the public problematization (with its emotional intensity) naturally leads to mobilization, rallying support for remedies. Emotions thus act as a catalyst that transitions the issue from mere discussion to the realm of decision-making and action.

Another important theoretical implication is how emotions interact with the "naturalization" cycle of representations. Emotions can periodically "interrupt" a settled state and reignite the cycle of problematization. A surge of collective outrage or fear in response to a new AI development can abruptly end a complacent period and put AI risks back on the agenda (as happened with the launch of ChatGPT, which suddenly made AI's capabilities tangible to millions, evoking both awe and alarm). These emotional surges ensure that AI's status as a public problem is periodically renewed and renegotiated. In effect, emotions are the heartbeat of the social representation cycle—each spike in emotional intensity sends a pulse that keeps the issue alive and evolving. This demonstrates that social representations are not static once naturalized; they can be re-energized by emotional publics, leading to iterative cycles of public problem construction.

These processes of framing AI as threat or manageable innovation—can be productively framed in terms of Rogers et al. (2019) diffusion of innovations framework. This model proposes that the general public adopts new technologies by progressing through a series of stages: knowledge, persuasion, decision, implementation, and confirmation. Emotionally charged media narratives in our data would thus seem to map onto the early stages of the process, that of 'knowledge' and 'persuasion,' where the general public's knowledge and attitudes are being actively constructed. Panic and fear may correspond to the process of resistance or uncertainty, in particular the case of late adopters or doubters, while stories about the economic potential of AI (e.g., Topic 5) speak directly to 'early adopters' as well as institutional champions. By employing this lens, we can better understand how problematization of AI not only reflects affective response, but can map onto other stages of the general public working through technological change, determining how and when AI comes to be understood as legitimate, desirable, or in need of regularization.

Toward an Integrated Model of Social Problems and Representations

Overall, our discussion affirms and enriches the theoretical model of representational naturalization (Negura and Plante 2021) used in this study. Our findings confirm that AI in national English-language Canadian newspapers discourse predominantly occupies the problematization stage, characterized by high issue salience and a focus on risks, which is exactly where we expected it to be. At the same time, we detected elements of the other stages—emergence, communication/mobilization, and resolution—overlapping in the discourse. This suggests that the process of constructing a public problem is not strictly linear, but rather concurrent and cyclical. Multiple phases of the cycle can operate simultaneously on different facets of the issue. For example, even as the media largely "problematize" AI, there are still emergent aspects (new developments being introduced with some optimism) and early resolution aspects (debates on regulatory solutions, hinting at the normalization to come).

We also account for the socio-cognitive and cultural underpinnings of this process. The representational naturalization model posits that a new phenomenon becomes taken-for-granted social reality through phases of collective sense-making (Negura and Plante 2021). In our discussion, we have seen how media narratives facilitated each of these phases for AI: raising initial awareness (emergence), fostering debate and conflict of interpretation (divergence leading to problematization), amplifying the issue and rallying responses (communication/mobilization), and hinting at stabilization through policy (resolution). The contribution of our study is to show that emotions structure and accelerate these representational phases. Emotions help drive the transition from one phase to the next and can even loop the process back to an earlier phase if a new shock occurs. In doing so, our research extends the representational naturalization model by adding a clearer affective dimension to it. We argue that social representations are "cognitive–emotional" constructs (to use a formulation consistent with Bouriche 2022), meaning that how people think about an issue is deeply intertwined with how they feel about it. The case of AI demonstrates that collective emotions are not just outputs of media representation but also inputs that shape the trajectory of the representation. For instance, fear of AI can both be a result of negative media coverage and a reason why media coverage continues to be negative (a self-reinforcing loop). Recognizing this bidirectional influence of emotion helps us better explain why certain issues (like AI) can dominate public discourse and policy agendas at one time and then recede at another.

Finally, this integrative approach underscores the importance of media in linking representation and action. The media acts as the bridge between society's symbolic processes and its concrete responses. Through our study, we observe that media discourse translated the abstract concept of "AI risks" into a socially recognized problem that people demand action on. This confirms a point often made in classic agenda-setting and social problems research: media attention is a precursor to policy attention (Hilgartner and Bosk 1988). Only after AI risks were widely represented and emotionally charged in the media did we see substantial mobilization (e.g., government task forces on AI, public petitions, industry pledges for AI ethics).

Our findings resonate with the agenda-setting theory (McCombs and Shaw 2007) in that the salience of AI in the media likely influenced its salience among the public and policymakers. But we go a step further by showing the content of that media attention—the social representations and emotional tone—shaped how stakeholders chose to respond. Calls for AI regulation, for example, were specifically tied to narratives of fear and moral indignation, illustrating that it is not just attention that matters, but the interpretation of the issue that media propagate.

In sum, this discussion has endeavored to root our empirical observations in a robust theoretical context. Moving forward, understanding the social construction of public problems like AI will require paying attention to this marriage of media and minds: how media discourse (with its frames, symbols, and sentiments) crystallizes certain understandings in the public, and how those understandings in turn drive social action and policy. Our study, grounded in the interplay of social representations and social problem theory, contributes to this broader project by illustrating a contemporary instance of these timeless processes at work.

Conclusion

The study contributes to an in-depth understanding of the construction in national English-language Canadian newspapers of AI as a public problem since the launch of ChatGPT in 2022. Using the representational naturalization model (Negura and Plante 2021) in the analysis of media narratives, we showed how the construction of AI as a public issue is a complex and cyclical process shaped by both socio-cognitive and affective dynamics.

Our results show that AI in national English-language Canadian newspapers is in the problematization stage, highlighting risks such as job replacement, privacy violation, and ethical dilemmas. These risks are amplified by strong emotions such as fear, anger and sadness, which not only shape public perceptions but also mobilize demands for intervention and regulation. Elements of other stages—emergence, communication and resolution, are also present in some topics, highlighting the non-linear and overlapping nature of this process. Thus, while positive narratives about the economic potential of AI reflect aspects of the emergent stage, persistent negativity in media discourse nevertheless suggests that public debate remains largely rooted in concerns about societal risks.

From a theoretical point of view, our study enriches the representational naturalization model by integrating the dynamic and structured role of emotions in the process of constructing public problems. Our results suggest that the stages of emergence, problematization, communication and resolution are not isolated but are often concurrent and iterative, influenced by technological advances, societal anxieties and media frameworks. Further studies are of course necessary to monitor the ongoing cycle, in order to understand how AI is constructed as a public problem through media discourse, as well as the dynamic of emotions, sensational narratives, and advocacy in shaping public perceptions and societal responses to emerging technologies.

References

Blumer, H. 1971. Social problems as collective behavior. Social Problems, 18(3): 298–306.

Bouriche, B. 2022. Social representations as cognitive-emotional processes: an integrative approach proposal. Papers on Social Representations, 31(2).

Brondi, S., A. Armenti, P. Cottone, B. M. Mazzara and M. Sarrica. 2014. Parliamentary and press discourses on sustainable energy in Italy: No more hard paths, not yet soft paths. Energy Research & Social Science, 2: 38–48.

Brondi, S., M. Pivetti, S. di Battista and M. Sarrica. 2021. What do we expect from robots? Social representations, attitudes and evaluations of robots in daily life. Technology in Society, 66: 101663. https://doi.org/10.1016/j.techsoc.2021.101663

Clark, D. B., E. E. Tanner-Smith and S. S. Killingsworth. 2016. Digital games, design, and learning: A systematic review and meta-analysis. Review of Educational Research, 86(1): 79–122.

Cools, H. and N. Diakopoulos. 2024. Uses of generative AI in the newsroom: mapping journalists' perceptions of perils and possibilities. Journalism Practice, 1–19.

Contarello, A., L. Fortunati and M. Sarrica. 2007. Social Thinking and the Mobile Phone: a Study of Social Change with the Diffusion of Mobile Phones, Using a Social Representations Framework. Continuum: Journal of Media & Cultural Studies, 21(2): 149–163. https://doi.org/10.1080/10304310701268687

Dandurand, G., F. McKelvey and J. Roberge. 2023. Freezing out: Legacy media's shaping of AI as a cold controversy. Big Data & Society, 10(2): 20539517231219242.

Ekman, P. 1992. Are there basic emotions? Psychological Review, 99(3): 550–553. https://doi.org/10.1037/0033-295X.99.3.550

Hansen, S. S. 2022. Public AI imaginaries: How the debate on artificial intelligence was covered in Danish newspapers and magazines 1956–2021. Nordicom Review, 43(1): 56–78.

Hilgartner, S and C. L. Bosk. 1988. The rise and fall of social problems: A public arenas model. American journal of Sociology, 94(1): 53–78.

Hu, X and H. Liu. 2012. Text analytics in social media. Mining text data, 385–414.

Jovchelovitch, S. 2002. Re-thinking the diversity of knowledge: cognitive polyphasia, belief and representation [online]. London: LSE Research Online. Available at: http://eprints.lse.ac.uk/2628

Korneeva, E., T. O. Salge, T. Teubner and D. Antons. 2023. Tracing the legitimacy of artificial intelligence: A longitudinal analysis of media discourse. Technological Forecasting and Social Change, 192: 122467.

Lemke, N., P. Trein and F. Varone. 2024. Defining artificial intelligence as a policy problem: A discourse network analysis from Germany. European Policy Analysis.

McCombs, M. E. and D. L. Shaw. 2007. The agenda-setting function of mass media. pp. 170–175. *In:* The Political Communication Reader. Routledge.

Moreira, L., J. C. Paiva and C. Morais. 2021. The social representations of the Internet: A systematic review of literature towards a groundbreaking research agenda. Papers on Social Representations, 30(1): 1.1–1.34.

Moreira, L., M. Sarrica, C. Morais and J. C. Paiva. 2023. More than a Common Place: A Social Representations Approach to the Internet. Papers on Social Representations, 32: 2.1–2.28.

Negura, L. and N. Plante. 2021. The construction of social reality as a process of representational naturalization. The case of the social representation of drugs. Journal for the Theory of Social Behaviour, 51(1): 124–144.

Nisbet, M. C., D. Brossard and A. Kroepsch. 2003. Framing science: The stem cell controversy in an age of press/politics. Harvard International Journal of press/politics, 8(2): 36–70.

Nguyen, D. and E. Hekman. 2024. The news framing of artificial intelligence: a critical exploration of how media discourses make sense of automation. AI & society, 39(2): 437–451.

Papacharissi, Z. 2015. Affective publics: Sentiment, technology, and politics. Oxford University Press.

Plutchik, R. 1980. A general psychoevolutionary theory of emotion. Emotion: Theory, research, and experience, 1.

Ratinaud, P. 2009. IRaMuTeQ: Implémentation d'une méthode d'analyse textuelle. [Software].

Reinert, M. 1983. Introduction à la lexicométrie. Un modèle statistique d'analyse. Langages, 67: 5–15.

Reinert, M. 1990. Alceste: Une méthodologie d'analyse des données textuelles et une application: Aurelia de Gérard de Nerval. Bulletin de Méthodologie Sociologique, 26: 24–54.

Reinert, M. 1993. Les "mondes lexicaux" et leur "logique" à travers l'analyse statistique d'un corpus de récits de cauchemars. Langage et Société, 66: 5–39.

Rogers, E. M., A. Singhal and M. M. Quinlan. 2019. Diffusion of innovations. pp. 432–448. *In:* An integrated approach to communication theory and research. Routledge.

Rössler, P. 2001. Between online heaven and cyberhell: The framing ofthe internet'by traditional media coverage in Germany. New Media & Society, 3(1): 49–66.

Sarrica, M., S. Brondi and L. Fortunati. 2020. How many facets does a "social robot" have? A review of scientific and popular definitions online. Information Technology & People, 33(1): 1–21.

Spector, M. and J. I. Kitsuse. 2017. Constructing social problems. Routledge.

Spyridou, P. and M. Ioannou. 2025. Exploring AI Amid the Hype: A Critical Reflection Around the Applications and Implications of AI in Journalism. Societies, 15(2): 23.

Sun, S., Y. Zhai, B. Shen and Y. Chen. 2020. Newspaper coverage of artificial intelligence: A perspective of emerging technologies. Telematics and Informatics, 53: 101433.

Taylor, G. and B. DeCillia. 2021. Canada: A strong foundation with an uncertain future. pp. 43–84. *In:* Trappel, J. and T. Tomaz (Eds.). The Media for Democracy Monitor 2021: How leading news media survive digital transformation (Vol. 2). Nordicom, University of Gothenburg. https://doi.org/10.48335/9789188855404-3

Von Scheve, C. and M. Salmella. 2014. Collective emotions: Perspectives from psychology, philosophy, and sociology. Oxford University Press.

Wagner, W., N. Kronberger and F. Seifert. 2002. Collective symbolic coping with new technologies: Social representations and the public sphere. British Journal of Social Psychology, 41(3): 561–578.

CHAPTER 3

From AlphaGo to ChatGPT

Mobilising Public Salience and Awareness of AI in the UK and South Korea, 2000–2023

*Jaesun Chae** and *Martin W. Bauer*

Introduction

This volume brings the task of Social Representation Theory (SRT) to the study of emergent technology. This raises two questions: how is AI represented in society? and what does AI represent when it generates narrative text? This chapter contributes answers mainly to the first question: how does AI become an object of representation? We consider this process to be one of feature selection, attention seeking and construction of meaning in response to a challenge.

Artificial Intelligence (AI) is recognised as a transformative force of the digital age, catalysing change across scientific and economic domains. The rapid spread of generative AI technologies—most notably demonstrated by the global influence of LLM (large language models) like ChatGPT after its launch in November 2022—has focused public attention (Qi et al. 2024).

Recent discourse, however, shows growing concerns, dismissing AI as overhyped with false promises (Kahn 2025). What is heralded as "revolutionary intelligence" may follow the patterns of previous technological hype cycles (The Economist 2025).

AI's history has been marked by cycles of grand visions followed by disappointment, creating alternating AI winters and springs (Butler 2016; Galanos 2023; Süerdem 2024). During spring periods, specific breakthroughs come to represent the technology at different times. For example, AlphaGo captured public

London School of Economics, UK.
* Corresponding author: j.chae1@lse.ac.uk; m.bauer@lse.ac.uk

DOI: 10.1201/9781003502753-6

attention and became a household name when it defeated a human Go champion in 2016 (Chae 2017). Similarly, ChatGPT catalysed interest in the entire Generative AI field after 2022.

Against this backdrop, our research argues that AI development is not an inevitable force of nature but rather the result of mobilization efforts by technology actors operating as a quasi-social movement. The study examines the movement surrounding AI through social and psychological lenses, drawing on social movement theory, which conceptualizes social movements as collective efforts to transform society (Blumer 1969) by mobilising symbolic and material resources to promote their objectives. Within this framework of resource mobilisation (McCarthy and Zald 1977; Tilly 1978), mass media serve the function of agenda-setting (Carroll 2015) and shaping public perception and understanding, which legitimizes and emboldens the technology movement.

Our study examines the dynamic relationship between media coverage and public discourse surrounding AI, addressing three research questions regarding the flow of attention, topicality and awareness: How has the media's salience of AI evolved? how have dominant topics of AI evolved? With salience and topicality flows, how has public awareness of AI changed?

Specifically, the study focuses on a comparative analysis of two technology nations: the UK and South Korea. Both countries are similar in their progress in the Western and Eastern parts of the global tech race, and they share a geopolitical stance as influential AI middle powers. Situated in a contradictory position where they must pursue AI-led economic growth while managing global tensions, these two nations hold on to a different significance in existing discourses centred on the US-China AI hegemony.

The study performs long-term observations of AI for the period 2000 to 2023. The period culminates in the global AI Safety Summits of 2023 in London and 2024 in Seoul. We are utilising two main approaches. First, a computer-supported content analysis of media articles on AI is conducted, focusing on major newspaper sources in each country to examine the salience and development of AI topics (attention and topic flow). Next, secondary analysis of public survey data is conducted to investigate how public opinion is shifting (flow of awareness).

The results reveal distinct focal characteristics of each country for the flow of attention, of topics and of awareness over the period 2015–2023. In conclusion, while AI development appears to be swept along by global imperatives, the study suggests that AI is manifest through distinct discourses tailored to each nation's unique geopolitical context. ChatGPT's rapid rise since 2022 has acted as a catalyst for the UK; however, there is no guarantee that this enthusiasm will not wane, as demonstrated by Korea's earlier trajectory following the AlphaGo event in 2016.

Taking these together, this study maintains that AI is a quasi-social movement that mobilizes attention, topicality and awareness around future opportunities. While real social movements seek to address real grievances, it remains opaque what problem AI is trying to resolve.

The Emergence of AI and Shifting Climates

AI can broadly be characterised as "a machine capable of reasoning like a human with language processing abilities" (Havenstein 2005). An ambition from the start was to create machines that imitate human common sense (Bauer and Schiele 2024). The field emerged in the 1930s–40s with foundational work in Logic by Gödel and Turing, leading to an official research program in 1956 (Cordeschi 2007). Despite recurrent enthusiasm, AI failed to meet expectations and entered several "AI winters" due to a lack of tangible achievements (Butler 2016; Galanos 2023). The field experienced a recent resurgence, starting in the late 1990s, following successes in robotics and linguistic translation (Kurzweil 2005), as well as achievements in gaming (Chess, Go), autonomous mobility and robotic assistants (Bundy 2017). The excitement reached unprecedented heights with OpenAI's ChatGPT in November 2022, Generative AI dramatically increasing public and policy attention.

Nevertheless, amidst all this excitement, concerns around AI have difficulties to be heard. Academics and writers have expressed scepticism, claiming that AI has yet to deliver on its promises, though global consultancies take notice: "few, if any, truly transformative changes within the next decade" (Goldman Sachs 2024). Economic historians argue that artificial intelligence [AI] is offering labour substitution, surveillance and propaganda, when it should deliver "useful machinery" [UM] that improves work and livelihoods, not lining the pockets of the few; what a better world needs is UM, not AI (Acemoglu and Johnson 2023). Beyond undue expectations, AI advancement has furthermore sparked discussions about potential catastrophic "existential risks" to humanity (Dafoe 2018). In 2015, scientists, including Stephen Hawking and Elon Musk, penned an open letter urging the development of AI safeguards. In March 2023, over 1,000 experts called for a six-month moratorium on giant AI systems, citing fatal risks to society. And in November 2023, the UK invited a global summit on AI Safety to London, and in 2024, this summit met again in Seoul. Might AI encounter its Asilomar moment? In 1975, a global meeting of leading microbiologists self-imposed a moratorium on risky recombinant DNA research in order to avoid powerful viruses to leak from the labs (see Bauer 2015).

Debates over how AI should be developed also intensified in the public arena. International surveys often reveal public concerns about job disruption and privacy protection, although perceptions remained ambivalent. While AI is expected to have a significant societal impact, concerns about potential harms often outweigh expected benefits (Kelley et al. 2021). Research also highlights how sociocultural contexts shape AI imaginaries, with East Asian perceptions being more positive due to favourable AI portrayals, whereas Western narratives lean towards more cautious depictions (Cave et al. 2018).

Technology as a Quasi-social Movement

Social movement, broadly defined as "collective enterprises to establish a new order of life" (Blumer 1969), encompasses dynamic networks of individuals cultivating beliefs and convictions to bring about societal change for the "better" (Tarrow 2011; McCarthy and Zald 1977; Tilly 1978). Unlike formal political parties or interest groups, movements operate as flexible structures of co-ordinating actions centred on four key variables: framing, networking, repertoire, and opportunities to address real grievances (Tarrow 2011).

Technology, generally characterised as the expertise used by humans to manage and control the world (Bridgstock 1998), also involves collective action to mobilise support and address challenges during periods of advancement. While modern technology arises from extending scientific enquiries, social movements seek to address grievances and suffering centred on labour, civil liberty, or gender relations that arise from ideology. It is, however, argued that technology also operates as collective actor-networks, i.e., as man-machine cyborgs (Latour 1987). However, tech movements often lack the grassroots grievances typical of social movements. While tech movements address abstract "grievances" such as "efficiency potentials not yet realised", these are not equivalent to the concrete pains of working conditions, low wages, racial discrimination or gender inequality. Hence, we characterise the AI tech movement as a "quasi-social movement" with the key variables project, attention, and context (Bauer 2015). This is consistent with the historiography of computing and information technology, which argues that its history is best written as the proliferation of solutions in search of problems (Ceruzzi 2003).

All movements, inherently reliant on and competing for resources, aim to reach a broader public; and so do tech movements. Emphasising resource mobilisation highlights the symbolic resources and the connection to institutional actors, stakeholders, and media outlets that cultivate a public imagination that grants legitimacy.

For resource mobilisation, communication plays a crucial role and has become a key area of research of public attention (Newig 2004). In this sense, the mass media serve as a prominent forum for the formation of public opinion, presenting both opportunities and challenges for activist groups (Kielbowicz and Scherer 1984). The aim of mobilisation is to expand media coverage, to set the agenda and to shape people's perceptions by controlling the narrative on a particular issue such as AI. Agenda-setting has been studied extensively, with empirical evidence demonstrating how media coverage influences public perceptions, for example, with regard to issue importance (Djerf-Pierre et al. 2025). It sets public priorities by highlighting certain issues over others, thus selecting "what to think about". This proves particularly influential for non-everyday issues, such as new technology, the state of the environment, or policy making, where the public relies on media for information (McCombs 2014; Windahl et al. 2009).

Geopolitics in the AI Tech Race: the UK and Korea

Our study compares the AI movement across the UK and South Korea. Both countries serve as influential middle powers in the global AI landscape, positioned between major AI superpowers while maintaining significant technological capabilities in their respective Western and Eastern contexts.

The United Kingdom of England, Wales, Scotland and Northern Ireland [UK]

Celebrating Alan Turing's pioneering WWII legacy, the UK is recognised for its concentration of AI startups, strong expertise in computing and outstanding research capacity (AI NOW Institute 2024), frequently ranking among the global leaders on international benchmarks.[1]

The competitive landscape began to shift after 2010, as advances in computing power and the availability of vast datasets intensified global rivalry (US Government 2016). The acquisition of British AI companies by major US corporations, such as Google's purchase of DeepMind (2014) and Microsoft's acquisition of SwiftKey (2016), underscored the need for the UK to develop strategies for attracting and retaining AI investments and expertise (Hall and Pesenti 2017).

In this context, the UK developed its national strategies, starting with the "2017 Digital Strategy", which identified AI as a "Grand Challenge" (Jones 2024). Following the successful 2016 Brexit referendum, which separated the UK once again from Europe, the government sought economic intervention to restore the nation's stature. The UK's AI commitment grew even stronger after OpenAI's success with ChatGPT in 2022, resulting in significant funding rounds and investments (Benaich and Hogarth 2023). The disruption in the global supply chain during the pandemic 2020–22 and the Russian military invasion of Ukraine on 24 February 2022 further reinforced the focus on AI sovereignty, including cyber defence (Jones 2024).

On the other hand, the increased capabilities of AI technology have raised safety concerns, with existential risk discussions gaining prominence in the press and among policymakers (Hall and Pesenti 2017). Nevertheless, while facing intensified competition from US investments, the UK identified an opportunity through AI safety leadership, positioning itself as an international hub for responsible AI development. This strategic approach culminated in hosting the first global "AI Safety Summit" in November 2023, which brought together major AI companies and facilitated agreements among nations, including the US and the EU (Poireault 2023).

South Korea

Korea has a relatively brief history of scientific development, establishing its computer industry in the 1970s and founding its first AI organisation in 1985 (Zhang 2016). Nevertheless, due to its rapid economic growth and advanced internet infrastructure, Korea has gained recognition as a testing ground for new technologies in Asia.

Driven by Korea's deeply rooted "sociotechnical imaginary" of national developmentalism (Jasanoff and Kim 2015), AI rapidly gained momentum around 2010 in Korea. The pivotal moment occurred in 2016, when DeepMind's AlphaGo defeated a Korean Go champion 4-1. This event, coinciding with the "Fourth Industrial Revolution" discourse in the Davos Forum, transformed the once-obscure tech into a mainstream phenomenon, setting the stage for a continuous influx of

[1] E.g., Standford HAI's Global AI Vibrancy Ranking, Oxford's Government AI Readiness Index.

news on AI over the subsequent years (Kim 2023). The public's astonishment at the champion's defeat generated significant attention, and the government responded with massive funding starting from 2017.

The subsequent COVID-19 pandemic further accelerated AI adoption. The "Korean New Deal" in 2020 proclaimed the nation's DNA (Data-Network-AI) as its guiding principle, emphasising AI-directed digital transformation to restore economic conditions after the pandemic (Roubini 2020). Corporate involvement also played a critical role. Korean enterprises actively lobbied to secure positions as key AI semiconductor suppliers, with Nvidia orders becoming regular news headlines. Startups added momentum to the sector, with Korea frequently ranking first in AI patents per capita (Maslej 2025).

Korea's commitment to maintaining its competitive edge in AI was also demonstrated through its active engagement in global AI governance discussions. Following the UK's pioneering AI Safety Summit in 2023, Korea co-hosted the subsequent global meeting in 2024, successfully facilitating international agreements.

These patterns show a notable strategic similarity between the two nations. Both the UK and Korea represent "strong middle powers" in the AI landscape, pursuing strategic autonomy while carefully navigating around dominant players. Their similar strategic positioning becomes particularly evident when examining key AI milestones: the symbolic moment when the UK's ambitious DeepMind project challenged Korea's Go champion through AlphaGo, and their consecutive stewardship of global AI Safety Summits in London 2023 and Seoul 2024. These interconnected developments reflect both countries' shared determination to carve out distinctive roles in global AI governance.

Public Attention and Awareness of AI

As reflected in the AI trajectories of both countries, AI technology appears to have once again entered a rejuvenating "spring phase" with "global heat" it has never experienced before. AI has effectively galvanised numerous successful events after a prolonged and exhausting "AI winter" (Benaich and Hogarth 2023; Greenberg 2017). Others fear another cycle of hype (Kahn 2025; LaGrandeur 2024; Süerdem 2023)when inflated expectations lead to disappointment and another cold period.

Motivated by these ambivalences, our study seeks to explore the AI tech movement through a cross-national comparison of the UK and Korea. Consequently, we pose the following research questions:

- How has the mobilisation of AI developed over the past decade in the UK and Korea?

This overarching question is addressed through three specific studies of the flow of public attention, topicality and awareness:

- How has the prominence of AI in media coverage evolved? (flow of media salience)

- How have the key themes surrounding AI shifted over time? (topic flow)
- How has public awareness of AI changed over time? (shifting awareness)

Data Sources

To investigate media salience, topic flow and public awareness, the research primarily employs two analytical approaches: computer-supported content analysis of news sources and secondary analysis of public opinion polling.

Firstly, our study focuses on legacy newspapers, and not broad social media sources. While social media platforms amplify diverse voices beyond the constraints of mainstream media (Lewis 2012), legacy news sources offer greater reliability and authority on the debate. As key gate-keepers for public opinion formation (Zaklama 2025), newspapers continue to frame the issues through influential voices while reflecting diverse stakeholder perspectives. Their coverage offers easily accessible data for tracing public debates on issues such as AI (Brennen et al. 2018).

Online newspaper coverage of four tech issues—nuclear, gene, climate change, and AI—from 2000 to 2023 was traced in both countries, with special focus on AI after 2015. For the UK, three national newspapers offer a balanced political representation: *The Telegraph* (conservative-right), *The Times* (centre-right), and *The Guardian* (liberal-left), accessible through the Nexis database with the keyword phrase "Artificial Intelligence." For Korea, three newspapers were also selected: *Dong-a Ilbo* (right-wing), *JoongAng Ilbo* (moderate-right), and *Hankyoreh* (left-wing). Articles were collected from *BIGKinds*, a Korea Press Foundation database, using "*In-gong ji-neung*" (Korean for "Artificial Intelligence").

The study employs basic media search and computational text analysis in R to examine news articles. Both data corpora were processed similarly to ensure methodological consistency. Specific data cleansing processes and pre-processing steps are detailed in the appendix.

Unlike Nexis in the UK, which provided full-text data, *BigKinds* in Korea only offered metadata of the items, including titles and full noun keywords. We did not reduce the UK corpus to match the Korean one, as the objective was to investigate relative changes within each country rather than direct structural comparisons. However, to test comparability, we replicated the UK analysis using noun-only features, which yielded similar results.

Secondly, our study also uses available public opinion data, which enables researchers and interested actors to monitor AI issues with different question-answer formats (Kiecolt and Nathan 1985). Media discourse and public perceptions are interconnected systems of meaning-making and public opinion formation, continually flowing, interacting and shaping one another (Gamson and Modigliani 1989). Understanding citizens' views on AI is operationally useful, but it is also required from a normative standpoint: the public is a stakeholder in the technology and not only a consumer market (Zhang 2024).

We examined nationally representative surveys on public attitudes towards AI conducted in the UK and Korea from 2015 to 2023. The corpus comprises seven national and international surveys covering approximately 55,000 respondents from both countries. These include government surveys such as the *Public Attitudes of Data and AI* from CDEI[2] https://www.gov.uk/government/organisations/centre-for-data-ethics-and-innovation, the *Public Attitudes Tracker* by BEIS,[3] and the *Public Understanding of Science* from KOSAC,[4] as well as surveys conducted by organisations including BRISTOWS, Delta Poll, Ipsos, and Hankook Research. Among them, both Delta Poll and Ipsos are multi-national surveys that include the UK and Korea.

Much of the available survey data is presented only in published reports as summary tables; raw data is only available partially and with great difficulties. We conduct descriptive analyses examining how public awareness changes over time, plotting key indicators across time to identify trend changes. Due to inconsistencies in time frames, varying item wording and response scales, direct comparisons UK-Korea require cautious interpretation. All our estimates are subject to the usual 2–3% sampling error for n = 1,000.

Media Salience—Floating Public Attention

As an initial step, we carried out a comparative analysis of news patterns from 2000 to 2023, examining four tech issues: artificial intelligence [AI], nuclear technology [nuclear], genetic engineering [gene], and climate change [climate], which encompass key strategic developments that have framed public debate on science and technology over the past decades (Bauer 2015). This baseline analysis revealed the relative attention given to AI compared to the concert of other significant technologies. We can thus assess the absolute salience (volume) and the relative salience of AI among the key science and technology news flow, as shown in Figures 1 and 2.

Figure 1 shows the level of media attention to AI in comparison to the other three issues in each country. The left axis and bar charts display the total number of articles on nuclear, gene, and climate change topics. The right axis and line graphs indicate AI's proportion of coverage relative to the total in percentages for all four issues.

While nuclear, genetics, and climate issues had salience all through the early 2000s, AI did not receive significant attention until around 2015/16, coinciding with tech events in the UK, such as the acquisition of AI start-ups by US firms and DeepMind's spectacle with the AlphaGo match in Korea. The rise in news media coverage of AI into 2023, and the relative decline of the other issues, reflects a reallocation of media attention and resources. Comparing AI with other debated issues shows how, after a prolonged period of dormancy, AI has gained momentum and emerged as an issue of comparable prominence in the 2020s.

---

[2] Centre for Data Ethics and Innovation.

[3] Department for Business, Energy and Industrial Strategy.

[4] Korea Foundation for Science and Creativity.

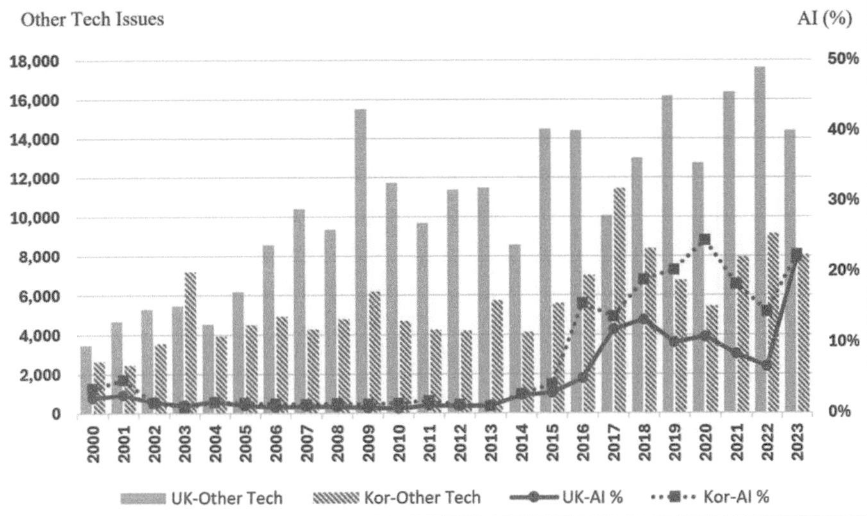

Figure 1: Media Coverage of AI Compared.

We concentrate our analysis on the 2015–2023 period to capture this trend more closely. Figure 2 shows the media flow of AI salience in the UK and Korea using an indexed scale, where 100 represents 2023, which is the highest point of coverage so far recorded over the timeframe.

While both countries exhibited overall upward trends in AI coverage, their temporal pattern characteristics differed. The UK started with an already relatively high baseline level but then experienced moderate, fluctuating growth until its peak in 2023—a roughly threefold increase from the previous year, and a 6-fold increase overall. Korea, by contrast, demonstrated a steep upward trajectory between 2015–2018, followed by a plateau and a second, smaller hike into 2023 of about 40% more coverage, while overall, equally with a 6 to 7-fold increase in AI coverage. Both countries demonstrated salience growth until 2018, followed by a temporary levelling-off period, ultimately setting the stage for the drama of 2023. Each country's attention is shaped by significant landmark events that captured the public's and media's attention. These events offered accessible entry points into AI discourse: the AlphaGo match captivated Korean audiences in 2016, while ChatGPT's arrival particularly resonated with UK audiences, heralding specific challenges in either context.

Topic Flow—The Changing Themes of AI

Having confirmed the significant rise in AI salience, our topic analysis gains deeper insight into the trajectory of AI-related themes across countries. The analysis employed computer-assisted text analysis using word frequency to identify key terms for every year over the 2015–2023 period. To ensure the identification of distinctive and contextually relevant keywords for each year, we utilized Term Frequency-

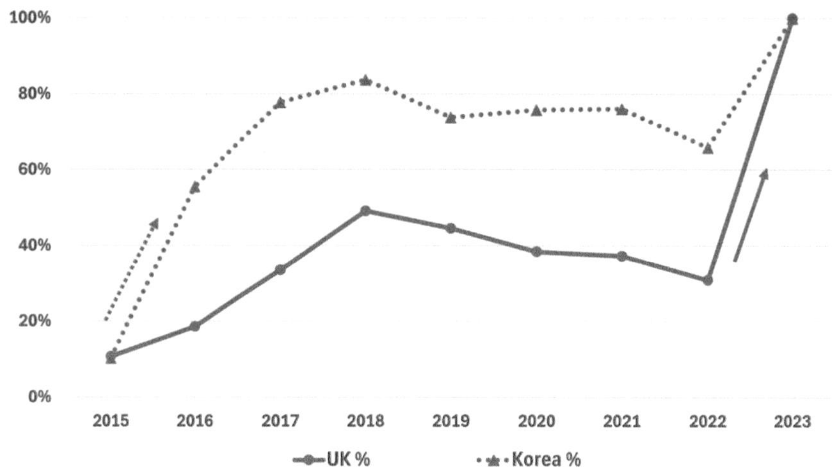

Figure 2: AI Salience.

Inverse Document Frequency (TF-IDF) weighting rather than relying solely on basic term frequency counts.

Table 1 shows the top 10 keywords for each year in the UK and Korea, ranked by their TF-IDF scores. The analysis reveals changing topicality patterns in AI discourse across both countries.

Shifting Focus in the UK

Table 1.1 shows the temporal distribution and progression in AI-related themes in the UK media. During the early period (2015–2017), AI discourse focused on terms such as "robot" and "machine", often contextualized through science fiction "films" [e.g., *Ex Machina 2015*] that captured the public imagination. This period also witnessed the emergence of applications, e.g., autonomous mobility and self-driving "cars" and AI-powered "games" gaining attention alongside major technology players such as "Google" and the growing local AI start-up industry.

The subsequent period (2018–2019) is marked by commercialization and capitalisation of Hi-Tech. Media coverage increasingly focused on "companies" and market dynamics, with terms like "business" and "fund" gaining prominence. This shift highlights the active capital market mobilization in the AI sector.

The 2020–2022 period presented the challenges of the COVID-19 pandemic, which redirected attention away from AI. The Virus now captured most of the news. However, "government" initiatives emerged as a key theme, reflecting policy efforts to maintain momentum in AI development. During this time, British AI companies such as "Sensyne" (healthcare AI) and "Arm" (semiconductor design) attracted attention in relation to the UK's AI capabilities and policy debates. In addition, in response to heightened public interest in health and disease, AI health applications also became prominent, with focused attention on "cancer" diagnosis and "patient" data management.

The most recent period (2023) witnessed a dramatic shift with "ChatGPT" becoming the dominant term, effectively reigniting widespread interest in AI across

Tables 1: Top 10 Keywords in the UK and Korea Over Time [ranked by TF-IDF]

Table 1.1: Top 10 Keywords in the UK (2015–2023).

	2015	**2016**	**2017**	**2018**	**2019**	**2020**	**2021**	**2022**	**2023**
1	robot	robot	robot	google	google	company	cancer	patient	company
2	car	google	job	company	company	patient	company	company	**chatgpt**
3	human	game	human	data	facebook	china	robot	cancer	Human
4	film	human	data	facebook	patient	google	arm	technology	business
5	google	facebook	company	robot	robot	government	china	data	google
6	data	company	google	patient	data	robot	patient	human	government
7	game	data	facebook	human	human	data	data	robot	technology
8	company	business	business	business	technology	human	human	business	use
9	facebook	app	car	china	million	technology	business	million	**microsoft**
10	machine	computer	technology	amazon	fund	sensyne	million	china	**risk**

Table 1.2: Top 10 Keywords in Korea (2015–2023, Korean terms translated into English).

	2015	**2016**	**2017**	**2018**	**2019**	**2020**	**2021**	**2022**	**2023**
1	computer	alphago	alphago	service	service	service	service	digital	**chatgpt**
2	google	lee sedol	service	samsung	data	corona	data	service	semicon
3	job	google	samsung	google	lee sedol	data	digital	data	service
4	service	computer	job	data	president	digital	platform	semicon	digital
5	aibo	service	naver	naver	digital	startup	startup	president	data
6	algorithm	job	google	speaker	bigdata	semicon	semicon	platform	openAI
7	data	president	s.phone	s.phone	program	online	corona	startup	president
8	s.phone	program	data	startup	global	smart	president	solution	naver
9	erica	deepmind	president	LG	samsung	platform	park jh	metaverse	platform
10	internet	data	speaker	bigdata	cloud	president	smart	program	startup

* Abbreviations: s.phone = smart phone, semicon = semiconductor, samsung = Samsung Electronics, LG = LG Electronics, lee sedol, park jh = names of Go-players, Lee Se-Dol and Park Jeung-Hwan

business and government sectors. ChatGPT, the name of a tool, now a metonymic symbol not only of LLM [large language model] processes that drive it, but of AI in its entirety: AI = ChatGPT. However, this surge in attention was accompanied by the emergence of "risk" as a significant keyword, indicating growing awareness of potential challenges and marking the beginning of a more nuanced discourse around AI governance and AI's societal implications.

Shifting Focus in Korea

Table 1.2 shows the top 10 keywords annually in Korea. During the early period (2015–2017), Korea's AI discourse similarly centred on robots and hardware, but with a more practical bent than the UK. Korean coverage emphasized specific, tangible applications. Robots developed in neighbouring Japan, such as "Aibo" and

"Erica," attracted attention, while discussions centred on concrete devices including "smartphones," "AI speakers," and "computers." Most notably, the Go-playing AI "AlphaGo" featured prominently in newspaper headlines, alongside its human opponent "Lee Sedol" and its London developer "DeepMind."

In line with this trend, major tech companies such as "Samsung," "Naver," and "LG" quickly became focal points of discussion, with emerging "startups" joining the conversation from 2018 onward. Unlike the UK, where government involvement became prominent only after 2020, Korea's government engaged much earlier, beginning in 2016, with the term "president" representing the administration in announcing numerous policy initiatives.

By 2020, "semiconductors" had emerged as a significant theme, as Korean IT giants such as Samsung and SK Hynix became key suppliers to global AI chip designers. This development further accelerated Korea's strategic efforts to position itself within the global AI ecosystem, as AI advancement became directly linked to national economic prosperity and stock market performance.

Amid Korea's persistent efforts to establish its presence in the global AI market, AlphaGo's social impact continued to resonate well beyond 2016. Lee Sedol's retirement from professional Go in 2019 and next-generation player Park Jung-Hwan's 2021 World Go Championship victory were both framed through the lens of the ongoing human-AI competition, demonstrating how deeply this narrative had resonated in Korean culture.

By 2023, Korea also witnessed significant attention towards "ChatGPT", which took centre stage as it did in the UK. The prominence of "semiconductors" consequently grew as well. However, AI-related concerns such as "risk," which appeared in British discourse, did not emerge in Korean conversations.

In Summary, we might say that the events of AlphaGo in 2016 in Korea and ChatGPT in 2023 in the UK took on signal functions, creating a societal alarm. However, the alarm signalled different dangers: in Korea, society was alerted to the risk of a "technological lag of the Nation", in the UK, the risk discourse pointed to a "technology out of control".

Public Perception—the Flow of AI Awareness

Having examined media salience and topicality of AI across the two countries, we also investigate public awareness trends through analysis of available survey and polling data. This analysis seeks to assess whether increased media coverage is followed by heightened public awareness, which would be consistent with the media agenda-setting hypothesis.

First, shifting public awareness of AI in the UK is indicated by five surveys with comparable questions and methodologies focused on AI knowledge: how familiar are people with AI? Figure 3 presents a combined index of AI awareness over time alongside our media salience [see Figure 2]. Note that the global Ipsos poll used a distinctly different question, so the Ipsos data is displayed using a secondary axis located on the right side of the graphs. The results demonstrate that AI awareness has grown in parallel with increased media coverage of AI topics. The results highlight a

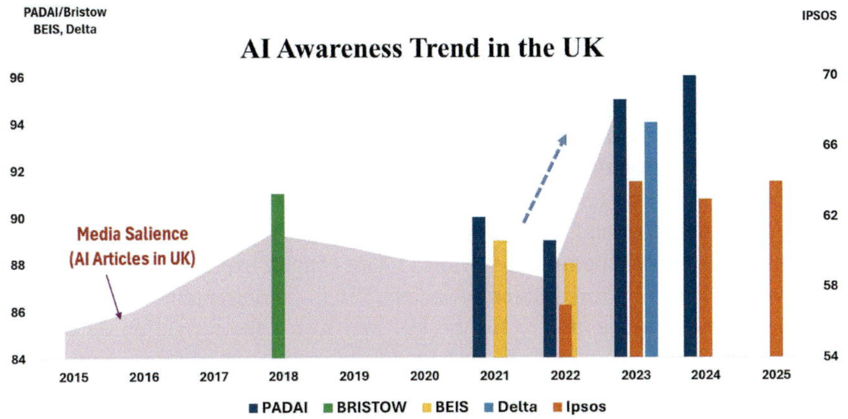

Figure 3: AI Awareness UK.

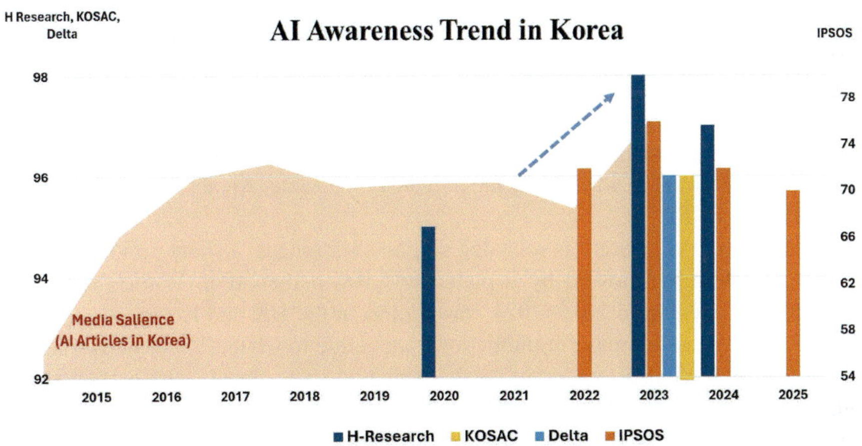

Figure 4: AI Awareness Korea.

spike in awareness in 2023, followed by a plateau thereafter. The time resolution of our data is not fine enough to assess the delay with which public awareness follows media salience, but clearly, the events of 2023 let public awareness surge in the UK.

The analysis of public AI awareness in Korea is indicated by four separate surveys with comparable questions about AI knowledge and understanding (Figure 4). The Korean data reveal a pattern consistent with UK findings, showing a pronounced surge into 2023 that significantly exceeded awareness levels recorded in 2020 and 2022; however, the surge is not as fast as in the UK and is abating. The surge coincides in both cases with intensified media coverage of ChatGPT's launch and the widespread adoption in use. The awareness level remains elevated in subsequent years in both cases, suggesting that 2023 has awakened public awareness of AI in significant ways.

Discussion and Conclusion

Social representation theory, the framework of this volume, seeks to elucidate a) how human common-sense deals with novelty—the unfamiliar, including emergent technologies, which is always challenging—and b) how the response of anchoring this challenge contributes to updating that very common sense. However, this process is far from being unidirectional and takes time. Under this lens, AI is both an object of representation and a humanoid subject-producer of representations on the other, and AI becomes a medium for doing so. Our chapter explored the first of these aspects, how AI is being represented in society. "Being represented" takes time, and this chapter traces comparatively for the UK and Korea the formation of a public opinion of AI. Our framing idea of a quasi-social movement that mobilizes attention and salience for AI thus elaborates a further question of SRT: How does the dance of subjects around an elusive object find its coordination in a project for the future? We might identify the answers as part of a shifting common sense.

Overall, our analysis of tech news showed a dormant period of AI between 2000 and 2015, masked by the other science and technology issues of nuclear, genetic and climate. Salience of AI increased in both countries after 2015, and was particularly sharp in 2023 following the launch of ChatGPT. However, a notable difference emerged in patterning: in the UK, with relatively low previous momentum, ChatGPT made a big difference in 2023. Korea had been primed; AI had previously raised the national alarm back in 2016 in the AlphaGo spectacle, so that ChatGPT added a smaller surge in 2023.

Analysis of AI awareness revealed similar differences: a sharp spike into 2023 for both countries, followed by a plateaued growth thereafter. While the overall increase over the period 2015–2023 was similar in the UK and Korea, the UK saw a more accentuated increase in public awareness due to ChatGPT, while Korea was already there, having been alerted by the AlphaGo story seven years earlier, which continued to resonate in the subsequent years. This increase in public awareness, in line with the flow of media salience, is consistent with the agenda-setting hypothesis: the media emphasise issues that afford public attention; they tell you what to think and talk about.

Salience Transfer

This paper demonstrates how public attention to AI is unfolding as salience transfers from media coverage to public perceptions. We showed a consistent upward trend in AI salience in both the UK and Korea, accompanied by corresponding growth in public awareness.

These findings are indeed consistent with agenda-setting: increased media coverage is followed by public attention with some lag, as public opinion develops. Our data demonstrate this media-driven impact on public awareness and interest: as AI-related news articles proliferated, public awareness of AI increased; and in comparing the UK and Korea, we can say that this increase is proportional. This pattern is especially evident in 2023, when ChatGPT's launch had generated widespread public excitement in both countries, triggering a public drama that continues in subsequent years.

However, we note the lack of specific historical data on public awareness during earlier pivotal moments, such as the 2016 AlphaGo match. This constraint stems from the fact that polling only championed dedicated AI questions from the early 2020s, having previously addressed AI within an emergent technology framework rather than as a distinct area of monitoring enquiry. The increasing presence of specific AI opinion polling is thus itself an indicator of salience transfer.

From Spectacle to Economic Imperative

Our keyword analysis reveals the evolution of dominant themes surrounding AI discourse over time. In both countries, earlier periods were characterised by discussions of emerging AI innovations—including robots and computers—often inspired by fictional narratives or high-profile events that captured public attention and fuelled imagination about an uncertain technological future.

The analysis of later years reveals a shift from AI-related innovations to their economic implications. This transition is indicated by the increasing prominence of AI companies and businesses in public discourse, along with associated products such as semiconductors.

Overall, the study shows a shift in AI discourse—from attention-grabbing "popular events" to promising "economic projects." This development indicates a sustained effort to leverage AI to enhance national competitiveness and economic prospects strategically.

However, even within the seemingly similar waves of change, subtle differences appear in the outcomes between the UK and Korea. These variations reflect each country's historical path and geopolitical context, shaping distinct thematic routes despite broadly comparable trends.

Contrasting Levels of Salience (2016~, Alphago Match)

The first difference between the UK and Korea is the salience of AI during the early stages, when the AlphaGo events occurred. Although both countries participated in the contest—with the UK as the AI developer and Korea with the human protagonist—the event had only a mild impact on AI salience in the UK. Several factors might explain this difference in resonance: there is limited familiarity with the board game Go in Western countries; DeepMind's achievements were shared with US-based Google as its parent company, and importantly, the UK's position as the inventor/innovator of the winning AI framed the outcome as an achievement rather than a threat.

In contrast, 2016 was undoubtedly a future shock in Korea, similar to the "Sputnik shock" which alarmed the USA back in 1957 about a potential USSR leadership in space technology. AlphaGo sparked intense interest and worries about domestic tech capabilities. Korea is a country with a history of relatively late-coming but fast-paced development, and it has an inherent obsession with "catching up" (Kim 2023) with advanced nations. The AlphaGo defeat came as a massive surprise and marked a National alarm: it demonstrated AI's groundbreaking capabilities, i.e., beating a human Go champion, and this came with the realization that the Nation risks being technologically left behind.

Korea suffered FOMO, fear of missing out. Overall, the event, combined with the complex philosophical nature of the ancient game of Go, positioned AI in Korean public discourse with a mixture of alarming fear and deference—AI viewed as an inevitable future to confront, other times as a powerful force of wisdom to embrace. This shift is evidenced by the media drama of AI-related coverage in 2016 and the event's continuing resonance in Korean AI discourse as a defining moment. The mobilisation effect of the AlphaGo event on public attention in Korea was the subject of an earlier study, which showed in a pre-post-comparison that increased media attention to AI had reduced awareness gaps across education and income (Chae 2017). The Go match became a recurring reference in Korean news long after its conclusion. For example, champion Lee Sedol's retirement match against a local AI Go program in 2019 reignited public attention and discussion about human-AI competition, which is a fascination Korea shares with Japan (Tanaka 2024).

Declining Attention During the Pandemic Crisis (2020~2022)

As strong middle powers in the global AI landscape, both the UK and Korea have demonstrated keen interest in national developments through technological shifts. The thematic analyses of earlier years exemplify this view; the UK maintained constant attention to US-led companies like "Google", "Facebook", and "Cars (Tesla)", while Korea brought neighbouring Japan's robot development into public discourse. However, after AI coverage had been on an upward trajectory during 2015–2018, the UK experienced a sudden drop in media attention during the COVID-19 pandemic; the Virus makes all the news. AI coverage reached a low point during the pandemic 2020-2022, and keyword analysis shows the rise of terms like "patient" and "cancer," which reflects AI mobilisation through integration with medical and health technology.

In contrast, Korea did not experience such a strong decline in AI salience during the same period. This can be partly attributed to Korea's milder pandemic experience: while Korea had nearly double the number of confirmed Covid cases than the UK, it had a six-time lower death rate (693 deaths per million population in Korea against 3,404 in the UK; source: *Wikipedia.org/wiki/Covid19_pandemic_death_rates_by_country [accessed 25 Sept 2025]*). This result was aided by extensive testing and digital contact tracing systems in a compliant community. The AI topic flow was also different during the pandemic. While pandemic-related themes such as "corona" similarly became prominent topics, Korea's ongoing focus on opportunity-driven sectors like "semiconductors" and "startups" maintained media focus on AI throughout this period.

Regaining Momentum with ChatGPT

In the UK, the media's waning attention to AI picked up rapidly after 2022. While the launch of ChatGPT was a key driver, the UK's struggle for economic growth post-Brexit and the need for recovery after COVID-19 also boosted this momentum. The quest for technological leadership grew crucial, indicated by terms such as "company," "business," and "government" becoming prominent in media discussions alongside "ChatGPT".

However, the rise of AI sparked a backlash, with the UK's AI-driven economic strategy facing growing frictions if not resistance. In 2023, "risk" emerged as a dominant theme of AI, including the discussion on existential risks for humanity, signalling mounting concerns, if not alarm, about its rapid advancement and marking a shift toward more critical discourse about a "technology out of control" [TOOC].

In contrast, while the release of ChatGPT reignited interest in AI, Korea experienced a more modest surge in 2023 than in the UK. Despite maintaining relatively strong emphasis on key economic narratives, the smaller increase in media coverage and public attention suggests waning momentum compared to the one witnessed over AlphaGo. Additionally, while UK media expressed concerns and apprehensions about AI's rapid advancement, Korea—having already faced disruption seven years earlier—showed no prominent negative themes and remained largely within an optimistic and opportunity-focused discourse.

A Hype Cycle at its Turning Point?

Given these observations, our findings illustrate how the public mobilisation of AI has developed across the UK and Korea, creating media salience and public awareness. The findings indicate that, although AI development is progressing as a global trend, it manifests with distinct national features.

Key elements of a tech movement functioned across both countries. Attention was catalysed by landmark events around AlphaGo and ChatGPT. Future projects materialized through national policies and funding initiatives driven by political and corporate actors pursuing economic interests. This mobilization effort operates within distinct geopolitical contexts—Korea's ambition to catch up, alerted by FOMO [fear of being missing out] [FOMO], and the UK's desire to maintain its position in the "AI arms race" (AI NOW Institute 2024), despite being alarmed by TOOC [a "technology out-of-control"].

The AI movement mobilized extensive resources, symbolic and material, through increased media coverage, thus setting the agenda and alerting public awareness. The movement, which manifested differently across nations, was also represented differently within each society: The UK experienced its alarm following ChatGPT's release, successfully capturing public attention under the theme of existential risk concerns about the technology. Conversely, ChatGPT prompted increased media visibility in Korea, but this occurred at a lower rate. The country's prior experience with the AlphaGo match of 2016, which had sparked unprecedented expectations about the social, economic and future implications of AI, appears to have moderated subsequent enthusiasm.

The history of computing has seen previous waves of enthusiasm for AI technology. Our findings raise questions about whether this round of mobilisation of AI will maintain its momentum or follow the pattern of a hype cycle? And might

2023 have been the peak? The greater the hyperbole surrounding expectations, the higher the risk of subsequent disappointment (Süerdem 2023).

We have shown that hyperbole can also signal alarm; only bad news is good news. But as history shows, alarming different issues: Korea was shocked in 2016 into fear that the Nation is losing out, the UK in 2023 was alarmed about a technology out-of-control. The AI challenge is the same, but the response reflects local dispositions.

Further Questions?

The questions of a turning hype cycle and the local responses to the global AI challenge warrant further research on a moving target. The AI story has clearly not ended, only just opened and maybe closing another chapter. These findings update our understanding of AI as a quasi-social movement mobilising around landmark events like AlphaGo and ChatGPT as temporary focal points.

We have shown agenda-setting at work, successfully telling people in the UK and Korea what to think about, namely, AI. However, we have not yet shown how the Nations are to think about AI, i.e., the question of how AI is framed? That is indeed our next task. Is AI framed in public discourse as a technological fait-accompli [TINA, there is no alternative] or is this an open development with societal choices [TAA, there are alternatives]. Acemoglu and Johnson (2023) have tried to show that we are indeed presented with a fait-accompli [AI] for the substitution of labour, omnipresent surveillance and effective propaganda, while there are potentials for useful machinery [UMs] to make work more efficient, create meaningful jobs and emancipate humans from drudgery and oppressive conditions in social networks in order to unleash the human capacity and better communication. It would be interesting to see how and how far these choices are part of the discourse or not?

Furthermore, any social movement has actors, and so do tech movements, and we might want to hear the voices of the AI protagonists, how they have contributed to the salience and topicality, and whether they are surprised by what came of it, and what they think of each other in the UK and Korea? For this continuing story, we are pleading: readers, please have patience, there is more to come!

Acknowledgements

Earlier drafts of this chapter were edited with the help of the platform claud.ai to refine the use of the English language; the final structure and terminology are human work. The first author would also like to thank the CIRS-25 Conference [Brest, 7-11 July 2025] and its organisers and participants, where the first author had the privilege of presenting this study.

References

Acemoglu, D. and Johnson, S. 2023. Power and progress: Our thousand-year struggle over technology and prosperity. Basic Books.

AI NOW Institute. 2024. 5. A Lost Decade? The UK's Industrial Approach to AI. AI Now Institute. https://ainowinstitute.org/publications/a-lost-decade-the-uks-industrial-approach-to-ai

Bauer, M. W. 2015. Atoms, Bytes and Genes: Public Resistance and Techno-Scientific Responses (1st edn, Vol. 126). NewYork, Routledge.
Bauer, M. W. and B. Schiele. 2024. AI and common sense: Ambitions and frictions. Routledge, Taylor and Francis Group.
Benaich, N. and I. Hogarth. 2023. State of AI Report. Google Scholar.
Blumer, H. (with Internet Archive). 1969. Principles of sociology. New York : Barnes and Noble. http://archive.org/details/principlesofsoci0000blum
Brennen, J. S., P. N. Howard and R. K. Nielsen. 2018. An Industry-Led Debate: How UK Media Cover Artificial Intelligence. Reuters Institute for the Study of Journalism. https://reutersinstitute.politics.ox.ac.uk/sites/default/files/2018-12/Brennen_UK_Media_Coverage_of_AI_FINAL.pdf
Bridgstock, M. 1998. Science, technology, and society: An introduction. Cambridge University Press.
Bundy, A. 2017. Preparing for the future of Artificial Intelligence. Springer London.
Butler, D. 2016. TOMORROW'S WORLD. Nature (London), 530(7591): 398.
Carroll, C. E. 2015. Mapping the Contours of the Third Level of Agenda Setting: Uniplex, Duplex and Multiplex Associations. In The Power of Information Networks. Routledge.
Cave, S., C. Craig, K. Dihal, S. Dillon, J. Montgomery, B. Singler et al. 2018. Portrayals and Perceptions of AI: Why They Matter. https://doi.org/10.17863/CAM.34502
Ceruzzi, P. E. 2003. A History of Modern Computing, second edition. MIT Press.
Chae, J. 2017. Going for AI – a study of techno-movement and hype-cycle through the AI Go-match in Korea [MSc Dissertation]. London School of Economics.
Cordeschi, R. 2007. AI Turns Fifty: Revisiting Its Origins. Applied Artificial Intelligence, 21(4–5): 259–279. https://doi.org/10.1080/08839510701252304
Dafoe, A. 2018. AI Governance: A Research Agenda. Centre for the Governance of AI, Future of Humanity Institute. https://cdn.governance.ai/GovAI-Research-Agenda.pdf
Djerf-Pierre, M., A. Shehata and B. Johansson. 2025. Media Salience Shifts and the Public's Perceptions About Reality: How Fluctuations in News Media Attention Influence the Strength of Citizens' Sociotropic Beliefs. Mass Communication and Society, 28(3): 459–484. https://doi.org/10.1080/15205436.2023.2299209
Galanos, V. 2023. Expectations and expertise in artificial intelligence: Specialist views and historical perspectives on conceptualisation, promise, and funding. https://doi.org/10.7488/era/3188
Gamson, W. A. and A. Modigliani. 1989. Media Discourse and Public Opinion on Nuclear Power: A Constructionist Approach. American Journal of Sociology, 95(1): 1–37.
Goldman Sachs. 2024. Gen AI: Too much spend, too little benefit? Goldman Sachs. https://www.goldmansachs.com/insights/top-of-mind/gen-ai-too-much-spend-too-little-benefit
Greenberg, P. 2017. Separating AI Reality from AI Hype. Customer Relationship Management (Malibu, Calif.), 21(2): 4–4.
Hall, W. and J. Pesenti. 2017. Growing the artificial intelligence industry in the UK. Department for Digital, Culture, Media and Sport and Department for Business, Energy and Industrial Strategy. Part of the Industrial Strategy UK and the Commonwealth.
Havenstein, H. 2005. Spring comes to ai winter. Computerworld, 39(7): 28–28.
Jasanoff, S. and S.-H. Kim. 2015. Dreamscapes of Modernity: Sociotechnical Imaginaries and the Fabrication of Power. University of Chicago Press.
Jones, R. 2024. The shifting sands of UK Government technology prioritisation. Soft Machines. Récupéré de. http://www.softmachines.org/wordpress/wp-content/uploads/2024/02/UK-Government-technology-priorities.pdf
Kahn, J. 2025. Is another "AI winter" coming? Here's what past AI slumps can tell us about what the future might hold. Fortune. https://fortune.com/2025/09/03/what-previous-ai-winters-can-tell-investors-and-executives-about-what-might-be-coming-next-for-ai/
Kelley, P. G., Y. Yang, C. Heldreth, C. Moessner, A. Sedley, A. Kramm et al. 2021. Exciting, Useful, Worrying, Futuristic: Public Perception of Artificial Intelligence in 8 Countries. Proceedings of the 2021 AAAI/ACM Conference on AI, Ethics, and Society, 627–637. https://doi.org/10.1145/3461702.3462605
Kiecolt, K. J. and L. E. Nathan. 1985. Secondary Analysis of Survey Data. SAGE Publications, Inc. https://doi.org/10.4135/9781412985796
Kielbowicz, R. B. and C. W. Scherer. 1984. The Role of the Press in the Dynamics of Social Movements.

Kim, S. Y. 2023. Development and Developmentalism of Artificial Intelligence: Decoding South Korean Policy Discourse on Artificial Intelligence. *In:* Cave, S. and K. Dihal (Eds.). Imagining AI: How the World Sees Intelligent Machines (p. 0). Oxford University Press. https://doi.org/10.1093/oso/9780192865366.003.0020

Kurzweil, R. 2005. The singularity is near: When humans transcend biology. Penguin Books.

LaGrandeur, K. 2024. The consequences of AI hype. AI and Ethics, 4(3): 653–656. https://doi.org/10.1007/s43681-023-00352-y

Latour, B. 1987. Science in action: How to follow scientists and engineers through society. Open University Press.

Lewis, S. C. 2012. The Tension Between Professional Control and Open Participation: Journalism and Its Boundaries. Information, Communication and Society, 15(6): 836–866. https://doi.org/10.1080/1369118X.2012.674150

Maslej, N. 2025. Artificial Intelligence Index Report 2025. Artificial Intelligence.

McCarthy, J. D. and M. N. Zald. 1977. Resource Mobilization and Social Movements: A Partial Theory. The American Journal of Sociology, 82(6): 1212–1241. https://doi.org/10.1086/226464

McCombs, M. E. 2014. Setting the agenda: The mass media and public opinion (Second edition.). Polity.

Newig, J. 2004. Public Attention, Political Action: The Example of Environmental Regulation. Rationality and Society, 16(2): 149–190. https://doi.org/10.1177/1043463104043713

Poireault, K. 2023. UK AI Safety Institute: A Blueprint for the Future of AI? Infosecurity Magazine.

Qi, W., J. Pan, H. Lyu and J Luo. 2024. Excitements and concerns in the post-ChatGPT era: Deciphering public perception of AI through social media analysis. Telematics and Informatics, 92: 102158. https://doi.org/10.1016/j.tele.2024.102158

Roubini, N. 2020, April. Ten reasons why a “Greater Depression” for the 2020s is inevitable. The Guardian. https://www.theguardian.com/business/2020/apr/29/ten-reasons-why-greater-depression-for-the-2020s-is-inevitable-covid

Süerdem, A. 2023. Hype Cycles in AI: Broken promises, empty threats in Bauer MW and Schiele B (2023) Science communication: Taking a step back to move *forward*. PUF.

Süerdem, A. 2024. The challenges and opportunities in large language models: Navigating the perils of stochastic and scholastic parrots in artificial understanding and common sense. *In*: Bauer, M. W. and Schiele, B. (eds.). AI and common sense: Ambitions and frictions. Routledge, Taylor and Francis Group. https://www.routledge.com/AI-and-Common-Sense-Ambitions-and-Frictions/Bauer-Schiele/p/book/9781032626178

Tanaka, M. 2024. Meanwhile in Japan: The possibility of techhno-animism for engaging deliberation for emerging technology. *In*: Bauer, M. W. and Schiele, B. 2024 (eds.). AI and common sense: Ambitions and frictions. Routledge, Taylor and Francis Group. https://www.routledge.com/AI-and-Common-Sense-Ambitions-and-Frictions/Bauer-Schiele/p/book/9781032626178

Tarrow, S. G. 2011. Power in Movement: Social Movements and Contentious Politics (3rd edn). Cambridge University Press. https://doi.org/10.1017/CBO9780511973529

The Economist. 2025. What if artificial intelligence is just a normal technology? The Economist. https://www.economist.com/finance-and-economics/2025/09/04/what-if-artificial-intelligence-is-just-a-normal-technology

Tilly, C. 1978. From mobilization to revolution. Random House.

US Government. 2016. Preparing for the future of artificial intelligence. National Science and Technology Council, Washington, D.C. 20502.

Windahl, S., B. Signitzer and J. Olson. 2009. Using Communication Theory: An Introduction to Planned Communication: Vol. Second edition. SAGE Publications Ltd.

Zaklama, S. 2025. Exploring the Foundations of Media Framing Theory. ResearchGate. https://doi.org/10.59573/emsj.9(1): 2025.7

Zhang, B. 2024. Public Opinion toward Artificial Intelligence. *In:* The Oxford Handbook of AI Governance. Oxford University Press. https://doi.org/10.1093/oxfordhb/9780197579329.013.36

APPENDIX

News Data Corpus and Preparation

UK Dataset

New items were obtained from *Nexis* using the search term "Artificial Intelligence" across *The Guardian, The Times, and The Telegraph* from 2015 to 2023. A total of 14,610 articles were first extracted for the basic review of the media coverage. For computational analysis of the keywords, the data were imported into R for data cleansing, which removed duplicates, articles that were too short or too long (n < 50 or n > 3,000 words), regional editions (Guardian US and Australia), and articles with date errors, yielding 9,155 articles for analysis. Pre-processing steps included tokenization, lowercasing, lemmatization, and removal of stopwords and publication-specific terms ("date," "published," "updated," "page," etc.).

Korean Dataset

Articles were retrieved from *BigKinds* using "*In-gong-ji-neung*" (Korean for "Artificial Intelligence") across *JoonAng Ilbo, Dong-A Ilbo*, and *Hankyoreh* from 2015 to 2023. A total of 14,568 articles were initially extracted for the basic review of the media coverage. For computational analysis of the keywords, the data cleansing process filtered out duplicates, unrelated contents (obituaries, photos of the day, HR announcement, etc), and articles that were too short or too long (n < 100 or n > 6,000 characters), with 14,330 articles remaining. Pre-processing was applied to the keywords through tokenization, lemmatization, and removal of stopwords and publication-specific terms ("newsletter," "subscription," etc.).

CHAPTER 4

The Political Impact of AI Social Representations

Gege Gatt, Joseph Borg* and *Mary Anne Lauri*

Introduction

Every new communication technology introduced in society gives rise to contrasting social representations and metaphors. Understanding these representations and metaphors is crucial, as they can shape our perceptions and actions both consciously and subconsciously. Artificial Intelligence (AI) is more than a new technology; it is a societal force reshaping politics and information ecosystems (Filgueiras 2022; Lee 2018). Its widespread adoption is transforming how information is created, shared, and consumed. As AI permeates deeper into societal frameworks, its implications for democracy and governance have become subjects of intense scrutiny (Kaye 2018). Understanding this interconnectedness is crucial for fostering a democratic environment (Doise et al. 2002).

Representations of AI reflect broader discourses about technology, progress, and human values (Staerkle et al. 2011). People who celebrate technological innovation as a sign of progress, have representations of AI that emphasize its transformative potential to improve efficiency, enhance convenience, and solve complex problems. However, there are also negative representations of AI resulting from concerns about privacy, replacing roles that humans do, job loss and harm to personal relationships (Kelley et al. 2021).

Emerging research suggests that public perceptions of AI (whether optimistic or sceptical) can influence electoral alignment, ultimately shaping the policy platforms adopted by political parties and governments (Jungherr et al. 2024). When, for

University of Malta.
Email: joseph.borg@um.edu.mt, ORCID: https://orcid.org/0000-0003-3983-307X,
Google Scholar: https://scholar.google.com/citations?hl=enanduser=UEEjCgoAAAAJ
Email: mary-anne.lauri@um.edu.mt; ORCID: https://orcid.org/0000-0001-5219-8010
* Corresponding author: gege.gatt@ebo.ai; ORCID: https://orcid.org/0009-0006-6635-8592
Google Scholar: https://scholar.google.com/citations?user=hP7ElToAAAAJ&hl=en

DOI: 10.1201/9781003502753-7

example, voters' perceptions of AI are predominantly depicted through positive metaphors (such as AI as a partner or guardian), political parties could be pressured to adopt different policies than if AI is viewed negatively (such as AI as a threat or job eliminator).

AI has been rapidly integrated into various facets of modern life including work environments, education, healthcare, justice system and politics. Society finds itself at the intersection of AI-driven realities, grappling with the profound socio-economic and cultural changes brought about by technological advancement. AI technology has emerged as a catalyst for societal transformation, revolutionising industries, economies, and everyday life (Dwivedi et al. 2023). From predictive algorithms to autonomous systems, AI's capabilities reshape how individuals interact with technology and each other.

This study explores whether there is a link between the social representations of AI and the way people vote in Malta. This small island nation in the Mediterranean has a population of approximately half a million citizens and has been a member of the European Union since 2004. The country operates a parliamentary representative democratic republic and features a historically entrenched two-party system dominated by the Partit Laburista (PL) and the Partit Nazzjonalista (PN). Voter turnout in Malta is among the highest in Europe, regularly exceeding 85%, reflecting a deeply embedded culture of electoral participation and partisanship. This context makes Malta a particularly compelling case for studying how emerging technologies intersect with political identity. The study also discusses the need of robust life-long media literacy programmes as the means of empowering people to consciously navigate safely in a new AI-driven environment.

Social Representations Theory and AI

Social Representations Theory suggests that people use socially constructed symbols, metaphors and representations to understand and communicate about new phenomena, offering a comprehensive framework for understanding the processes of meaning-making within social groups (Moscovici 1961, 1984). These representations help individuals make sense of novel phenomena by anchoring them within culturally shared meanings (Jovchelovitch et al. 2001; Brondi et al. 2021). When confronted with new and innovative phenomena, individuals engage in a process of collective coping by anchoring unfamiliar concepts or innovations to existing knowledge and objectifying them into tangible symbols or metaphors they understand (Wagner 2020). Analysing such metaphors is essential, as they are fundamental to human cognition (Lakoff and Johnson 2003). These metaphors range from utopian framings of AI as a societal cure-all to dystopian portrayals of it as a threat; echoing broader technophilic and technophobic worldviews embedded in cultural discourse.

Historical perspectives illustrate these contrasting views on every new communication technology (Orben 2020). Ancient Greek philosopher Socrates, in Plato's *Phaedrus*, criticised the invention of writing, arguing that it weakens memory and creates the illusion of knowledge without true understanding. The invention of printing by moveable type in the fifteenth century ushered in the Gutenberg Galaxy. This did not only change the way we perceive the world, but also brought about a radical cultural, religious, political and scientific revolution (Ong 1967; Eisenstein 1983). Centuries later, Arnheim (1971) lamented that television would reduce people to pathetic isolated hermits. Similar concerns were voiced about the internet (Kraut 1998), despite later studies that link internet use to higher well-being (Vallance and Wain 2024). Postman (1987) warned that television was leading to cultural decline while McChesney (1999) feared that the information highway fosters isolation, atomisation, and marginalisation. On the other hand, for McLuhan (1964) the electronic media enhances connectivity, functioning as a universal nervous system. Reactions to AI are passing through a similar process.

Representations of AI are reworked through media discourse. Media narratives contribute to the formation of new representations or the consolidation of existing ones. The study by Kelley et al. (2021) with 10,000 persons from eight countries in six continents found that social media posts, TV reports, movie portrayals and online articles were the four top sources of learning about AI. Media content circulates through various channels influencing societal understanding and acceptance of AI. Thus, media serves as a conduit for the transmission and reinforcement of some representations over others, exerting significant influence on collective perceptions. Through news coverage, TikTok content, opinion pieces, and entertainment, the media frames narratives about AI's potential benefits and risks. Often media representations of AI lack nuance (Caves et al. 2018) with sensationalist portrayals of AI in popular culture, like the movies "Her" in 2013 and "Ex Machina" in 2015, which contribute to the dissemination of myths and misconceptions further complicating public understanding.

Media, when leveraged by entities with political and economic power become even more powerful in shaping representations. Political entities and interest groups strategically utilise AI-driven algorithms to amplify certain narratives while suppressing others, tailoring content to specific audiences to shape public opinion and political behaviour (Settle 2018). Social media algorithms, powered by AI, personalise individual users' content feeds, reinforcing pre-existing beliefs and limiting exposure to diverse perspectives, thereby fostering echo-chambers that distort societal pluralism. AI-generated deepfake technology further complicates the media landscape, blurring the lines between truth and fabrication and raising concerns about misinformation (Kreps et al. 2022). Media acts as a creator, a mirror and a manipulator of AI narratives. For this reason, it is important for societies to remain resilient against algorithmically curated realities which influence political information and voting behaviour (Filgueiras 2022; Lee 2018). Media literacy, within an appropriately regulated environment, can act as a critical factor in mitigating the spread of disinformation. Individuals with high levels of media literacy may be empowered to navigate the challenges and opportunities presented by AI in an informed and responsible manner.

Politics and the Representations of AI

Representations of AI can influence collective decisions and voting outcomes by framing political debates and shaping public opinion (Kelley et al. 2021). Political actors harness AI-driven tools for voter targeting, campaign messaging, and sentiment analysis, seeking to gain a competitive edge in electoral contests. AI can be harnessed to change the way people vote, reinforce their political voting preferences or shape attitudes towards political parties (Yi et al. 2023). Concerns about the weaponisation of AI for disinformation campaigns and voter manipulation have raised alarms about the integrity of democratic elections. Moreover, AI's role in surveillance and predictive policing raises questions about civil liberties and the erosion of privacy rights in democratic societies. Political actors may leverage representations of AI to advance their agendas, framing issues in ways that resonate with voters. Candidates who advocate for policies to regulate AI and protect workers may appeal to voters concerned about the societal implications of technological advancement, while those who champion AI-driven innovation and deregulation may attract support from voters who prioritise economic growth and competitiveness.

Research in the United States, the United Kingdom and several European countries has demonstrated a relationship between how people view AI, their political orientations and voting habits. Borwein et al. (2025) studied the political implications of the perceived threat of technological change in 15 European countries, including France, Germany, Italy, and Ireland. Their study showed that subjective perception of personal automation exposure relates to increased support for left parties and decreased support for populist-right parties. Concern over threats to workers' friends and family members increased support of the populist right (Borwein et al. 2025).

Yi et al. (2023) researched the role media plays in shaping public opinion on AI by analysing over 7,500 articles from both liberal and conservative media. The results showed that the liberal media depict a more negative sentiment towards AI compared to conservative media. This negativity is partly attributed to concerns that AI may amplify existing societal biases, such as those related to race, gender, and income inequalities.

Political beliefs drive opinions on AI (Zaheer 2023). Different political groups have different core beliefs and priorities for society; consequently, adherence to a political party or ideology shapes the perception of different groups about the role of AI. Republicans in the United States are polarised between those who desire minor regulation and those who want extreme regulation of AI while Democrats were more consistent in desiring moderate regulation (Zaheer 2023).

A UK-wide study by PublicFirst (Dupont Wride and Ali 2023) found that the public is largely curious about AI, with similar levels of excitement and concern across the sample. By contrast, the British Social Attitudes Survey presents a more politically nuanced picture. In this survey, Margetts et al. (2025) found that left-leaning voters were more optimistic about AI's impact on economic efficiency, and more likely to associate AI with positive emotions such as curiosity, excitement and hope.

Republicans in the United States are most concerned about the effects of AI on personal privacy, national security, and elections (Polarization Research Lab 2024). While Democrats are also concerned about these issues, they are divided over whether AI may make national security worse or better.

A significant recent contribution to the field is Jacobs' (2024) model of AI-induced political polarisation. This draws on economic and cultural dimensions to explain how individuals' labour market positioning in relation to automation influences their ideological alignment. Jacobs finds that those deemed automation-susceptible tend to be economically left-leaning but culturally conservative, whereas those whose roles are complemented by AI tend to adopt culturally liberal and fiscally conservative positions. In practical terms, those feeling personally threatened by AI gravitate toward protectionist or populist positions, whereas those benefiting from AI adopt more libertarian or status-quo attitudes.

Additionally, Jacobs found that Republicans are more inclined than Democrats to perceive AI systems as biased against men and White individuals. Conversely, Democrats are somewhat less likely than Republicans to acknowledge the existence of such biases in AI systems.

This brief literature review shows that while the relationship between one's political beliefs and AI is complex, there is a distinction between the views of left-wing and right-wing voters. This is to be expected as several aspects of the worldview of both groups of voters are different. Taken together, these theoretical frameworks (particularly Jacobs' model of AI-induced political polarisation and Moscovici's theory of social representations) inform the rationale for this study's empirical design and method. It investigates how the public's metaphorical framings of AI intersect with voting behaviour in Malta.

Method

This study employed a cross-sectional survey from a representative sample of people living in Malta (n = 409) aged 18 and over, to find out whether there is a relationship between social representations of AI and voting patterns.

Data collection was carried out by Misco International between 27 June and 10 July 2024, using a quota system to represent the age and gender distribution of the Maltese population based on: "Demographic Review 2014" National Statistics Office, Valletta, 2016. The questionnaire was administered using CATI by trained personnel and took between four and five minutes to complete. Participants were told about the nature of the study and assured that only aggregate data would be used when presenting the research. Data was cleaned and the responses were then transferred to SPSS for analysis.

The questionnaire began with two demographic items capturing respondents' age and gender. These were followed by an open-ended prompt inviting participants to articulate, in their own words, their understanding of the term *Artificial Intelligence*. The final item asked participants to indicate their chosen party in the June 2024 European Parliament elections.

Table 1 summarises the ideological positioning, affiliations, and vote share of political parties that contested the 2024 European Parliament elections in Malta with official electoral outcomes obtained from Malta's Electoral Commission.

Results

When asked to define Artificial Intelligence in their own words, one third of the sample (n = 131) said that they did not know what AI is. This is interesting especially in the context of understanding the social representations of AI. The participants'

Table 1: Political parties in Malta and voting patterns in European Parliament Elections 2024.

Exponents	Political Position	Description of Political Orientation	Membership	Exponent's share of votes (%)
PN	Centre	A political entity that seeks to balance market-driven economic policies with social welfare provisions, advocating for moderate social reforms, pragmatic governance, and consensus-driven policymaking.	Member of the European People's Party (Christian Democrats)	42.02%
PL	Left	A political organisation with traditionally left-leaning policies that supported state intervention in the economy, yet with a present spate of market-driven economic activities.	Member of the Progressive Alliance of the Socialists and Democrats in the European Parliament	45.26%
Volt	Left	A Eurofederalist and progressive pan-European political party and movement, which advocates for greater European cooperation across Europe.	Member of Volt Europa	0.11%
ABBA	Right	A political organisation that emphasises free-market economics, limited government intervention, national sovereignty, and traditional social values, often prioritising law and order, lower taxation, and private enterprise.	N/A	0.20%
Imperium Europa	Extreme Right	Focused on preservation of the white European race and opposition to liberal democracy.	N/A	2.62%
AD+PD	Green	A political organisation that prioritises environmental sustainability, social justice, and grassroots democracy, advocating for ecological policies, renewable energy investment, climate action and social equity through progressive economic and social reforms.	Member of European Greens	1.19%
Independent	Independent	A set of independent candidates not grouped into a political party and subscribing to one or more of the groupings above	N/A	8.59%

verbatim responses were analysed using Thematic Analysis, following Braun and Clarke's (2006) six-step framework. The objective was to identify, analyse, and report patterns or themes within data. Two researchers conducted the analysis independently, and their results were subsequently compared to ensure consistency and reliability.

Eight themes were identified (Table 2). A deeper examination of these themes revealed divergent perspectives on AI. While some participants expressed highly optimistic views, others perceived AI as a significant threat.

Table 2: Thematic Analysis of the responses to the open-ended question asking 'How would you define Artificial Intelligence (AI) in your own words?'

Themes	%	Description of theme	Examples of Responses *Verbatim*
Makes life easier	16%	Describes AI as useful in tasks, simplifying life, or improving daily functions.	• A useful tool • Positive/helpful tool • It helps with health-related issues • Personal benefit
Technological help	15%	Refers to AI as a technological enabler of daily tasks.	• Like ChatGPT • Next-Gen assistance • New technology • A version of google
Computerised human/ robot	9%	Suggests AI could aid human roles, emphasising automation and robotics.	• It could create life-style changes • A work-related digital tool • Editing software • Performing and simplifying tasks
Threat to Humanity	7%	Expresses fear that AI could harm humanity, take jobs, or pose risks.	• We are not yet prepared for it • Lack of privacy • AI could be the cause of destroying the cyber system • Need to be controlled to protect the data protection
Useful tool for research and gathering information	5%	Indicates AI as a collaborative tool aiding research and problem-solving.	• Digital intelligence/assistance • A mission driven system • Decision making technology
Will create problems	3%	Suggests AI could have unintended negative consequences, affecting various sectors.	• Decrease in jobs in the film industry • Could cause problems for the older generation • A means of making people lazy
Is the future	3%	Perceives AI as a force driving technological advancement and future progress.	• Evolving • A technologic invention that will change the world • New technology
Artificial/fake	2%	Reflects skepticism, believing AI is not real or is misleading.	• Something negative • No need for it • Mind blocker
The unknown	32%	Respondents have no opinion or knowledge about AI or sees it as mysterious	• Something related to technology • Unpredictable • Not interested • Don't know

Upon further analysis of the open-ended responses and the eight initial thematic codes, five overarching social representations of AI were derived.

These representations encapsulate a spectrum of interpretive positions. At one end, some respondents viewed AI as a *threat to humanity*, reflecting existential or ethical anxieties. At the other, AI was characterised as *beneficial and supportive*, denoting high perceived utility. Between these poles, two nuanced perspectives emerged: AI as a *transformative force* associated with progress and innovation, and AI as a *stabilising force* linked to institutional reinforcement and societal order. A fifth and distinctive category, *AI as the Unknown*, captured responses marked by ambiguity, uncertainty, or disengagement. Together, these five representations offer a typology of how AI is socially constructed within the Maltese electorate.

Table 3 presents the outcome of a second-level interpretative coding process, where the five overarching social representations identified in Table 2 were further

Table 3: Social Representations of AI and Associated Metaphorical Framings.

Social Representation of AI		**%**	**Themes**	**Metaphors derived from Verbatims**	**Description of Social Representation**
1	**AI as a helper/ assistant**	31%	Helper/ Makes life easier Technological help	Invisible assistant	This reflects the view that AI enhances daily life by automating tasks and simplifying processes, making technology an indispensable tool for efficiency.
2	**AI as a stabilising force**	14%	Computerised human/ robot Useful tool for research and gathering information	A well-calibrated tool	AI is seen as supporting structure for governance and security, reinforcing societal order and aiding in decision-making processes.
3	**AI as a threat to humanity**	12%	Threat to Humanity Will create problems Artificial/fake	A monster	Encompasses fears about AI replacing human jobs, violating privacy, or leading to harmful consequences for individuals or society.
4	**AI as a progressive tool**	3%	Is the future	A second brain	Represents AI as a driver of innovation, shaping the future through technological advancements and fostering economic and social development.
5	**AI as the unknown**	32%	Do not know	Fog	A category representing individuals who do not have a defined understanding of AI, express disinterest, or remain uncertain about its role and implications in society.

analysed through the lens of metaphor theory (Lakoff and Johnson 2003; Schmitt 2005). Drawing on the principles of anchoring and objectification as well as Lakoff and Johnson's (2003) theory that metaphors structure how people cognitively frame and emotionally relate to abstract phenomena, we identified dominant metaphorical framings that respondents used, explicitly or implicitly, to make sense of AI as a social object.

Representations of AI with Declared Voting Patterns

In June 2024, almost 73% of the population in Malta voted in the European Parliament elections to fill six seats representing Malta (Electoral Commission, Malta - https://electoral.gov.mt/ElectionResults/MEP). Thirty-nine candidates contested the elections: Eight of these candidates were affiliated with the PN, currently the political party in opposition, while nine were candidates for the PL, currently the party in government. There were 13 independent candidates, four candidates for the Green Party, two for Imperium Europa, two for ABBA, and one candidate for Volt. The elected candidates were three from PN and three from PL.

When results of how participants voted were cross-tabulated with their representations of AI, a significant relationship was found (Chi-square = 37.44, p-value = 0.039) suggesting that there is an association between how respondents perceive AI and voting patterns. Table 4 displays the standardised residuals resulting from the cross-tabulation of AI representations and declared voting behaviour for major parties. Positive residuals indicate an over-representation of a given voting group within a particular social representation, while negative values reflect under-representation relative to expected frequencies.

The standardised residuals presented in Table 4 indicate which social representations of AI were mentioned more (or less) frequently than expected under the assumption of no association between voting behaviour and representation type.

The three representations cited more frequently than would be expected under conditions of no association between voting behaviour and AI representations were: (i) 'AI as a Helper or Assistant' among respondents who declined to disclose their voting preference, (ii) 'AI as the Unknown' among voters who supported PL, and (iii) 'AI as a Stabilising Force' among those who reported not voting. Conversely, two representations occurred less frequently than expected: (i) 'AI as a Helper or Assistant' among PL voters, and (ii) 'AI as a Stabilising Force' among PN voters.

The sparsity of the data suggested however that this value is to be interpreted with caution. Subsequently a Correspondence Analysis was carried out with the aim of obtaining a visual representation of the relationships between the two levels in a contingency table, that of the representations of AI and voting patterns.

Correspondence Analysis is mostly descriptive in essence, an aid to interpretation and theory construction rather than model testing (Hammond 1993). In Correspondence Analysis, the different levels of the two categorical variables are given scores on one or more dimensions, in such a way that levels that are more alike will get similar scores. Therefore, if the scores are plotted as graphs with the dimensions as axes, levels that are alike appear close to each other, whereas levels that are dissimilar appear far apart. The analysis was carried out using the R package FactoMiner.

Table 4: Standardised Residuals From the Contingency Table of Representations of AI by Voting Behaviour.

Representation	PL	PN	Independent	Right	Green	Not voted	Refused to answer
AI as a helper/ assistant	–2.815	0.041	1.375	0.181	0.511	–1.893	3.063
AI as a stabilising force	–1.185	–2.620	–0.413	0.862	1.365	2.606	0.095
AI as a threat to humanity	1.361	1.015	–0.077	0.005	–0.548	0.173	–1.892
AI as a progressive tool	–0.106	0.038	–0.042	0.003	–0.302	–0.316	0.410
AI as the unknown	2.776	1.219	–0.977	–0.842	–1.029	–0.175	–1.945

Discussion

A key contribution of this study lies in its identification and interpretation of the metaphors underpinning the public's social representations of AI. As shown in Table 3, each representation is anchored in metaphorical language that reflects deeper cognitive and affective schemas. For instance, AI is conceptualised as *"a human robot that never sleeps"* when framed as a helper, or as *"a monster"* when imagined as a threat to humanity. These metaphors are not merely rhetorical flourishes; they function as cognitive anchors that help individuals simplify and objectify complex technological phenomena. Their use reveals the ways in which AI is emotionally coded and narratively situated in the public imagination, with significant implications for trust, adoption, and political response.

The Correspondence Analysis conducted in this study reveals structured associations between social representations of AI and declared voting preferences. The map shows that these associations are not random but align along two latent dimensions (Dim 1 = 63.33%; Dim 2 = 34.71%) that jointly explain over 98% of the variance. Social representations are not politically neutral: they align with ideological identities in ways that reflect deeper cultural, economic, and informational divides. Notably, voters affiliated with the PL clustered near representations of AI as a *Threat* or as *Unknown*, reflecting ambiguity and concern. By contrast, those aligned with right-leaning parties and independents tend to cluster closer to representations like *Helper/Assistant*, *Progressive Tool*, or *Stabilising Force*, suggesting a more instrumental or optimistic framing of AI.

This spatial distribution suggests that AI metaphors are serving as symbolic stand-ins for broader political worldviews. This echoes McLuhan's (1964) view of media as an extension of the human nervous system, shaping how people interpret their environment, and in this context, constructing the symbolic frameworks through which citizens anchor meaning. Each representation listed below appears to crystallise a distinct socio-political orientation, reflecting how individuals project their values onto the perceived role of AI within society.

AI as a Helper or Assistant

This representation emphasises AI's capacity to simplify daily life and perform functional tasks. It clusters near 'Independent' and 'Refused to Answer' voters. These groups indicate an ideologically diffuse orientation. The proximity suggests that individuals who hold pragmatic, tool-based views of AI are less ideologically polarised. They may value utility over ideological framing, viewing AI as an extension of existing technologies and a pragmatic enhancement to progress.

AI as a Stabilising Force

This category demonstrates a moderate association with 'Not Voted' and 'Right-of-Centre' voters and lies in the upper-right quadrant of the Correspondence Analysis map (Figure 1). The placement implies that those with conservative tendencies may view AI as a mechanism for control, predictability, and institutional continuity. This interpretation aligns with political preferences that value law, order, and stability; attributes which AI embodies when framed as a governance tool to enhance security.

A notable paradox emerges in the Correspondence Analysis: the social representation of AI as a Stabilising Force aligns most closely with individuals who chose to abstain from voting altogether. This raises an interpretive challenge. One possible explanation is that these respondents exhibit technocratic trust but political disillusionment, viewing AI as capable of stabilising systems more reliably than electoral politics. In this sense, their abstention may not reflect apathy, but rather a post-political mindset wherein technology is seen as a more effective mechanism for governance than democratic participation. This insight reflects the depoliticising effect of techno-optimism (Morozov 2013).

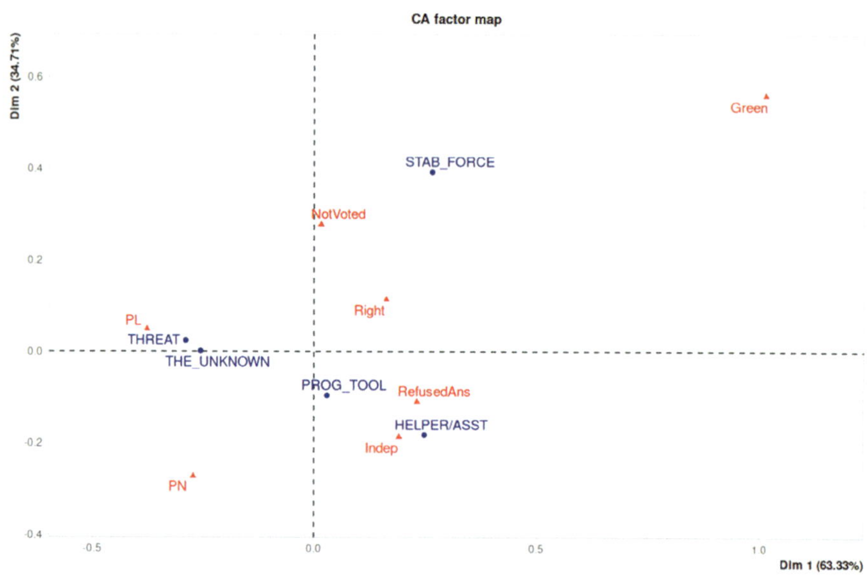

Figure 1: Correspondence Analysis of Social Representations of AI and voting patterns.

Greens, in the top-right quadrant are not close to any major AI representation, other than a weak association to 'AI as a Stabilising force.' Their distance suggests that voters identifying with the Green Party may hold AI views not strongly captured in the five dominant social representations, or it could also be that the sample size of Green voters which captured 1.19% of the 2024 European Parliament election result was too small to register strong clustering.

AI as a Threat

Positioned closely to PL, this representation frames AI as a source of societal disruption, job loss, and privacy erosion. This interpretation resonates with narratives that emphasise social equity, labour protections, and scepticism toward unregulated technological advancement. The spatial proximity of PL and the representation of 'AI as Threat' reflect a leftist critique of techno-capitalism, where innovation is not inherently positive but must be scrutinised for ethical and socio-economic consequences. PN voters are located in the bottom-left quadrant, but distanced from any of the five main representations of AI. Its relative proximity to 'AI as a Threat', 'AI as the Unknown' and 'AI as a progressive tool' but not overlapping with them, suggests ideological detachment or diffusion. Thus, PN voters may not express a dominant or unified view of AI possibly indicating fragmentation within its support base regarding AI's societal role.

AI as a Progressive Tool

This representation appears near 'Refused to answer' and 'Independent' voters, much like the representations of AI as Helper/Assistant view. AI is linked to societal advancement, innovation, and problem-solving potential. The detachment from mainstream political groups might indicate that those who embrace this representation prefer visionary thinking and are less anchored in traditional partisan lines.

AI as the Unknown

The most populous category is closely associated with PL and also, yet more distantly, to the respondents who abstained from voting or voted for parties to the right. This position signals a significant knowledge gap or uncertainty about AI. The correspondence suggests that even within traditionally left-leaning groups (such as the PL), digital and AI understanding may be uneven. Alternatively, it may reflect cognitive dissonance (Festinger 1954) between support for progressive social policies and discomfort with fast-moving technological change.

Extending Jacobs' Model of AI Polarisation

Building on the foundational hypotheses proposed by Jacobs (2024), this study offers empirical validation that social representations of AI are not only polarising forces in the labour market but also carry tangible electoral consequences. Jacobs suggests that individuals deemed 'automation susceptible' (those whose roles are most at risk) tend to be economically left-leaning and culturally conservative. Conversely, individuals in roles with high AI exposure (those whose work is complemented rather than replaced by AI) are more culturally liberal and economically right-leaning.

This study provides empirical confirmation of these dynamics within a specific electoral context. In Malta, left-leaning voters (especially those aligned with the PL), were more likely to express social representations of AI as a "threat" or as the "the unknown", mirroring Jacobs' category of AI 'losers'. Meanwhile, respondents identifying with right-leaning parties exhibited a stronger association with the representation of AI as a "Progressive Tool", and a secondary, though less pronounced, association with "AI as a Stabilising Force" echoing Jacobs' characterisation of AI 'winners', those whose professions are structurally complemented by automation. This consistency across very different contexts strengthens the generalisability of Jacobs' model.

Where Jacobs proposed a theoretical connection between automation anxiety and political realignment, this study offers a significant extension: a direct link between AI social representations and declared voting patterns. By applying Correspondence Analysis, the results show how AI's symbolic role maps meaningfully onto political identities. This mapping evidences a phenomenon Jacobs could only infer; that social representations of AI shape electoral behaviour. This finding suggests that polarisation through technological framing is not solely a function of large-party systems or labour demographics but may be a broader consequence of how digital innovation is socially constructed.

Furthermore, while Jacobs focused on large-scale macro-democracies like the United States, this study offers a fine-grained look at a microstate democracy, Malta, where ideological divides are similarly reflected in how AI is understood and how representational uncertainty (particularly among PL voters) may undermine democratic engagement despite high electoral participation. In this context representational uncertainty refers to a state in which individuals within a political group lack a stable or single coherent social representation of AI. It is important to distinguish this from political disengagement especially since many of these individuals are electorally active, yet exhibit unclear representations of AI which may render them more susceptible to populist appeals. This concept aligns with Moscovici's theory of social representations, which suggests that when societies encounter novel or disruptive phenomena without pre-existing metaphors or culturally shared schemas, uncertainty and ambiguity tend to dominate collective sense-making.

Building on this, our findings confirm that AI is not politically neutral and subsequently reinforce the importance of embedding media literacy and civic education within traditional political discourse.

Implications for Political Strategy and Public Engagement

The results of this study underscore that social representations of AI are not neutral abstractions but symbolically loaded perspectives that align closely with political identity. These findings hold actionable implications for both political strategy and civic engagement.

One of the most salient findings emerging from the Correspondence Analysis is the strong spatial proximity between the social representation of "AI as the Unknown" and respondents who reported voting for the PL. This alignment suggests a meaningful correspondence between representational uncertainty and left-leaning electoral affiliation within the Maltese context. Rather than signaling political disengagement, this grouping suggests that even among electorally active, left-leaning individuals, there exists a substantial level of uncertainty, ambiguity, or lack of clarity regarding what AI is and how it functions. This may reflect gaps in media literacy, a broader skepticism about emerging technologies, or a disconnect between political ideology and technological understanding.

The association challenges assumptions that disengagement from AI discourse is confined to politically passive or marginalised groups and instead highlights the need for targeted education and communication strategies, even within established voter bases.

Forging the Centre. Shaping a Cohesive Narrative

Despite capturing 42.02% of the national vote in the 2024 European Parliament elections, the voters within the PN appear statistically unanchored to any singular representation of AI within the Correspondence Analysis map. Its spatial isolation from the core cluster of AI, especially when contrasted with the PL's strong association to 'AI as Threat' and 'AI as the Unknown', suggests that PN voters do not coalesce around a shared narrative of AI's role in society, a signal of ideological diffusion or representational ambiguity. In democratic terms, such a vacuum leaves a significant segment of the electorate potentially under-engaged in the normative debate about technology and power. As discussed, political meaning is often constructed through metaphors, and where metaphors are absent or fragmented, public policy risks becoming technocratic, detached from civic resonance. Strategically, PN's electoral weight positions it uniquely to forge a cohesive, centre-ground narrative on AI; grounded in institutional trust, regulatory responsibility, and participatory citizen empowerment. Doing so could revitalise democratic engagement not by polarising views, but by giving shape and substance to a currently unaligned middle.

The Moral Neutrality Myth: Public Trust Must Be Earned, not Assumed

The Correspondence Analysis reveals that different political constituencies anchor AI to distinct metaphors ranging from viewing it as a 'threat', to that of being a 'tool for progress or stability'. This dispersion across the factor map confirms that technology is not perceived as neutral, but rather as deeply embedded in societal structures, shaped by (and shaping) ideology, media narratives, and lived experience. One clear result is that for example, the left-leaning voters, particularly those aligned with PL, are closely associated with the representations of 'AI as the Unknown' suggesting discomfort or mistrust rooted in uncertainty. Meanwhile, right-of-centre voters align more closely with 'AI as a Progressive Tool' showing that even positive associations are ideologically conditioned.

This makes one thing clear. Neutrality in technology is a myth, and political messaging should respond accordingly. Leaders who fail to speak transparently about AI's ethical dimensions such as its role in surveillance, labour displacement, or algorithmic bias, risk being perceived as complicit in opaque systems. Conversely, those who acknowledge both the risks and the promise of AI, and who propose concrete regulatory safeguards (e.g., transparency standards, independent audits, ethical use and bias mitigation tools), are better positioned to build trust across divergent ideological groups. The data supports a growing imperative for political actors to engage the public in conversations about the values embedded within AI systems, not just their functionality.

Democratic Resilience Through Media Literacy

Artificial Intelligence now shapes governance and electoral dynamics (Jungherr 2023) making regulatory safeguards alone insufficient for preserving democratic integrity. Equally vital is the cultivation of voter competencies that enable citizens to interpret AI-driven representations and make informed choices. When individuals possess the analytical tools to traverse the intertwined domains of technology, media, and politics, democracy itself gains resilience.

Digital and information literacy provides these tools by fostering critical analysis of complex content streams. Learners trained in these competencies can evaluate the credibility of sources, identify misleading narratives, and uncover underlying biases in media portrayals. The European Commission (2023) emphasizes that such proficiency empowers people to navigate digital services, exploit emerging communication channels, and shield themselves and their families from harmful or illicit material.

Other European Union bodies echo this call, highlighting that literacy initiatives are indispensable for undermining disinformation and enhancing civic decision-making (European Digital Media Observatory 2024). Although these programmes cannot eradicate falsehoods single-handedly, they significantly bolster society's capacity to resist manipulative content in concert with regulatory measures.

Yet the promise of information literacy confronts formidable obstacles in today's online landscape. The swift expansion of social platforms and the democratization of content creation have fragmented audiences into echo chambers dominated by uniform viewpoints (Chan et al. 2023). Simultaneously, the deluge of misinformation erodes trust in established news outlets, complicating efforts to foster critical engagement. These challenges also open pathways for innovation. By integrating AI-powered teaching platforms and interactive modules, educators can create immersive learning environments that adapt to individual needs and sustain learner interest.

In an era where algorithmic curation precedes every political message, equipping the electorate with information literacy is not a pedagogical luxury but a democratic necessity. Such programmes must move beyond abstract theory: they should be deeply rooted in local contexts, woven into national curricula, and reframed as "citizen literacy." This broader framing emphasises not only the mechanics of AI but its repercussions for power structures, labour markets, and public accountability, thereby fortifying communal resistance not just to disinformation but to civic disempowerment.

The results from this survey illustrate the urgency of this approach. Respondents categorised under "AI as the unknown" were predominantly supporters of the PL. Their political engagement suggests that low conceptual understanding of AI does not necessarily correlate with disengagement from democratic processes. Instead, it reveals a media literacy gap among active voters, highlighting the need for targeted literacy education.

Grasping AI's impact on society requires more than general awareness; it necessitates fluency in the technological narrative increasingly shaping political discourse. The findings suggest that some voters remain tethered to traditional party loyalties while lacking familiarity with how those parties engage with emerging technology. This disconnect underscores the need to embed media literacy within broader civic education strategies particularly for electorally engaged but digitally under-informed populations.

Conclusion

The findings of this study provide validation for the claim that AI is not only a technological or economic artefact, but a culturally mediated political phenomenon. As AI continues to determine how information is produced, disseminated, and interpreted, it becomes increasingly central to the democratic process; both as a tool of governance and as an object of political representation. In this context, the politicisation of AI should not be seen as incidental, but as an inevitable outcome of how democratic societies cope with emerging technologies. Hence media literacy emerges as a critical safeguard against the erosion of democratic values. This study contributes to the mapping of social representations of AI against declared voting preferences using Correspondence Analysis. In doing so, it offers an evidence-based extension of Jacobs' (2024) model of AI-induced polarisation, grounding abstract hypotheses in a specific electoral and cultural context; that of a microstate democracy within the European Union. While Jacobs theorised that automation anxiety may shape political attitudes, this study goes further, showing how metaphors of AI may become embedded in social representations and serve as symbolic proxies for broader ideological cleavages.

By equipping individuals with the skills to navigate AI-driven media environments critically, societies can uphold democratic principles, foster informed citizenship, and mitigate the risks posed by misinformation and disinformation. In the pursuit of a more democratic future, media literacy must remain a cornerstone of educational and civic engagement efforts.

Furthermore, the results challenge the enduring assumption that political disengagement stems solely from apathy. Instead, it suggests that representational uncertainty (a situation in which electorally active citizens lack a coherent metaphorical framework to process technological change) may be a central driver of democratic vulnerability. Citizens who cannot articulate what AI is are less likely to hold their political leaders accountable for how it is governed. In this vacuum, narratives can be co-opted by partisan media or populist actors, further fragmenting public trust.

It is being suggested that media literacy programmes for the digital age should approach the subject holistically. They have to "become more agile and responsive to technological and indeed philosophical developments to information and media" (Tiernan et al. 2023, p. 906). Hristovska (2023) emphasises the need for a multi-pronged approach involving education, technology development and media responsibility to effectively address the impact of AI generated content on media literacy and public perception of AI. Moreover, partnerships between governments, educational institutions, and civil society organisations can expand access to media literacy education and foster a culture of critical inquiry and civic engagement.

A cross-sectional survey, while useful for understanding prevalence and associations between factors, has several shortcomings. The study acknowledges limitations including non-response bias, survey fatigue and a non-random sample. Moreover, politics in Malta is very polarised and is always on the public agenda. Respondents would often try, by their answers, to put the party they support in good light by giving socially desirable answers. Despite these limitations, this survey remains useful for understanding population characteristics at a given time. Future research should explore longitudinal changes in these representations, investigate how media exposure conditions them, and examine how they evolve across different democratic contexts.

This study contributes to the growing body of literature demonstrating that AI is not simply a computational force, but a cultural and political phenomenon. It demands not only technical regulation, but socio-political interpretation and ethical accountability. The gap seen in politically engaged citizens who may still perceive AI as 'unknown' stresses further the need of media literacy. This gap is not benign; it risks eroding informed political engagement, enabling technocratic policymaking to proceed without democratic scrutiny, and deepening inequalities in knowledge and power. As such, media literacy must now be treated as a prerequisite for democratic participation.

Acknowledgments

The authors would like to express their gratitude to MISCO (Malta) for their professional execution of the national survey. Their methodological expertise and support were invaluable to this research.

References

Arnheim, R. 1957. 8 re-print 1971. Film as Art. University of California Press.

Borwein, S., B. Bonikowski, P. J. Loewen, B. L. Lee-Whiting and B. Magistro. 2025. Perceived technological threat and vote choice: Evidence from 15 European democracies. West European Politics, 48(3): 534–561. https://doi.org/10.1080/01402382.2023.2297601

Braun, V. and V. Clarke. 2006. Using thematic analysis in psychology. Qualitative Research in Psychology, 3(2): 77–101. https://www.researchgate.net/publication/235356393_Using_thematic_analysis_in_psychology.

Brondi, S., M. Pivitelli, S. Di Battista and M. Sarrica. 2021. What do we expect from robots? Social representations, attitude and evaluations of robots in daily life. Technology in Society, 66(2): 101663. DOI: 10.1016/j.techsoc.2021.101663

Bauer, M. and G. Gaskell,. 1999. Towards a Paradigm of Social Representation. Journal of the Theory of social Behaviour, 29(2): 163–186.

Caves, S., C. Craig, K. Dihal, S. Dillon, J. Montgomery, B. Singler et al. 2018. Portrayals and perceptions of AI and why they matter. The Royal Society. https://doi.org/10.17863/CAM.34502

Demographic Review 2014. National Statistics Office, Valletta, Malta.

Chan, C.K.Y.A. 2023. Comprehensive AI policy education framework for university teaching and learning. Int. J. Educ. Technol. High Educ. 20, 38. https://doi.org/10.1186/s41239-023-00408-3

Doise, W. and C. Staerklé. 2002. From social to political psychology: The societal approach. pp. 151–172. *In:* Monroe, K. (Ed.). Political psychology. Lawrence Erlbaum. Mahwah, NJ.
Dwivedi, Y. K., A. Sharma, N. P. Rana, M. Giannakis, P. Goel and V. Dutot. 2023. Evolution of artificial intelligence research in Technological Forecasting and Social Change. Research topics, trends and future direction. Technological Forecasting and Social Change. 192 (July) 122579 DOI: https://doi.org/10.1016/j.techfore.2023.122579
Eisenstein, E. 1983. The Printing Revolution in Early-Modern Europe. Cambridge: Cambridge University Press.
Electoral Commission Malta. 2024. Electoral Commission Malta: https://electoral.gov.mt/ElectionResults/MEP
European Commission. 2023. Communication from the Commission: Guidelines pursuant to Article 33a(3) of the Audiovisual Media Services Directive on the scope of Member States' reports concerning measures for the promotion and development of media literacy skills.
European Digital Media Observatory. 2024. EDMO Guidelines for Effective Media Literacy Initiatives.
Festinger, L. 1954. A theory of social comparison processes. Human Relations, 7(2): 117–140.
Filgueiras, F. 2022. The politics of AI: Democracy and Authoritarianism in developing countries. Journal of Information Technology and Politics, 19(4). DOI 10.1080/19331681.2021.2016543
Hammond, S. 1993. The descriptive analysis of shared representation. pp. 205–222. *In:* Breakwell, G. M. and D. V. Canter (Eds.). Empirical approaches to social representations. Oxford: Claredon Press
Hristovska, A. 2023. Fostering media literacy in the age of AI: Examining the impact on digital citizenship and ethical decision making. KAIROS: Media and Communication Review, 2(2): 39–59. https://nbn-resolving.org/urn:nbn:de:0168-ssoar-93927-0
Jacobs, J. 2024. Artificial shock and socio-political polarization. Technological Forecasting and Social Change. 199: 123006. https://doi.org/10.1016/j.techfore.2023.123006
Jacobs, J. 2024. Automation and Identity: How Labour Market Disruption Shapes Political Ideology. Technological Forecasting and Social Change, https://doi.org/10.1016/j.techfore.2023.123006
Jovchelovitch, S. 2001. Social representations, public life and social construction [online]. London: LSE Research Online. Available at: http://eprints.lse.ac.uk/2649 Available in LSE Research Online: August 2007.
Jungherr, A. 2023. Artificial Intelligence and Democracy: A Conceptual Framework. Social Media and Society, July- September 1–14.
Jungherr, A., A. Rauchfleisch and A. Wuttke. 2024. Deceptive uses of artificial intelligence in elections strengthen support for AI ban. August 26: 2024
Kaye, D. 2018. Report on artificial intelligence technologies and implications for freedom of expression and the information environment. United Nations Human Rights Office of the High Commissioner. https://www.ohchr.org/EN/Issues/ FreedomOpinion/Pages/ReportGA73.aspx
Kelley, P. G., Y. Yang, C. Heldreth, C. Moessner, A. Sedley, A. Kramm et al. 2021. Exciting, useful, worrying, futuristic: Public perception of artificial intelligence in 8 countries. AIES'21 Conference Proceedings.
Kraut, R., M. Patterson, V. Lundmark, S. Kiesler, T. Mukopadhyay and W. Scherlis. 1998. Internet paradox. A social technology that reduces social involvement and psychological well-being? American Psychologist, 53(9): 1017–31. DOI: 10.1037/0003-066x.53.9.1017. PMID: 9841579
Kreps, S., R. McCain and M. Brundage. 2022. All the news that's fit to fabricate: AI-generated text as a tool of media misinformation. Journal of Experimental Political Science, 9(1): 104-117. DOI: 10.1017/XPS.2020.37
Lakoff, G. and M. Johnson. 2003. Metaphors we live by. Chicago and London: The University of Chicago Press.
Lee, K. 2018. AI superpowers: China, Silicon Valley, and the new world order. Houghton Mifflin Harcourt.
Margetts, H., T. Sippy, R. Modhvadia, O. Bruff and A. Dutta,. 2025. Artificial intelligence (BSA 42). London: National Centre for Social Research.
McChesney, R. W. 1999. Rich media, poor democracy: Communication politics in dubious times. Urbana and Chicago: University of Illinois Press.
McLuhan, M. 1964. Understanding media: The extensions of man. Cambridge, MA: MIT Press.
Morozov, E. 2013. To Save Everything, Click Here: The Folly of Technological Solutionism. London : Allen Lane.

Moscovici, S. 1961. La Psychanalyse : Son image et son public. Etude sur la representation sociale de la psychanalyse. Paris : Presses Universitaires de France.

Moscovici, S. 1984. The phenomenon of social representations. pp.3–69. *In:* Farr, R. M. and S. Moscovici (Eds.). Social Representations. Cambridge: Cambridge University Press.

Ong, W. J. 1967. The Presence of the Word. New Haven and London: Yale University Press.

Orben, A. 2020. The Sisyphean Cycle of Technology Panics. Perspectives on Psychological Science, 15(5): 1143–1157. https://doi.org/10.1177/1745691620919372

Plato. (ca. 370 BCE). Phaedrus (B. Jowett, Trans.). Retrieved from https://faculty.georgetown.edu/jod/texts/phaedrus.html

Polarization Research Lab. 2024. Partisan animosity and America: The path to the 2024 presidential election. Dartmouth College, University of Pennsylvania, and Stanford University. Retrieved from https://prlpublic.s3.amazonaws.com/reports/January2024.html

Postman, N. 1987. Amusing ourselves to death: Public discourse in the age of show business. London: Methuen.

PublicFirst. 2023. What does the public think about AI? UK. Authors: Jonathan Dupont, Seb Wride and Vinous Ali.

Schmitt, R. 2005. Systematic metaphor analysis as a method of qualitative research. The Qualitative Report, 10(2): 358–394.

Settle, J. 2018. Frenemies: How Social Media Polarizes America. DOI: 10.1017/9781108560573

Staerklé, C., A. Clémence and D. Spini. 2011. Social Representations: A Normative and Dynamic Intergroup Approach. Political Psychology, 32(5): 759–768.

Tiernan, P., E. Costello, E. Donlon, M. Parysz and M. Scriney. 2023. Information and media literacy in the age of AI: Options for the future. Education Sciences, MDPI.

UC Office of the President. 2024. The politics behind AI: How political beliefs drive opinions on artificial intelligence in California. Retrieved from https://escholarship.org/uc/item/1tn9z87k

Vallance, C. and P. Wain. 2024. Internet access linked to higher wellbeing, study finds. Retrieved from https://www.bbc.com/news/articles/c89z2v0pjl3o

Wagner, W. 2020. Social representation theory: An historical outline. *In:* Oxford Research Encyclopedia of Psychology, Social Psychology, History and Systems of Psychology. DOI: 10.1093/acrefore/9780190236557.013.606

Xiao, X., Y. Su and D. K. L. Lee. 2021. Who consumes new media content more wisely? Examining personality factors, SNS use, and new media literacy in the era of misinformation. Social Media + Society, 7(1). https://doi.org/10.1177/2056305121990635

Yi, A., S. Goenka and M. Pandelaere. 2023. Partisan media sentiment towards artificial intelligence. Social Psychological and Personality Science, 15(6). https://doi.org/10.1177/19485506231196817

Zaheer, A. 2023. The politics behind AI: How political beliefs drive opinions on artificial intelligence in California. University of California, Davis, POL 195: Special Studies in American Politics.

Work and Education Transformation

CHAPTER 5

Artificial Intelligence and World of Work

Social Representations of AI in the Context of Job Displacement and Transformations

Mircea Vultur

Introduction

Artificial intelligence (AI) is a broad term, and its definition can be flexible. Generally, AI refers to any system that implements mechanisms resembling human reasoning; any machine capable of reproducing behaviors associated with humans, such as reasoning, planning, and creativity. Essentially, the work of any machine that appears intelligent can fall within the realm of AI. This technology aims to overcome the limitations of human intelligence, its uncertainties, and its errors (Russell and Norvig 2020).

In recent years, AI has emerged as a powerful driver of change in the world of work, automating tasks in industries from manufacturing to finance. This has led to significant labor market transformations. Currently, artificial intelligence impacts most sectors of the world of work, ranging from the automation of simple and repetitive tasks to the execution of specific and highly complex functions that humans find difficult to perform. Artificial intelligence is thus on its way to becoming the norm in many companies. The integration of artificial intelligence into the workplace has evolved over decades, driven by advancements in computing, data availability, and algorithmic innovation (Vendramin and Valenduc 2023; Bubeck et al. 2023).

Institut national de la recherche scientifique Québec, Canada.
Email: mircea.vultur@inrs.ca

DOI: 10.1201/9781003502753-9

The interplay between technological advancement and world of work is not new; it has played out over several centuries. Each wave of innovation, from the steam engine to electricity to computers, has brought both transformative opportunities and societal anxieties (Susskind 2021). In the late 19th and early 20th centuries, for example, electricity and new industrial machines (like the assembly line famously introduced by Henry Ford in 1913) further transformed labor. Repetitive tasks became easily automatable. This era created unprecedented economic growth and many new types of jobs (from electrical engineers to automobile mechanics), but not without periods of social strain. For instance, as factories became more efficient, fewer manual laborers were needed in certain roles, contributing to labor movements and demands for worker protections. Each wave of automation also increased productivity significantly, which in the long run tended to lower the cost of goods, stimulate demand, and eventually create different jobs, a pattern that economists call the "compensation effect" (technology displaces some jobs but creates others through growth) (Grossman and Helpman 1991).

By the mid-20th century, a new form of automation emerged, one that targeted cognitive work. The invention of digital computers in the 1940s and 1950s meant that machines could not only assist muscle power but also perform calculations and data processing. Many routine clerical tasks (like bookkeeping, file sorting, and later, data entry) were now done by computers or simple software. This was sometimes called the "Information Technology Revolution" (ITR) (Pal 2008): many routine, rule-based tasks became automated by software; for example, sorting records or calculating payroll could be done far faster by a computer program than by a human clerk.

The emergence of AI in the world of work is part of this process. Its origins can be traced to the mid-20th century, when early concepts of machine learning and automation began to emerge (Susskind 2021). During the 1950s and 1960s, the development of simple expert systems demonstrated AI's potential to replicate human decision-making in specific domains, such as medical diagnosis and logistics. The 1980s marked a significant shift with the proliferation of computers in workplaces, enabling AI systems to support decision-making processes in industries like manufacturing, finance, and healthcare. Technologies such as rule-based systems and robotics began automating repetitive tasks, laying the groundwork for efficiency gains and productivity improvements. The advent of big data and advanced machine learning in the 2000s accelerated AI's role in the workplace (Vendramin and Valenduc 2023). With large-scale data processing and predictive analytics, AI transitioned from automating tasks to enhancing strategic decision-making. This period saw AI's applications diversify into fields like supply chain optimization, customer relationship management, and risk assessment. More recently, the rise of deep learning and natural language processing has brought transformative applications such as chatbots, intelligent virtual assistants, and autonomous systems. These tools have reshaped industries by streamlining operations, enhancing personalization, and creating new job categories focused on AI development and oversight (Garneau 2021; Vultur 2024). Actually, AI is deeply embedded in the workplace, influencing areas like hiring practices, performance analytics, and collaborative environments (Deranty and Corbin 2024). While it has created opportunities for innovation and growth, its history also reflects ongoing debates about its implications for the world of work. These debates about the central place occupied by AI in the world of work owes a great deal to the impressive technological advances achieved, particularly since 2010, which have made Big Data and deep learning accessible; but the debate is also about the representation of technology as a tool for progress. The influence exerted by so-called "expert thinking", which developed mainly in the United States, has been decisive in the emergence of AI. This thinking conveys a representation of technology as a "problem-solving tool". Behind the development of AI, there is therefore also an ideological representation of technology as a vector for solving problems: *the technicist representation of the world of work.*

Theoretical Frameworks on Social Representation of AI in the Word of Work

Debates about AI and word of work often oscillate between optimism about AI-assisted growth and anxiety about "technological unemployment"[1] and other harmful impact of this new technology (Hawking 2016; Aghion 2020). In understanding these debates, the concept of social representations introduced by Moscovici (1984) appears as a heuristic element. The theory of social representation refers to the systems of values, ideas, and practices shared by a group that enable individuals to make sense of social phenomena (Negura, Plante and Lévesque 2019). When applied to artificial intelligence, social representations emerge as collective frameworks through which societies interpret the role, significance and implications of AI in various contexts including the world of work. As a transformative technology, AI is integrated into these representations by individuals and groups, influencing how it is perceived and adopted (Vaast 2007).

In the workplace, as in society at large, social representations of AI often revolve around themes of opportunity and threat. For instance, AI may be represented positively as an enabler of efficiency and innovation, emphasizing its role in automating repetitive tasks and enhancing productivity. Conversely, it may be represented negatively as a source of job insecurity, perpetuating fears of automation-induced unemployment and creating social anxiety over jobs displacement (Susskind 2021). The shared representations of AI also reflect broader societal dynamics, including power relations and cultural norms.[2] For example, in industries where AI is embraced as a strategic asset, it is often represented as a neutral and objective force driving progress. However, in communities where AI adoption is perceived as imposed or disruptive, it may symbolize inequality, surveillance, or a loss of human agency. Social representation theory also suggests that how workers and the public at large conceptualize AI will influence their responses to it. Different social groups may develop different representations of AI (Gal and Berente 2008; Bao and al. 2022). Tech enthusiasts often represent AI as a powerful tool or partner that can augment human capabilities. This can lead to behaviors like early adoption of AI tools and a

[1] This term was famously used by classical economist John Maynard Keynes in 1930 to describe job loss driven by new technology.

[2] See the chapter written by Lilian Negura and Noëlle Tchatchou in the present book.

focus on learning new skills to work alongside AI.[3] Skeptics workers, by contrast, might represent AI as a threat, a soulless machine coming for their jobs, which can lead to resistance, fear, or calls for halting AI deployments. These representations are continuously fed by social communication, including media stories, government statements and scientific literature in the field of AI (McKelvey and Roberge 2023). Understanding social representations of AI is important for navigating its integration into society, as these shared perceptions shape public acceptance, policymaking, and the ethical frameworks guiding AI deployment. Recognizing the diversity of these representations helps address concerns and foster inclusive discussions about the role of AI in shaping the future of work.

The social representations of AI in the workplace can be understood through various perspectives and sources that reflect how people perceive, understand and interact with AI in professional environments. For example, we can analyze how AI is depicted in news, movies, and literature about work. Social media platforms and forums reveal also common sentiments about AI in work settings. Social sciences explore studies on the societal impact of AI, its role in redefining jobs, and its influence on professional identities. In a socioeconomic perspective, research investigate how AI is perceived in terms of job creation, displacement, and skill requirements. Each of these perspectives influences representations of AI oscillating between positive ("AI as a productivity enhancer") and negative ("AI causing job loss and other harmful") narratives.

The aim of this chapter is to present representations of AI in relation to various aspects of the world of work, using social representation theory as an analytical framework and literature on contemporary changes in work under the impact of AI as an empirical basis. We will refer to social representation that are diffused throughout the scientific literature in the field of socio-economics of work considering Moscovici's "late" perspective on social representation (Moscovici and Hewstone 1984; Moscovici 1993), that scientific knowledge acts as a social representation when it fulfills the conditions of being collectively negotiated, dynamically influenced by group interactions, and inherently embedded in the practices of communication, persuasion, and consensus within scientific communities (Jodelet 2013; Negura and Lavoie 2016). As highlighted by Negura, Sarrica, and Rizzoli in this book, various players, including experts, contribute to shaping and interpreting the meanings associated with emerging technological realities. These interpretations are expressed and circulated through public debates and media articles, which often emphasize either the positive or negative aspects of these technologies. These viewpoints evolve into widely accepted representations, frequently grounded in scientific literature. In this context, socio-economic research is the source of these representations and serves as a crucial medium for communicating and disseminating these representations.

Based on literature treating about the impact of AI in the word of work, the chapter will first present the *positive representations of AI* referring to (a) innovation and productivity growth, (b) AI as a new source of value and (c) AI as a tool improved Human Resource Management (HRM). Secondly, the chapter looks at the

[3] On this topic see the chapter written by Emmanuelle Vaast in the present book.

negative representations of AI, which refer to (a) the loss of jobs, (b) the worsening of socio-economic inequalities, and (c) the increased possibilities for monitoring and controlling workers. As mentioned above, from a methodological point of view, the analysis will be based on scientific literature in the field of socio-economics of work, mainly books, articles and research reports (including reports produced by consultants and trade unions), analyzing AI and work.

Positive Representations of AI

AI as a Driver of Innovation and Productivity Growth

Among the most widely shared social representations of AI is the view that AI serves as a powerful driver of innovation and productivity. This positive perspective is particularly prevalent among researchers, who emphasize AI's potential to accelerate technological advancement. This representation is supported by concrete examples, such as the rapid improvements in AI's learning capacities. For instance, facial recognition technology is frequently cited as evidence of AI's growing effectiveness, with error rates dropping from 4.1% in 2014 to just 0.08% in 2020 according to the National Institute of Standards and Technology (NIST) (Crumpler 2020). Another prominent representation portrays AI as a catalyst for transforming the workplace, notably through the development of robots with increasing physical autonomy (Wisskirchen 2017). This perspective emphasizes AI's contribution to productivity gains, enabled by autonomous robots equipped with sensors that minimize the need for human intervention. These robots can now adapt in real-time, learning from their environment, a notion that reinforces the social belief in AI's capacity to bridge the gap between simulated and real-world learning. High-profile examples, such as Boston Dynamics' Atlas robot performing parkour or the Spot robot conducting industrial inspections, have helped solidify this image in the public imagination. The emergence of the Internet of Things (IoT) is also closely tied to these representations. The increasing adoption of IoT technologies, from 13% of companies in 2014 to over 25% in 2023 (Vendramin and Valenduc 2023), illustrates the growing perception of AI as a pervasive and valuable asset across industries. Finally, deep learning, often described as mimicking the human brain, is at the center of another widely circulated representation: that AI systems is capable of self-learning without human supervision. This representation highlights AI's ability to autonomously process large datasets, detect patterns, and perform complex tasks such as object recognition, language translation, and decision-making. AI is not merely an assistant but is increasingly seen as capable of replacing certain human expert tasks, contributing to competitive advantage (David 2020; Coté and Su 2021).

By optimizing processes, enhancing efficiency, and driving innovation, AI offers unprecedented opportunities to strengthen economic growth and improve business competitiveness. The increased automation enabled by AI reduces production costs while simultaneously improving product quality, thus boosting companies' competitiveness on global markets (Brynjolfsson and McAfee 2016). In the service sector, AI is revolutionizing professional practices. AI algorithms enhance decision-making by rapidly analyzing massive volumes of data. In scientific discours, several social representations of AI portray it as a powerful enabler of efficiency, accuracy, and economic growth across various sectors. In financial services, for exemple, a dominant perspective positions AI as a superior tool for detecting fraud and predicting market trends, enabling faster decision-making and enhanced business outcomes. This positive representation extends to the healthcare sector, where AI is widely perceived as a transformative force that improves the speed and accuracy of medical diagnoses (Holzinger et al. 2023). Studies highlight how AI-driven medical imaging tools, such as MRI, CT scans, and retinal scans, are now capable of diagnosing diseases like cancer and diabetic retinopathy with accuracy comparable to, or even exceeding, that of expert radiologists (Agrawal et al. 2018). This fuels the social belief that AI has life-saving potential while also contributing to the reduction of healthcare costs.

Another recurring representation frames AI as a driver of innovation and market expansion. From this viewpoint, AI is not only improving existing processes but also creating new products, services, and previously untapped markets. In logistics, for example, AI is socially constructed as a key to transforming supply chain management, offering predictive tools that optimize demand forecasting, reduce delays, and minimize waste, ultimately meeting consumer expectations with unprecedented efficiency (Boute and Udenio 2022).

Moreover, a widely shared scientific narrative emphasizes AI's democratizing effect on technology. This perspective suggests that AI empowers small and medium-sized enterprises (SMEs) by providing access to advanced technologies that were once reserved for large corporations. Through affordable, scalable AI solutions, SMEs are now able to automate tasks, perform sophisticated data analyses, and compete more effectively in their respective markets (McKinsey 2018).

At a macroeconomic level, AI is also represented as a major engine of future global economic growth. According to PricewaterhouseCoopers (2018), AI could contribute up to $15.7 trillion to the world economy by 2030, an estimate that reinforces the perception of AI as a transformative economic force. This growth is projected to come from both productivity gains and increased consumer demand, supporting the prevailing belief in AI's potential to reshape industries and societies on a global scale.

AI as a New Source of Value

AI is increasingly perceived as a source of new value in the workplace. It is represented as a tool that automates repetitive and sometimes hazardous tasks, such as data entry and assembly line manufacturing, thereby reshaping how work is imagined and organized. Within this perspective, AI-enhanced jobs are often viewed as opportunities for workers to engage more in creativity, strategic thinking, and initiatives that require human judgment (Lane and St-Martin 2021). Across various sectors, AI is portrayed as a key driver of value creation. For example, in financial services, JPMorgan Chase presents the use of an AI coding assistant as a means to allow engineers to concentrate on higher-value projects like data analytics, reinforcing a narrative where AI contributes to improving efficiency and innovation (Franey 2025).

Another dominant social representation emerging around AI is its association with new high-skilled employment opportunities. Institutions like the World Economic Forum (2020) contribute to this representation by disseminating forecasts that emphasize growing demand for specialized roles such as data analysts, machine learning experts, robotics engineers, and cybersecurity specialists. These discourses anchor AI to themes of economic growth, digital transformation, and future employability, shaping public perception and guiding career aspirations. Similarly, the U.S. Bureau of Labor Statistics projects significant growth in data science jobs by 2026 (Schroeder 2021), reinforcing the social image of AI as a creator of desirable, future-proof professions.

Lane and Saint-Martin (2020) identifies the emergence of three distinct categories of jobs supporting AI development: trainers (who train AI systems), explainers (who interpret AI-generated results) and sustainers (individuals who ensure AI systems function as intended). Other AI-related roles will also emerge, including scientists designing algorithms, less-skilled workers who train AI systems with data and correct errors, professionals who explain AI-driven solutions to executives, and those in both the private and public sectors who regulate AI usage.

Beyond jobs related to AI development, social and emotional skills will also become increasingly important. By 2030, demand for social and emotional skills is expected to increase by 26% in the U.S. and 22% in Europe across all industries (Bughin et al. 2018). While some of these skills, such as empathy, are innate, others, like communication, can be developed through training. Entrepreneurial spirit and initiative are expected to be the most sought-after skills. Additionally, the demand for leadership and personnel management will grow significantly. The need for complex cognitive skills such as creativity, critical thinking, decision-making, and processing complex information is also expected to rise by 2030 (Coté and Su 2021).

AI as a Factor in Improved Human Resource Management

From the perspective of social representations, the AI in Human Resource Management (HRM) is not simply viewed as a technological tool, but rather as a socially constructed object whose meaning is shaped by organizational discourse and expert narratives. AI is often represented as a modernizing force that brings efficiency and enhances employee engagement in HRM practices (Van Esh et al. 2019).

A central representation of AI in HRM focuses on its role in recruitment. AI is commonly presented as a solution to the inefficiencies of traditional hiring processes. According to PriceWaterhouseCoopers (2018), this representation is particularly dominant in large international companies, especially those based in the United States, where 40% of HR functions already incorporate AI in recruitment. Through this lens, AI is anchored to the idea of speed and precision: it is perceived as capable of filtering resumes rapidly and objectively according to predefined criteria, selecting the most suitable candidates from vast application pools (Upadhyay and Khandelwal 2018).

This perspective is further reinforced by the widely circulated notion that AI liberates human recruiters from repetitive, low-value tasks, such as resume screening and data entry, which traditionally consume up to 40% of recruiters' working time. AI is represented as performing these tasks in seconds, allowing human recruiters to focus on higher-order activities like interviewing and negotiating contracts with top talent (Windley 2021). Through this discourse, AI is objectified as an enabler of strategic human interaction rather than as a replacement for it.

Further representations of AI position it as a decision-support system for managers, especially in tasks such as candidate evaluation, employee productivity tracking, and communication improvement (Carlos 2021). The increasing adoption of AI-powered virtual assistants, or chatbots, within HRM is also framed within a social representation that emphasizes cost reduction, improved user experience, and rapid response capabilities (Hupfer 2020). These chatbots are symbolically linked to continuous learning, adaptability, and efficiency, reinforcing AI's image as a dynamic partner in organizational processes. By analyzing real-time employee performance data, AI is seen as an early-warning system capable of detecting declines in engagement toward work. This enables organizations to intervene proactively, with the goal of improving retention and workplace satisfaction.

In the context of gig work,[4] AI create a new archetype for HRM. Workers are guided by technologically mediated HR practice or "algorithmic HRM" conceptualized as "the use of software algorithms that operate on the basis of digital data to augment HR-related decisions and /or automate HRM activities" (Meijerik al. 2021, p. 2457). Algorithms can manage large numbers of gig workers simultaneously without human intervention. Tasks like assigning work, scheduling, and monitoring can be automated. For instance, Uber uses algorithms to match drivers with passengers instantly and dynamically set prices based on demand. Automating HR functions also reduces the need for human managers and administrative overhead. Instead of hiring a large HR team, a food delivery platform can rely on an app to manage onboarding, payments, and dispute resolution. Algorithms can provide workers with instant feedback on performance, helping them adjust behaviors in real time (Vultur and Wartel 2024). A courier app might alert drivers when they're deviating from optimal routes or driving too slowly. AI in the context of gig work can foster mutually beneficial exchange of information between drivers and passengers, provide emotional support and compagnonship, and lead to instrumental helping behaviours as well. Such bonds occasionally become strong, as when drivers build independent relationship and friendship with customers ouside the app (Kameswaran et al. 2018). These social representations of AI as a beneficial tool is bolstered by these outcomes: evidence of greater productivity, new quality of jobs, better management can feed a narrative in which AI have a positive representation in society. Throughout history, productivity gains from technology have been the cornerstone of rising living standards. If managed properly, AI has the potential to continue that trajectory: reducing drudgery, expanding creative opportunities, and enabling humans to achieve more in less time. However, as argued by Jean-Marie Seca and Matthieu

[4] Gig work refers to a type of employment where individuals are paid for completing short-term, flexible tasks or projects, rather than holding a traditional full-time job with a long-term contract. This work is typically arranged through digital platforms or apps (like Uber, Fiverr, or Upwork), and workers are usually classified as independent contractors rather than employees.

Seca in their chapter of this book, social representations of AI across various sectors are significantly shaped by institutional policies, as well as underlying economic and ideological frameworks. In the field of education, for instance, organizations such as UNESCO and the OECD tend to promote a highly favorable narrative around AI, emphasizing inclusion, pedagogical innovation, and support for teachers. However, the authors suggest that these institutional discourses are often idealized and closely aligned with broader industrial and strategic agendas, including national development goals and global competitiveness.

Negative Representations of AI

Job Losses

In the "Parable of Horseshit" Elisabeth Kolbert discusses the decline of horses due to technological developments, particularly the invention of the internal combustion engine (Kolbert 2014). This new technology transformed a creature that, for millennia, had played a central role in economic life, on farms, in fields, and in cities, by relegating it to the sidelines. Kolbert recounts how, in the 19th century, horses were essential to urban life. They pulled carriages, plowed fields, and were critical for transport and logistics. As cities grew, so did the "horseshit problem", tons of manure cluttering urban streets became a massive public health issue. The anticipated crisis over how to manage horse waste, however, never materialized, because of the rise of the automobile and technological innovation. By the early 20th century, horses were quickly replaced by cars, tractors, and trains. Within a few decades, horses were no longer economically viable, their labor became obsolete.

Using this parable as a starting point, in a series of articles written in 1980, Wassily Leontief made an important claim in socioeconomic analysis. He argued that what technological progress had done to horses, it would similarly do to human beings: drive them out of work. Just as cars and tractors replaced horses, computers, robots, and other products of AI would displace human labor (Leontief 1983). Progress may benefit society as a whole but leave specific groups behind, in this case, horses, or certain categories of workers.

The portrayal of AI as a destroyer of jobs is a prominent representation in research on the world of work (Gymrek et al. 2023). Many researchers question whether the creation of new jobs by AI will be sufficient to offset the loss of jobs. Responses to the issue of job displacement caused by AI range from alarmism (Frey and Osborne 2013; Westlake 2014) to neutrality (Calo 2016; Frey 2019), and even optimism (Brynjolfsson and McAfee 2016; Danaher 2020). In a study still widely cited as evidence of the imminent decline of skilled employment in the face of AI, Frey and Osborne (2013), argue that 47% of American jobs are at high risk of automation by the mid-2030s (Frey and Osborne 2013). The authors sought to identify the jobs most at risk based on tasks most vulnerable to automation. They point out that many other factors will need to be considered before full automation of jobs within organizations. This will depend, among other things, on context, cost and regulatory concerns. In this perspective, AI is often represented as both a source of disruption and a generator of opportunity in the labor market. For example, McKinsey's widely cited report contributes to the social representation of AI as a driver of large-scale job transitions, projecting that by 2030, automation could require between 400 million and 800 million individuals globally to move into new roles. This scenario feeds into a collective narrative of unavoidable transformation where adaptation becomes a central expectation for workers. However, alternative social representations circulate alongside this one, offering a more optimistic framing of AI's long-term impact on employment. PricewaterhouseCoopers (PwC) advances the perspective that although AI-driven automation may lead to job losses, these will be largely offset by the emergence of new jobs created by AI technologies (Hawksworth et al. 2018). This reinforces a socially accepted image of AI as a technology that simultaneously disrupts and regenerates the labor market. Within this narrative, PwC explicitly downplays the risk of mass technological unemployment, instead anchoring AI to ideas of economic balance and job evolution. Further social representations are shaped by the International Labour Organization (ILO), which emphasizes that globally, the number of jobs susceptible to enhancement through AI far exceeds those at risk of replacement. This collective framing positions AI not as a simple threat, but as a tool that can elevate human work in many sectors. According to the ILO, approximately 13.4% of jobs are likely to be augmented by AI, compared to only 5.1% that may be replaced, thereby reinforcing a perspective that AI is more about transformation than elimination.

On the plan on job losses, social representations surrounding AI also intersect with issues of gender inequality. The ILO and researchers like Gymrek et al. (2023) highlight that the risks and benefits of AI are unevenly distributed across genders. In developed countries, women are often represented as both more exposed to job displacement by AI and more likely to occupy roles that could be enhanced by AI technologies. This duality constructs a social image where women are simultaneously positioned as vulnerable and as potential beneficiaries in the evolving labor landscape shaped by AI. These perspectives introduce complexity to the dominant representations, emphasizing that the consequences of AI are not uniformly experienced across populations.

The concept of "creative destruction" introduced in 1939 by Austrian economist Joseph Schumpeter, serves as a powerful analytical framework for understanding the impact of AI on employment. Schumpeter argued that economic growth is inherently dynamic, characterized by the continuous cycle of job destruction and job creation. As technological innovations emerge, outdated roles and industries decline, making way for new jobs driven by novel production techniques and business models. This ongoing transformation, fueled by the innovative reconfiguration of economic factors, leads to productivity gains and economic expansion. In this context, the displacement of certain jobs is not merely a loss but a crucial enabler of new opportunities in emerging sectors and more efficient enterprises.

The Growth of Inequality

In the socioeconomic litterature, a widely circulating social representation positions AI as a potential amplifier of existing inequalities, particularly when its integration into the labor market is perceived to disproportionately benefit specific social groups

(Acemoglu and Restrepo 2020; Susskind 2021). This representation emphasizes that highly skilled workers are commonly seen as the primary beneficiaries of AI-driven transformations. Scientific narratives often portray these workers as those best positioned to capitalize on technological shifts, whether through rising wages, enhanced job security, or new entrepreneurial opportunities within the AI economy. Conversely, another dominant social representation frames low-skilled workers as the most vulnerable group in the face of AI adoption. Within this perspective, low-skilled jobs are frequently depicted as the first to be automated, subject to restructuring, or degraded into precarious, low-paying positions (Christiaens 2023). This representation contributes to the collective image of a labor market where low-skilled workers face increasing competition, reduced bargaining power, and deteriorating working conditions.

The distinction between beneficiaries and those at risk is also anchored to representations of education and skill acquisition. Workers with advanced education or technical competencies are often depicted as complementary to AI systems (AI developers, data analysts, and professional) who can strategically leverage AI in their roles are socially represented as experiencing increased demand and wage growth. In contrast, workers who perform routine, repetitive tasks and who lack access to advanced education are collectively perceived as replaceable and more likely to face downward mobility or exclusion from high-quality employment opportunities.

This dichotomy reinforces a broader social representation of AI as a force driving labor market polarization, a pattern previously observed in earlier waves of automation. Referencing studies such as those by Susskind (2021), the narrative suggests that AI may intensify this polarization by accelerating the decline of middle-skill jobs, positions such as bookkeeping, assembly line work, and basic customer service are often symbolically identified as prime targets for AI-driven automation. This widely circulated view contributes to the social understanding that displaced middle-skill workers may be pushed into lower-paid service roles or face significant barriers in retraining for high-skill occupations.

AI can also concentrate wealth by reducing labor's bargaining power and share of production in different communities and countries: as some communities (or countries) thrive by adopting AI, others may be left behind (Lehdonvirta 2016). For example, many rural or deindustrialized regions could see job opportunities dwindle if AI enables companies to centralize operations or if local businesses invest in job-eliminating automation to cut costs (Kirov and al. 2022). This spatial disparity can contribute to political and social divisions, as has been observed in past instances of industrial decline. On a societal level, if large numbers of people feel "left out" of the AI-driven prosperity, it can fuel populist backlashes or public distrust in technology and institutions. There is also a concern that AI might enable a form of "winner-takes-all" economics. Companies that successfully harness AI could dominate markets with lower costs and higher productivity, potentially squeezing out competitors. This market concentration could lead to fewer but larger employers, which can negatively impact workers' leverage (fewer choices of employer, possibly more monopsony power keeping wages down). Furthermore, AI could shift income from labor to capital, owners of AI systems (the companies and their shareholders) might reap significant profits, while the average worker sees relatively less benefit, thus increasing wealth inequality.

Increased Workplace Surveillance and Control

One of the most negative representations of AI concerns workplace organization. A frequently overlooked consequence of new technologies is the increased *monitoring and control* of workers (Bales and Stones 2020; Aloisi and De Stefano 2022; Cousineau et al. 2024). This goes beyond the use of digital tools and extends into AI-driven algorithmic management. The term "algorithmic management", as noted previously, refers to situations where algorithms take over some supervisory functions, relying on the collection of digital data about employees, such as geolocation, individual performance metrics, biometric data, and more.

Initially developed within the platform economy, algorithmic management has since expanded into other industries, including logistics, manufacturing, and retail (Delfanti 2021). On one hand, this leads to the automation of managerial functions, dictating what workers must do, in what order, and within what time frame. Although human managers still oversee operations, workers now receive instructions directly from AI-driven systems. For instance, in food delivery applications, algorithms assign tasks and dictate optimal delivery routes (Baril and Vultur 2025). In warehouse environments, algorithmic management functions similarly: wearable devices and connected tools track workers' performance. In warehouses like those operated by Amazon, voice guidance systems are reported to be used in approximately 70% of shipment preparation tasks. These systems direct workers by instructing them on which aisle to go to in order to retrieve a parcel. In such scenarios, humans effectively operate under the control of machines, serving their instructions. This shift raises concerns about work intensification, as well as the potential loss of professional expertise and a sense of dispossession in one's role. AI creates a despotic and paternalistic workplace environment (Altenried 2020). Supervisors have access to this data and can intervene if an employee's performance drops. These performance indicators are often used to make decisions about an employee's future within the company. This approach is particularly common in platform-based work, but it is also widely applied in call centers, where employees are monitored in real time using voice recording and analysis software. The people are attracted to the plateformes because of its facilities of acces. Over time, however, the platform transforms itself, enacting an "invisible cage", "an algorithmically enforced system of control that is objective, efficient, and scalable" (Rahman 2024, p. 180). There is an asymmetry in the way organizations use of algorithms benefits actors. These algorithmic systems are not the only elements that are opaque to workers. Because platform, contracts, controls, and tools constantly change without warning or explanation, workers are subject to cognitive exhaustion as a result of the need for continual adaptation.

An other dominant social representations presents AI as a pervasive system of data collection, where even collaborative robots (designed to assist rather than supervise) are perceived as mechanisms of surveillance (Cousineau et al. 2024). Through this lens, AI is associated with the generation of detailed, granular performance data, such as idle times and machine downtimes, which feeds a broader societal concern about invasive monitoring. Within this shared understanding, AI-based surveillance is no longer seen as merely operational but as potentially encroaching on personal and behavioral aspects of employees' lives, fostering stress and threatening well-being (Lane and St-Martin 2021).

These representations contribute to a social narrative that frames AI as transforming workplace governance. Traditional employer-employee relationships, once personified by identifiable authority figures, are now seen as replaced by an impersonal algorithmic system. This system is socially constructed as an invisible, anonymous force that exercises control without human presence or emotional capacity. Algorithms, in this view, are objectified as neutral yet uncontrollable agents that eliminate the subjective elements of human negotiation and compassion (Zarka 2016). Through this representation, AI is depicted as reshaping managerial practices by introducing a form of power that is neither visible nor negotiable, but automated and seemingly unquestionable.

The expansion of AI-driven management is further framed by social representations that evoke comparisons to Kafkaesque systems of domination, where bureaucratic logic becomes overwhelming, absurd, and fundamentally inhuman. Within this perspective, AI is not merely a tool but a symbol of modern alienation (Casilli 2019), an omnipresent force that creates environments where individuals feel powerless, trapped in opaque structures governed by rules they cannot understand or influence. Additionally, there is a growing social discourse that connects AI to broader risks to individual freedom and democratic accountability. In this representation, algorithmic management is no longer confined to efficiency gains but is seen as a mechanism that erodes both workplace autonomy and wider political freedoms (Christiaens 2023). This framing suggests that as AI's influence extends across industries and governance structures, the line between corporate surveillance and social control becomes increasingly blurred.

Conclusion

The rise of AI in the word of work is a double-edged sword, bringing transformative changes that carry both promise and peril. The rise is not simply a technological evolution but a social phenomenon deeply embedded in representations that shape how societies perceive, accept, and resist change. Based on scientific literature in the field of socio-economics of work, mainly books, articles and research reports, analyzing AI and work, this chapter has shown that AI is simultaneously constructed as a promise of progress and a source of disruption. We traced the historical continuum from past automation waves to the current AI revolution and discussed theoretical frameworks of social representations. From an analytical standpoint, what emerges is that the social representations of AI are not fixed; they are dynamic, and distributed in two ways, positive and negative. The positive representations, AI as a driver of innovation, as a creator of new value, and as a modernizer of human resource management are anchored in scientific discourses that emphasize productivity, growth, and the optimization of work. These representations often serve to legitimize and accelerate the adoption of AI by focusing on its benefits for businesses and skilled workers, while marginalizing concerns about inequality and labor degradation. Conversely, the negative representations, the losses of job, the exacerbation of socio-economic inequalities, and the intensification of workplace surveillance, highlight the risks of reinforcing social exclusion, eroding worker autonomy, and creating algorithmically driven environments that reduce human agency. These representations are particularly salient in the experiences of low-skilled workers, gig economy participants, and those operating in tightly controlled digital platforms, where AI is less a tool of empowerment and more a vector of domination.

From a socio-economic perspective, the future of AI in the workplace is not predetermined; it is profoundly contingent on the collective choices we make, the regulatory frameworks we build, and the inclusiveness of the social dialogue surrounding these technologies. If policymakers, businesses, and civil society fail to address the asymmetries embedded in current trajectories, such as job polarization, rising inequality, and pervasive surveillance, the negative social representations risk becoming self-fulfilling prophecies. Conversely, if AI is deployed with safeguards, inclusive strategies, and a commitment to distributing its benefits broadly, the more optimistic representations may materialize.

Looking toward the future trajectory of AI in the labor market, we can consider a spectrum of scenarios (Moddy 2018). In a best-case scenario, AI becomes a general-purpose technology that boosts economic growth, creates enough new jobs (or reduces work hours with stable incomes) such that overall employment and quality of life remain high. Work could become more engaging as drudgery is automated. Society might shorten the workweek, using productivity gains to allow more leisure and learning (some have suggested AI could enable a 4-day workweek or similar, spreading work among more people). In this optimistic future, social representations of AI would likely shift to predominantly positive, AI as an empowering collaborator and an engine of prosperity, analogous to how we now view electricity or the internet as largely positive forces.

In a worst-case scenario, AI adoption outpaces our ability to adapt. Many people could find themselves unemployed or stuck in low-wage gig jobs, with a wealthy few benefiting from AI-driven profits. Inequality could reach untenable levels, and social unrest or backlash against technology could grow. In this future, the representation of AI would be quite negative, perhaps as a malevolent force or the culprit of social breakdown. Extreme narratives, like the idea of a dystopia of mass unemployment, would gain prominence. It's worth noting that both extremes are speculative; the reality will likely be somewhere in between and will be mediated by the choices we make now.

Ultimately, social representations of AI in the world of work are essential analytical lenses to understand not just what AI is, but what AI means to different actors. These representations do not simply reflect an objective reality; rather, they contribute to the construction of that reality. Consequently, the implications of AI in the world of work extend beyond technological considerations and encompass significant social, ethical, and political dimensions. The trajectory of AI's integration into the workplace will ultimately be shaped by the representations we adopt and the actions we take in response to them.

References

Acemoglu, D. and P. Restrepo. 2020. Robots and Jobs: Evidence from US Labor Markets. Journal of Political Economy, 128(6): 2188–2244.

Aghion, P., C. Antonin and S. Bunel. 2020. Le pouvoir de la destruction créatrice, Paris. Odile Jacob.

Agrawal, A., J. Gans and A. Goldfarb. 2018. Prediction Machines: The Simple Economics of Artificial Intelligence. Harvard Business Review Press.

Aloisi, A. and V. De Stefano. 2022. Your Boss is an algorithm. Artificial intelligence, platform work and labour. Hart Publishing.

Altenreid, M. 2020. The platforms as a factory: crowdwork and hidden labour behind Arificial Intelligence. Capital & Class, 44(2): 145–158.

Bales, R. A. and K. Stone. 2020. The Invisible Web at Work: Artificial Intelligence and Electronic Surveillance in the Workplace. Berkeley Journal of Employment and Labor Law. UCLA School of Law, Public Law Research Paper No. 19.

Bao, L., N. M. Krause, M. N. Calice, D. A. Scheufele, C. D. Wirz, D. Brossard et al. 2022. Whose AI? How different publics think about AI and its social impacts. Computers in Human Behavior, 130: 107–182.

Baril, E. and M. Vultur. 2025. Navigating streets, restaurants and algorithms: a study of young immigrant food delivery couriers in Montréal and Toronto. Labour/Le travail, 95.

Boute, R. N., and M. Udenio. 2022. AI in Logistics and Supply Chain Management. pp. 46–65. *In:* Merkert, R. and K. Hoberg (Eds.). Global Logistics and Supply Chain Strategies for the 2020s: Vital Skills for the Next Generation. Springer International Publishing, https://doi.org/10.1007/978-3-030-95764-3_3.

Brynjolfsson, E. and A. McAfee. 2016. The Second Machine Age: Work, Progress, and Prosperity in a Time of Brilliant Technologies. New York : W. W. Norton.

Bubeck, S., V. Chandrasekaran, R. Eldan, J. Gehrke, E. Horvitz, E. Kamar et al. 2023. Sparks of Artificial General Intelligence: Early Experiments with GPT-4. arXiv. http://arxiv.org/abs/2303.12712.

Bughin, J., E. Hazan, L. Lund, P. Dahlström, A. Wiesinger and A. Subramaniam. 2018. Skill shift: Automation and the future of the workforce, McKinsey, 23 mai, https://www.mckinsey.com/featured-insights/future-of-work/skill-shift-automation-and-the-future-of-the-workforce

Calo, R. 2016. Robot Law. Cheltenham, Edward Elgar.

Carlos, J. 2021. Drishti : AI-powered Human Production against the Machines, Harvard Business School Digital initiative, 21 avril, https://digital.hbs.edu/platform-digit/submission/drishti-ai-powered-human-production-against-the-machines/.

Casilli, A. 2019. En attendant les robots. Enquête sur le travail du clic. Éditions du Seuil.

Christiaens, T. 2023. Digital working lives. Worker autonomy and the gig autonomy. Rowman & Littlefield.

Côté, A.-M. and Z. Su. 2021. Évolutions de l'intelligence artificielle au travail et collaborations humain-machine. Ad Machina, 5: 144–160. https://doi.org/10.1522/radm.no5.1413

Cousineau L., A. Ollier-Malaterre, X. Parent-Roxheleau and É. Charbonneau. 2024. La surveillance éléctronique au travail : nouvelles réalités et enjeux. pp. 123–136. *In:* Paren-Rocheleau, X. and Ollier-Malaterre, A. (Eds.). Le management à l'ère numérique : Nouvelles pratiques, réalités et régulations. Presses de l'Université du Québec.

Crumpler, W. 2020. How Accurate are Facial Recognition Systems – and Why Does It Matter, Center for Strategic & International Studies, 14 avril, https://www.csis.org/blogs/technology-policy-blog/how-accurate-are-facial-recognition-systems-–-and-why-does-it-matter

Danaher, J. 2020. Welcoming Robots into the Moral Circle: A Defence of Ethical Behaviourism, Science and Engineering Ethics, 26(4): 2023–2049, https://philarchive.org/archive/DANWRI

David, E. 2020. How The Future Of Deep Learning Could Resemble The Human Brain, Forbes Technology Council, 11 novembre, https://www.forbes.com/councils/forbestechcouncil/2020/11/11/how-the-future-of-deep-learning-could-resemble-the-human-brain/

Delfanti, A. 2021. The Warehouse. Workers and robots at Amazon. Pluto Press. Northampton.

Deranty, J.-P. and T. Corbin. 2024. Artificial Intelligence and Work: A Critical Review of Recent Research from the Social Sciences. AI and Society, 39(2): 675–691. https://dl.acm.org/toc/aiso/2024/39/2

Franey, J. 2025. JP Morgan, credits AI coding assistant for boosting engeneer's efficiency by up to 20%. New York Times, March 14.

Frey, C. B. 2019. The Technology Trap: Capital, Labour, and Power in the Age of Automation. Princeton. Princeton University Press.

Frey, C. B. and M. A. Osborne. 2013. The Future of Employment: How Susceptible Are Jobs to Computerisation?. Oxford Martin School Working Papers. 17 septembre, https://www.oxfordmartin.ox.ac.uk/downloads/academic/future-of-employment.pdf.

Gal, U. and N. Berente. 2008. A social representations perspective on information systems implementation: Rethinking the concept of "frames". Information Technology & People, 21(2): 133–154.

Garneau, J. M-E. 2021. La numérisation de l'économie et les mondes du travail: principals avancées technologiques et enjeux. pp. 21–46. *In:* Bernier, J. (Ed.). L'intelligence artificielle et les mondes du travail. Québec. Presses de l'Université Laval.

Grossman, G. M. and E. Helpman. 1991. Innovation and Growth in the Global Economy. MIT Press.

Gmyrek, P., J. Berg and D. Bescond. 2023. Generative AI and jobs: A global analysis of potential effects on job quantity and quality, ILO Working Paper 96 (Geneva, ILO). https://doi.org/10.54394/

Hawking, S. 2016. This is the most dangerous Time for Our Planet. Guardian, 1 December.

Hawksworth, J., R. Berriman and S. Goel. 2018. Will robots really steal our jobs?: an international analysis of the potential long term impact of automation, *PricewaterhouseCoopers*, https://www.pwc.co.uk/services/economics-policy/insights/the-impact-of-automation-on-jobs.html.

Holzinger, A., K. Keiblinger, P. Holub, K. Zatloukal and H. Müller. 2023. AI for life: Trends in artificialintelligence for biotechnology. New Biotechnology, 74: 16–24.

Hupfer, S. 2020. Talent and workforce effects in the age of AI, *Deloitte,* 3 mars, https://www2.deloitte.com/us/en/insights/focus/cognitive-technologies/ai-adoption-in-the-workforce.html.

Jodelet, D. 2013. La rencontre des Savoirs. Papers on Social Representations. 22: 91–92.

Kameswaran, V., L. Cameron and T. R. Dillahunt. 2018. Support for social and cultural capital developmentin real timeridesharing services. Proceedings of the 2018 CHI conference in human factors in computing system. ACM, avril.

Kameswaran, V., S. Pal, T.C. Veinot, R. Brewer, A. Parameshwar, Y. Vidhya et al. 2018. We can go anywhere' understanding independence through a case study of ride-hailing use by people with visual impairments in metropolitan India Proceedings of the ACM on Human-Computer Interaction, Volume 2, Issue CSCW, No.: 85, Pages 1-2 https://doi.org/10.1145/3274354.

Kolbert, E. 2014. The real reason nobody buys electric cars. The New Yorker. https://www.newyorker.com/magazine/2014/02/24/there-goes-the-sun. February 24.

Kirov, V., B. Anđelković, T. Jakobi and M. Kosheva. 2022. Is there decent work in the online food delivery business? Case studies of Bulgaria and Serbia. Sociological Problems, 54(1): 297315.

Lane, M. and A. Saint-Martin. 2021. The impact of Artificial Intelligence on the labour market: What do we know so far? OECD Social, Employment and Migration Working Papers. 256, https://dx.doi.org/10.1787/7c895724-en.

Lehdonvirta, V. 2016. Algorithms that divide and unite : delocalisation, identity and collective action in microwork. pp. 53–80. *In:* Flecker, J. (eds.). Space, place and global digital work. Palgrave Macmillan.

Leontief W. 1983. Technological advance, Economic Growth and the Distribution of Income. Population and Development Review 9(3): 403–410.

McKinsey and Company. 2018. AI, Automation, and the Future of Work. McKinsey Global Institute.

Meijerik, J., M. Boons, A. Keegan and J. Marler. 2021. Algorithmic Homan Resource management. Synthesaizing developments and cross-disciplinary insight on digital HRM. The International Journal of Human Resources Management 32(12): 2545–2562.

McKelvey, F. and J. Roberge. 2023. Recursive power : AI gouvernmentality and technofutures, in Handbook of critical studies of artificial intelligence. Edward Elgar Publishing, Northampton. pp. 21–32.

Moody, K. 2018. High tech, Low Growth : Robots and the Future of Work, Historical Materialism. (26) 4: 3–34.

Moscovici, S. 1984. Le domaine de la psychologie sociale. *In:* Moscovici, S. (ed.). La psychologie sociale. Paris. Presses Universitaires de France.

Moscovici, S. and M. Hewstone. 1984. De la science au sens commun. pp. 539–566. *In:* Moscovici, S. (ed.). La psychologie sociale. Paris. Presses Universitaires de France.

Moscovici, S. 1993. Toward a social psychology of science. Journal for the Theory of Social Behaviour, 23: 343–374.
Negura L, N. Plante and M. Lévesque. 2019. The role of social representations in the construction of power relations. Journal Theory Social Behaviors, 1–17. https://doi.org/10.1111/jtsb.12213.
Negura, L. and C. Lavoie. 2016. La pensée sociale et professionnelle dans l'action. pp. 11–29. *In:* Negura, L. (Ed.). L'intervention en sciences humaines: l'importance des représentations. Québec. PUL.
Pal, S. K. 2008. 21st Century Information Technology Revolution, Ubiquity, juin.
PriceWaterhouseCoopers. 2018. Artificial Intelligence in HR: a No-brainer, https://www.pwc.nl/nl/assets/documents/artificial-intelligence-in-hr-a-no-brainer.pdf.
Rahman, H. 2024. Inside the Invisible Cage: How Algorithms Control Workers, Oakland. University of California Press.
Russell, S. J. and P. Norvig. 2020. Artificial Intelligence: A Modern Approach (4th ed.). Pearson.
Schroeder, B. 2021. The Data Analytics Profession And Employment is Exploding – Three Trends That Matter, Forbes, https://www.forbes.com/sites/bernhardschroeder/2021/06/11/the-data-analytics-profession-and-employment-is-exploding-three-trends-that-matter/?sh=1b15bade3f81.
Susskind, D. 2021. A world without work. Technology, Automation and How We Should Respond. Penguin Books.
Upadhyay, A. K. and K. Khandelwal. 2018. Applying artificial intelligence: implications for recruitment. Strategic HR Review. doi:10.1108/shr-07-2018-0051.
Vaast, E. 2007. Danger is in the eye of the beholders: Social representations of Information Systems security in healthcare. Journal of Strategic Information Systems, 16(2): 130–152. https://doi.org/10.1016/j.jsis.2007.05.003.
Van Esch, P., S. Black and J. Ferolie. 2019. Marketing AI recruitment: The next phase in job application and selection. Computers in Human Behavior, 90: 215–222, https://doi.org/10.1016/j.chb.2018.09.009.
Vendramin, P. and G. Valenduc. 2023. Comprendre le phénomène des plateformes dansa la transition numérique : une perspective européenne. *In:* Vultur, M. (ed.). Les plateformes de travail numériques. Polygraphie d'un nouveau modèle organisationnel. Presses de l'Université Laval. 19–40.
Vultur, M. 2024. La reconfiguration du monde du travail par l'émergence des plateformes de travail numériques : origines, développement et impacts. Organisations et Territoires, 33(3).
Vultur, M. and L. Wartel. 2024. Sécurité, liberté et sens du travail dans l'économie des plateformes, Qu'en pensent les jeunes travailleurs d'Uber et UberEats qui opèrent au Québec? Ad Machina, 8.
Westlake, S. 2014. Our Work Here Is Done: Visions of a Robot Economy. London: Nesta. https://www.nesta.org.uk/report/our-work-here-is-done-visions-of-a-robot-economy/
Windley, D. 2021. Is AI The Answer To Recruiting Effectiveness, Forbes Human Resources Council, 16 juin, https://www.forbes.com/councils/forbeshumanresourcescouncil/2021/06/16/is-ai-the-answer-to-recruiting-effectiveness/.
Wisskirchen, G., B. Thibault Biacabe, U. Bormann, A. Muntz, G. Niehaus, G. J. Soler et al. 2017. Artificial Intelligence and Robotics and Their Impact on the Workplace, https://www.ibanet.org/LPD/Human_Resources_Section/Global_Employment_Institute/Projects.aspx.
World Economic Forum. 2020. The Future of Job. octobre, http://www3.weforum.org/docs/WEF_Future_of_Jobs_2020.pdf.
Zarka, Y-C. 2016. Métamorphose du monstre politique. Paris. Presses universitaires de France.

CHAPTER 6

Navigating the Future With AI

Sociotechnical Imaginaries and Social Representations in the Work of AI Developers

Emmanuelle Vaast

Introduction

Recent developments in Artificial Intelligence (AI), particularly in machine learning and deep learning, hold the promise of profound transformations for individuals, organizations, and societies. While AI is frequently associated with positive potentials, its development has also provoked widespread anxieties and debates (Teodorescu et al. 2021). AI's capacity to learn and act autonomously introduces uncertainty and opacity into socio-technical systems, as its consequences may evolve in inscrutable and unpredictable ways.

To date, there has been a tendency within scholarship to separate investigations of AI development from studies of the consequences associated with its deployment and use (Bailey and Barley 2020). However, such a separation overlooks how sociotechnical imaginaries—collectively held visions of AI's societal roles and futures (Jasanoff and Kim 2009)—permeate the day-to-day work of AI developers. These imaginaries shape expectations regarding AI's capabilities and risks, framing how developers conceive the future societal impact of the technologies they help build. Simultaneously, social representations—the symbolic constructions, narratives, and metaphors through which developers make sense of AI (Moscovici 1984; Vaast 2007)—mediate how these broader imaginaries are enacted in practice.

Desautels Faculty of Management, McGill University.
Email: emmanuelle.vaast@mcgill.ca

DOI: 10.1201/9781003502753-10

Understanding how AI developers approach their work and interpret the possible future consequences of the technologies they develop is crucial, particularly given AI's growing embeddedness in domains such as labor, security, and healthcare. The consequences of AI use are already generating mixed and, at times, troubling outcomes (Marabelli et al. 2021). To unpack the tension between AI's promises of societal benefit and the realities of disappointing or harmful results, we must examine how AI developers negotiate between dominant sociotechnical imaginaries and their own social representations of AI.

This is especially salient given growing ambivalence among AI developers regarding the societal implications of their work. Some developers have surfaced issues of algorithmic discrimination based on ethnicity and gender (Buolamwini and Gebru 2018), while others have raised alarms about the risks posed by AI systems (Kellogg et al. 2020). These concerns have, in some cases, led to collective actions. These developments point to a workforce that is actively reflecting on the futures their work helps to shape, futures that are infused with utopian and dystopian imaginaries, and mediated through the representational practices of developers themselves.

Accordingly, this chapter asks: *How do AI developers make sense of the future consequences of their work?* By focusing on how developers' social representations are informed by, and feed into, sociotechnical imaginaries, this chapter seeks to provide a more nuanced understanding of how AI futures are imagined and enacted at the site of technological production.

Theoretical Foundations

The Work of AI Developers as Technology Development

The rise of AI holds transformation potential for society and organizations (Makridakis 2017). AI has shown promise of positive outcomes in multiple domains, yet the transformative potential of AI has lingered on the cusp of being fulfilled (Canals and Heukamp 2020). Alongside its promises, AI introduces risks such as biases in facial recognition systems and discrimination in hiring and policing.

The increasing centrality of AI and the growing demand for AI developers' work have generated new uncertainties and ethical dilemmas (Griffin et al. 2024; Malmio 2023). There are growing signs of awareness among AI developers and organizations regarding the ambivalent consequences associated with AI. Professional associations and technology companies have adopted codes of conduct that set ethical boundaries for AI development (Washington and Kuo 2020). Ethical discussions have become a feature of leading AI conferences.

As AI systems increasingly rely on deep learning and neural networks, training and refining AI have become even more complex. AI's semi-autonomous and autonomous capabilities further exacerbate challenges as users and sometimes developers cannot anticipate the systems' outcomes. Yet, these outcomes are still shaped by AI developers' work. The properties of algorithms, the data they rely on, and the models that underpin them emerge from developers' decisions.

AI developers—computer scientists, data scientists, machine learning researchers, engineers, etc. (Piorkowski et al. 2021)—bring their domain expertise to bear in various organizational and institutional settings. Their hybrid arrangements across academia, corporations, and startups reflect the intertwining of AI as scientific domain and engine for innovation, but also foreground the ethical tensions that developers navigate (Orr and Davis 2020).

Within developer communities, social representations play a crucial role in how developers make sense of their work, tools, and responsibilities (Markova 1996; Moscovici 1984; 2001). Developers co-construct what AI "is" and "ought to be" through these shared representations, often privileging technical over social considerations. Such representations also shape boundary work (Gieryn 1983), as developers demarcate their roles relative to others.

Sociotechnical Imaginaries in AI Developers' Work

The development of AI takes place within broader sociotechnical imaginaries—collectively held, institutionally stabilized, and publicly performed visions of desirable futures that are attainable through advances in science and technology (Jasanoff and Kim 2009). These imaginaries shape the meanings, priorities, and ethical considerations embedded in AI development, influencing both the communities of practice and the symbolic content of social representations.

Sociotechnical imaginaries are particularly salient in the case of AI, a technology poised to shape future societies across multiple domains (Makridakis 2017). These imaginaries infuse AI development with a dual orientation toward both the present and imagined futures. Developers draw upon and negotiate with imaginaries of AI as a force for societal progress or disruption, embedding these visions into their practices and identities.

AI developers do not merely produce technical systems; they also engage in ongoing sensemaking regarding the anticipated consequences of their work (Borup et al. 2006; Brown and Michael 2003). These visions are not purely technical but are shaped by social representations—the circulating metaphors, narratives, and categories that developers use to make complex technologies intelligible (Wagner et al. 1999). For example, developers may conceptualize AI as an "objective tool," a "disruptive innovator," or a "risky black box," and these representations mediate how they act within dominant sociotechnical imaginaries.

Social representations operate at the intersection of the developers' community practices and broader societal discourses. These representations provide the symbolic resources developers draw upon to anchor abstract and uncertain technological futures into recognizable frameworks (Moscovici 1984). Representing AI as an "intelligent assistant" rather than as an "autonomous agent" may affect how developers perceive their responsibility for downstream consequences, influencing choices about transparency, accountability, and risk mitigation in their design practices (Jasanoff 2015).

Moreover, social representations function as cognitive shortcuts, shaping how developers communicate about AI within their teams, to non-technical stakeholders, and to the broader public. Such representations stabilize meaning around AI's role and potential impacts, but they also reproduce dominant power dynamics and cultural values embedded within sociotechnical imaginaries (Irani 2015). For example, the representation of AI as a neutral, scientific instrument aligns with imaginaries that emphasize technological determinism and progress, while obscuring human agency and ethical complexity (Eubanks 2018).

Yet, sociotechnical imaginaries are not merely external influences imposed on developers—they are also actively co-produced by developers themselves (Jasanoff 2015). As developers articulate their visions of what AI should do and who it should serve, they contribute to the reproduction or contestation of broader societal narratives around AI. This interplay reveals a recursive relationship between developers' micro-level sensemaking and macro-level imaginaries, where collective visions are continually anchored and objectified through the everyday discourses and practices of technology creators.

Near and Distant Futures in Sociotechnical Imaginaries

The distinction between near and distant futures (Augustine et al. 2019) further enriches our understanding of how sociotechnical imaginaries unfold in AI developers' work. Near futures often relate to incremental developments and short-term societal impacts of AI, focusing on achievable applications such as automation of tasks, customer service optimization, or algorithmic decision support systems. These projections tend to align with institutional imaginaries that emphasize economic efficiency, productivity, and market competitiveness (Bareis and Katzenbach 2022).

Conversely, distant futures are infused with imaginaries that are more speculative and discontinuous from the present, often linked to grand narratives of AI as a revolutionary force or even an existential risk. These visions evoke profound uncertainty, as they imagine AI fundamentally reshaping societal institutions, labor markets, or democratic structures (Crawford et al. 2019). For developers, navigating between these temporal horizons requires reconciling organizational pressures to deliver immediate technical solutions with an awareness of potential long-term consequences embedded in broader sociotechnical imaginaries.

Social representations are crucial in managing this tension. For example, developers oriented toward near-future imaginaries may represent themselves as problem-solvers or scientific experts working within known boundaries of technical feasibility. In contrast, those engaged with distant-future imaginaries might adopt social representations of themselves as guardians, critics, or ethical agents, tasked with mitigating speculative risks such as bias, surveillance, or loss of human autonomy (Moss and Metcalf 2020; Shilton 2018).

Developers may switch between these temporal framings depending on project scope, organizational culture, or external pressures (Roux-Rosier et al. 2018). This dynamic underscores how sociotechnical imaginaries are not static but actively shaped and re-shaped by developers' localized social representations and work practices.

Methods

Research Design and Data Collection

This chapter relies upon a qualitative design that suited its purpose because it enabled to dig deeply onto how some AI developers worked and what perspectives they held regarding AI and its consequences. This design was also conducive of theoretical elaboration by bringing inspiration for new ideas. In a typical manner for qualitative research (Miles and Huberman 1994), I engaged in the collection of diverse data to examine the research question. For one, I worked with a research assistant to conduct 63 semi-directed interviews. Among these interviews, 55 were with AI developers, i.e., with practitioners actively involved in the development of AI. The other eight interviews were with actors who were part of the AI ecosystem without being actively involved in developing AI.

Semi-directed interviews were useful for this study because they revealed important ways in which AI developers saw their work and its consequences. We reached AI developers by at first going through McGill University AI resources and Faculty and then expanding contacts via snowballing mechanisms. We engaged in theoretical sampling to get a variety of perspectives among AI developers. We strove to get diversity in terms of stage in their career (from finishing training to very established), gender, country of work, and ethnicity.

The AI community involves people with specific training and types of daily work and jobs (e.g., theoretical researchers, machine learning researchers, computer vision specialists, software engineers, etc.). During the interviews, respondents discussed their background, career path, and current work. Respondents talked about the challenges and rewards they associated with their work of developing AI. They discussed the ways in which others perceived AI and whether they considered these perceptions accurate. Interviewees reflected upon the situation of their domain as fast-growing and characterized by statistical imbalances in terms of, e.g., gender or ethnicity (Crawford 2016).

I also collected archival data from publicly available social media posts and discussions (from Twitter, reddit, Facebook, blogs, and GitHub primarily) among AI developers. The rationale for collecting these archival social media data was that they provided an important, and distinct, perspective on AI developers and on the ways in which they talked about their work and its consequences. Archival data from online sources can be particularly revealing of delicate matters as these discussions occur organically among people (Vaast and Levina 2015). I collected 125 social media discussions, with more than 2,500 associated posts.

I also engaged in two years of participant observation in the Observatory on the societal impacts of AI and digital technology, henceforth Observatory on AI. It is an organization that brings together Academics (computer science and social sciences), members of Industry (large technology companies and start-up companies) and non-for-profit organizations dedicated to overseeing developments associated with AI.

I had access to email communications, newsletters, reports, and other documentation. I also had access to events organized fully or partly by the Observatory on AI. This participant observation helped gain an understanding of how AI developers thought about the potential of their work and interacted with different actors on the challenges and opportunities of AI development.

Data Analysis

The data analysis process followed guidelines for qualitative grounded theorizing (Charmaz 2006; Feldman 1995; Strauss and Corbin 1998) and relied upon different stages of coding of the data. Analyses started as the data collection unfolded. They proceeded in stages, with interrelated steps of qualitative coding, memo writing, and gradual theory elaboration (Charmaz 2006).

First, as the interviews and social media data collection were engaged, I went through the AI documentation and reviewed the work of the Observatory on AI to become better versed in AI techniques and vocabulary as well as in the AI community. This was helpful to converse with the AI developers in interviews and to make better sense of their social media discussions.

Then, I undertook an open coding of the interviews that involved identifying key themes associated with the background, work, and consequences of one's work among AI developers. On the basis of these early codes, I wrote initial memos that summarized different perspectives that AI developers in the community adopted about their work, as recommended in Locke et al. (2008). The early analyses of the interviews suggested that AI developers considered in two different manners the consequences of their work.

Then, I turned to an open coding of the social media data to examine further these initial insights. The open coding of the interviews had brought light to the social media data. Moreover, the social media discussions among AI developers illuminated that the question of the future consequences of the work of developing AI was closely associated with the definition of expertise for some AI developers and disconnected from it for other AI developers. This insight led to the analytical elaboration of two models of expertise in the AI community, i.e., ideal-typical ways in which AI developers viewed their work and considered its consequences.

Having labelled these two models tangential and integral, I went back to the entire data set to see how the two models contrasted and manifested in work in the AI community. This led to specify differences associated with existing concepts in scholarship.

Then, I searched, in the entire dataset, for ways in which the two models of expertise were connected to each other. I discovered how the AI community resolved some of the tensions between the two models and how this affected AI developers' work. This led me to unpack two ways in which AI developers connected the two models.

Imagined Futures in AI Development

Tangential and Integral Models of Expertise

The tangential model of expertise considered that the future consequences of AI developers' work were peripheral to the work of AI development. It emphasized that AI developers were scientists creating new possibilities for others to implement and with which to deal. The integral model of expertise highlighted that the future consequences of AI were connected to the work of developing AI. The AI developers who leaned towards the integral model acknowledged the importance of recognizing not only the potential for positive innovations but also for negative consequences associated with the work they accomplished. This section contrasts these models of expertise in terms of their imagined futures, the identity of experts, and boundary work.

Imagined Futures

Imagined futures correspond to the futures that people imagine and that explain in part what people do or want to do. The tangential model of expertise was associated with the imagined future as distant future. The integral model was associated with the near future.

The imagined future of the tangential model was "*distant*" (Augustine et al. 2019) in that it was expressed in abstract terms, with a disconnect between the present (i.e., the current situation and the work practices of AI developers) and the future (i.e., what new AI developments would enable in an indeterminate future). For instance, an event organized by the Observatory on AI was dedicated to deep learning. A keynote speaker, a pioneer of deep learning, explained that:

"With all the improvements we have seen in recent years, some people think we are done and just need to scale up what we have already learnt to wider topics and problems, however, I think there are many pieces in the puzzle which are missing."

Such a quote exemplifies how, in a tangential model, AI was making promising advances but that it was difficult to foresee exactly what the distant future would be as well as how it would be reached as seen in the reference to "*many pieces in the puzzle which are missing*." The tangential model of expertise emphasized that AI was progressing in leaps and bounds rather than linearly.

The integral model of expertise focused on the "*near*" future (Augustine et al. 2019), that is, on a more concrete future connected to the present. The imagined future of the integral model was practical and associated with existing practices and contexts. In this model, the work of AI developers was to enhance what already existed and to minimize existing limitations, rather than to create unknown possibilities. The model emphasized the concreteness and direct connection between the present work in AI and the near future. An interviewee was a Faculty at a large University and worked on deep learning and neural networks. He explained the continuity between current AI developments and past technological progress:

"It's important to recognize that at this particular stage of AI, the way that [it] works right now, it's not actually that fundamentally different from any new technology in that what it allows us to do is just accomplish certain tasks more rapidly at higher volume with, you know, less human effort involved, much as almost any other technology we've invented over time." (Ryan, interview)

This interviewee viewed developing AI as akin to developing prior technologies and as mostly building and improving upon existing foundations.

Identity as AI Developers

In the tangential model, AI developers articulated their identity as that of scientists who contributed to scientific progress and followed rigorous methods whereby they first developed assumptions and then tested and refined them to generate new AI. For instance, an interviewee who was a director of AI and data science in a technology company explained that, to him: "*there's no disagreement in science, there's just proving right and proving wrong*." This interviewee emphasized the primacy of the scientific method and of the identity of AI developers as scientists.

The scientist identity underlined that, in this model, AI developers considered that they engaged in science to create a better future for many. An interviewee who was a research engineer and project manager in a technology company illustrates this:

"People need to realize we do science, because we need/we want to solve problems and with that, the future will be something that you will be enjoying as well." (Alejandro, interview)

This interviewee considered that AI developers engaged in science to generate a distant future that would benefit society. By contrast, the integral model of expertise articulated an identity for AI developers whereby they were scientists *and* citizens engaged in society. This model presented AI developers as in part activists whose work strove to identify and limit causes of discriminations in the algorithms they and others were building. The mix of expertise and activism was visible in this social media post by the collective "*@Data4BlackLives,*" self-presented as "*a movement of scientists and activists. Data as protest. Data as accountability. Data as collective action*:"

"What is our duty as Black researchers, academics, and community organizers to collect data and tell our stories? Algorithms are not new; they are manifestations of old systems that devalue and dehumanize our people." (@Data4BlackLives, Tweet, 01/2019).

In the integral model of expertise, AI developers were scientists engaged in society. For instance, an interviewee worked on developing AI at a large university whose focus was on technology engineering and design. She considered that her gender, and the fact that she was in a social minority within her department, made her more aware than most of her peers of potential negative future consequences of AI.

Boundary Work

The two models of expertise were also distinct in their boundary work. They differed in terms of how AI developers demarcated the boundaries between their domain of work and others (Gieryn 1983). The tangential model separated domain of work and responsibilities of AI developers. The integral model was more extensive in its consideration of the domain of work and of the responsibility of AI developers.

The tangential model emphasized the positive potentials associated with scientific progress in AI. Some AI developers articulated a conviction that they worked towards creating new systems that would yield positive innovations and change. They foresaw benefits for people and societies associated with the development of new working algorithms. AI developers who leaned towards the tangential model cautioned against reining in the development of AI. Limiting progress in AI would ultimately stifle innovations that could bring many positive outcomes to society. An interview respondent who worked in AI in a finance company commented:

"I think that the people who develop algorithms have a big responsibility to document their work. (…) I think a lot of responsibility falls on the people who are using the tool to try and use it correctly. Yeah, it's a tough balance because you wouldn't want to kind of stifle innovation." (John, interview)

In the tangential model, AI developers were responsible for developing new working algorithms but not for the potentially negative consequences. This amounted to delegate to others the tasks of dealing with the potential for negative consequences associated with AI development. To limit negative potentials of AI, the tangential model of expertise emphasized that others who were not AI developers but who understood the implications of AI should engage with actors interested in using AI. AI developers from the tangential model appealed to others to explain to the larger public how to use, and not to use, new algorithms.

The integral model of expertise included within the domain of work the importance of recognizing not only the potential for positive innovations but also for negative consequences associated with the work they accomplished and the algorithms they participated in developing. An interviewee who was a senior research scientist in AI at a University noted:

"I do believe that we are held responsible for what we create... We should all be held accountable for the work that we do, and the decisions that we made. And we should have the responsibility to think through those implications. And what damage we could be doing, or what good we could be doing." (Sandra interview)

The integral model also emphasized that it was within the purview of AI developers' work to engage via communication with the general public and organizations adopting AI systems. In this model, AI developers expected to translate some of the technical complexity of AI developments in terms that non-experts could understand. An interviewee who worked in developing AI for a large technology company for instance expanded on the importance of communicating with others:

"Communication... we want to make sure that our product team understands limitations because all the algorithms developers, the AI researchers know their limitations. They never thought that the product team or the business development team could ever ignore those limitations. But that [has] happened quite often."

(Bao interview)

This interviewee explained how, to be successful, he viewed his work as involving not only being technically proficient, but also being able to explain the intricacies of AI to others.

In sum, the two models represented distinct ideal types associated with different imaginations of the future, conceptions of AI developers' identity, and boundary work. AI developers had to find ways to deal with tensions between the two models. The remainder of the section unpacks how AI developers did so by integrating the tangential model of expertise and by blending the two models.

Integrating the Tangential Model of Expertise

Because of the prestige of the potential for scientific advances associated with AI, some AI developers leaned more towards the tangential model but started integrating some aspects of the integral model onto their expertise and work. They accomplished this by expanding the definition of expertise and by expanding the boundaries of their work.

Expanding the Definition of Expertise in AI Development

Some AI developers expanded the definition of expertise in AI development by considering not only scientific and technical skills but also competences arising from social identities, demographics, and experiences. This amounted to incorporating a tenet of the integral model onto the tangential model.

A controversy and the ways in which some AI developers responded to it illustrates this expansion of the definition of AI expertise. In June 2020, a team of AI developers released the PULSE algorithm, a computer vision model developed by Duke University researchers that created deepfake pictures from pixelated ones. AI developers soon recognized that the algorithm yielded biased results. The high-resolution pictures it produced were overwhelmingly Caucasian, regardless of the ethnicity of the original picture. AI developers discussed what made the algorithm so obviously biased in its results. Yann LeCun, a prominent figure in AI, Turing Award winner, Professor at NYU, and Chief AI scientist at Facebook, reflected the tangential model of expertise by asserting the prominence of the training data as source of bias:

"ML [Machine Learning] systems are biased when data is biased. This face upsampling system makes everyone look white because the network was pretrained on FlickFaceHQ, which mainly contains white people pics."

(Yann LeCun Twitter post, 06/2020)

This comment led AI developers to debate whether bias could only come from the data. LeCun then offered his perspective on AI developers as objective scientists, as in the tangential model:

"I think people like us should strive to discuss the substance of (…) questions in a non-emotional and rational manner. It's hard sometimes. But that's why we can call ourselves scientists." (Yann LeCun Tweet, 06/2020)

Timnit Gebru, a well-known researcher on algorithmic bias and data mining and co-lead of the Ethical AI team at Google at the time, criticized LeCun's explanation of the bias:

"Yann, I suggest you watch me and Emily's [Denton] tutorial or a number of scholars who are experts in this are. You can't just reduce harms to dataset bias. For once listen to us people from marginalized communities and what we tell you. If not now during worldwide protests not sure when." (Timnit Gebru Twitter post, 06/2020)

Gebru explained that her expertise stemmed not only from her formal training, which included a Ph.D. in Electrical Engineering from Stanford University, but also from her being an Ethiopian-American woman working in a domain where these demographics were underrepresented. Her perspective exemplified the integral model. Following the exchange, many AI developers discussed what expertise in AI development entailed and where it came from. Some AI developers who leaned toward the tangential model acknowledged that they needed to expand their definition of expertise. This was the case for Thomas Dietterich, Emeritus Professor of Intelligent Systems at Oregon State University, who first stated:

"Everything you say here [LeCun contending that an algorithm is biased when the data is biased] makes sense, but I think it is missing the broader point. When we deploy software (whether built with ML or by other means), we must consider the possibility that it has unknown failure modes that may harm people."
(Thomas Dietterich tweet 06/2020).

This comment was typical of the tangential model since Dietterich emphasized the "*unknown failure mode that may harm people*" that could only become revealed once the new technology is deployed. Future consequences of the new technology could only be fully understood once implemented and in use. Dietterich however continued:

"Both anticipating and detecting harms requires a diverse team that brings a variety of conceptual models and experience to the table. Engaging such a team from the start is essential to anticipating harms in advance." (Thomas Dietterich tweet, 06/2020)

With this comment, Dietterich expanded the definition of expertise to include "*a variety of conceptual models and experience*." He explained that this was important for AI developers to be able to foresee future negative consequences of new AI developments.

Moreover, some AI developers reflected upon who they were as scientists, as in the tangential model, but also questioned how their gender and/or ethnicity participated in shaping their perspective, as in the integral model. For instance, an interviewee was a researcher in AI focusing on usability and security. He explained that he was proud of the scientific progress that his work in facial recognition was making (as consistent with the tangential model) but that he was uncomfortable with the fact that his work could contribute to discriminatory systems (as consistent with the integral model). He was African-American and aware that facial recognition systems tended to yield adverse results for him and many others. He commented:

"I don't like facial recognition algorithms because [they] don't recognize me well, but you find them in most airports these days… We're developing [these algorithms] for a wide society of people, which includes a wide range of people. That includes both men and women and everybody else. So if we are not inclusive of everybody, then [we] fail at the start, right? [We]'re not achieving the objective."

(Kofi interview)

Such a quote illustrates how some AI developers reflected upon how being part of a social minority within the AI community made them recognize risks with AI developments that others were less likely to be aware of right away. They brought together their social identities and their identity as scientists. They considered that demographics were part of their expertise by making them more or less close to future negative consequences of AI.

Expanding the Boundaries of the Work

Some AI developers who leaned toward the tangential model expanded the scope of what they considered to be part of their work. They selectively expanded the boundaries of their work either by reflecting upon how to deal with data bias or by engaging with others to discuss potential future consequences of AI development. Doing so, they incorporated some of the characteristics of the integral model onto the tangential one.

For instance, an interviewee was Head of a Machine Intelligence research group at a large technology company. In his work, he oversaw AI-based R&D for the company. His outlook was associated with distant future possibilities of AI for new technological developments. He explained that, at first, he and his team had not been involved in the selection of datasets to train new AI models. This was consistent with the tangential model. However, given current developments in AI at his company and in other contexts, he had come to consider that ensuring proper datasets was important to his work. He noted:

"Be bold, but also be responsible. Right? (…) How do we make sure that a responsible decision is made based on data? Because, somehow, we believed that [with] enough data, our algorithms can reflect the real world. And this is obviously not true."

(Reinhold interview)

This interviewee further explained how he and his team had recently developed and implemented a system for "*Responsible AI*" that included selecting, cleaning, testing, and refining data for training models. This AI developer had started including in his and his team's work what used to be done by others to limit negative future potential consequences of AI, which was consistent with the integral model.

Moreover, some AI developers included within the domain of work not only making fundamental and applied contributions to AI but also communicating with others to influence policy and, more generally, to address potential consequences of the work. For instance, an interviewee was prominent in AI for his fundamental advances in the domain of deep learning. He explained how he now considered it part of his work not only to contribute to progress in AI with his research, but also to interact with others who were non-experts in AI to bring more awareness to the potential negative consequences of some AI applications. He commented:

> "What I'm doing right now [is] talking to people who care about those social issues [associated with AI]. Talking to politicians, talking to the media, I think there needs to be an involvement of scientists. Even though their own work may not be specifically targeting these kinds of ethical issues, questionable applications, it could be used for this... So I think I have a responsibility to get involved in the democratic debate and discussion."
>
> (Joseph interview)

This AI developer mostly followed the tangential model of expertise as he considered himself a scientist first and foremost and was particularly invested in the distant future. Yet, he had included as part of his work a tenet of the integral model, that of communicating with non-experts in AI.

The engagement of AI developers with non-experts was also evident in an Observatory on AI event in which some AI developers who had worked on the development of contact tracing instruments to respond to the COVID-19 pandemic convened with social scientists, legal scholars, the media, and public servants. These AI developers presented their work and discussed with others its consequences in terms of controlling the spread of the virus and of threatening privacy. Their participation in such an event was a manifestation that they considered engaging with others part of their work. It illustrated how some AI developers worked with others to grapple with questions related to what their work would do in terms of positive (e.g., AI-enabled contact tracing could work and bring public health relief) and negative implications (e.g., for privacy).

Integrating the tangential model brought some elements of the integral model onto the tangential one. However, it did not involve changes in the ways in which AI developers imagined the future. Changes in imagined futures were associated with blending the two models.

Blending the Two Models

AI developers' changing imaginations of the future cultivated a blending of the two ideal-typical models. Blending the two models involved re-imagining the future and bridging near and distant futures.

Re-imagining the Future

The imagined futures associated with AI were not fixed. They changed when AI developers engaged in (re-)thinking about the future that could emerge as AI developed and became more sophisticated. Re-imagining the future was still about speculating on what the future technology would do, but it involved doing so while taking into account what past AI developments had accomplished and how the AI community and the broader societal context was changing. AI developers re-imagined the future in an ongoing manner.

An interviewee, a research scientist working at a research institute and a university, commented upon how little AI developers who worked on prior generations of AI (here, recommender systems) used to imagine negative consequences of their work and how much things had changed:

"Around maybe 2010, there was a lot of excitement in machine learning around the so-called recommender systems. (...). So I think a lot of people were working on recommender systems back then but had no interest at all to think about the ethical implications. Thinking that their work was shielded from possible misuses. And now it's pretty clear that actually all of our work has like really strong implications on the way that people think and even in interact with other people." (Jacques interview)

This interviewee remarked that changes in AI and happenings in society had made it necessary for current AI developers to think about what positive and negative implications new AI developments could bring. To him, re-imagining the future had become essential for AI developers. As another example, an interviewee was a research scientist in Natural Language Processing in a research institute dedicated to making fundamental and applied progress in AI. He explained how the AI community had started reflecting upon recent past and current developments of AI to direct new efforts and limit potential negative consequences in the future:

"There are some people [who] want to push the boundary of their research, which per se is a good thing. (…) I start to see a lot of these workshops or even conferences that start to talk about more of: "Okay, let's stop for a second. Let's make the point: Where are we going? Where, how much have we already done and how much of this is dangerous?"" (Milenko interview)

This interviewee pointed at efforts taking place among AI developers at conferences and workshops. These efforts were taking stock of current AI developments to reassess what the future consequences of AI could be. This interviewee also alluded to differences among AI developers regarding "*push*[ing] *the boundary of their research*," which emphasized the difficulties and need for the AI community to discuss these issues. AI developers recognized that imagining the potential consequences of their work was challenging and would always be incomplete but that it was necessary. AI developers contemplated new possibilities that opened once progress in AI had been made. They redefined what they saw as the future of AI based on its current developments. An interviewee who was a professor of computer vision and deep learning explained:

"Predicting the future… is really super difficult, right? Some people in the field are [of] the opinion that we're like five years away and some people in the field are saying we're like 100 years away… What we don't see is: what are the problems after that? And this notion of AI is always this receding horizon."

(Adam interview)

This interviewee commented upon the importance and the challenge of re-imagining the future in an ongoing manner because working in AI involved working with a "*receding horizon*." Moreover, another interviewee referred to the NeurIPS conference organizers' decision, in 2020, to require a "*statement on the potential broader impacts of their* [submitters'] *work*" (call for papers, NeurIPS 2020) for all submissions, i.e., a discussion of the potential future consequences of the work. He observed that this decision reflected the growing recognition among AI developers of the need to think and rethink the possible future consequences of AI development:

"The outside world should not underestimate the amount of conversation that happens within the field, whether it [a new AI development]'s bad, which topics to tackle, how to tackle them, what are the implications? There was a lot of discussion like: "Should this be needed? Lack of purpose and of liberty? What should we favor? Should we be more inclusive? Should we not be more inclusive?" This is a moving thing then…" (Tony interview)

This comment and the new requirement from this conference illustrated how, given ongoing progress in AI, the AI community was re-imagining the future and its own work.

Bridging Near and Distant Futures

Some AI developers worked on bridging the gap between the promises and risks of the distant future related to AI and the near future associated with its current developments. They did so by keeping a perspective of the potential for new progress associated with AI in the distant future while working on realistic improvements for the near future. It was challenging to do so because AI was difficult to comprehend, even for its experts. An interviewee explained how he struggled with the deep learning algorithms he relied upon to make progress in Natural Language Processing:

"My big problem with the deep learning algorithms, it's the huge interconnection of nonlinear operations that makes it close to impossible for a human to understand what is happening… It's starting to be hard to get an intuition of what can happen. (...) So, let's start to get a feeling of how the model could behave, which also means how the model could negatively behave. It's still very, very hard... The community is definitely looking at this aspect." (Milenko interview)

This AI developer emphasized the need for him and the AI development community to grasp better what the AI they developed could do. Bridging near and distant futures was a community effort because it was challenging for AI developers to envision the eventual consequences of the AI they developed. As an example of how AI developers bridged the near and distant futures, during an event organized in part by the Observatory on AI, a computer science professor who was also affiliated with a large technology company presented her perspective on advances in reinforcement learning, an area of machine learning associated with how intelligent agents should make decisions to maximize cumulative rewards. She noted that the (distant future) potential of reinforcement learning was high but far from accomplished and that there was no way to know when it would be achieved. She explained how current efforts from AI developers in industry and Academia worked on incremental improvements (near future) in reinforcement learning as constituting what she called "*Lego bricks*" that would bring the distant potential into realistic realms.

Bridging near and distant futures also entailed that AI developers consider the connections among past, present, and future of AI. AI developers reflected upon their current practices and what they could mean for the near and distant future. A presentation during an event organized in part by the Observatory on AI illustrates this point. The presentation gave insights on few-shot learning, a new technique that aimed at developing machine learning from only a few examples ("few-shot"). This presentation by a researcher at a large technology company focused on the "*Past, present, and future of few-shot learning*." The researcher situated few-shot learning within the history of AI and explained how past and current practices shaped its near and distant future consequences.

Bridging the near and the distant futures was ongoing and oriented the work of AI developers. For instance, an interviewee specialized in Natural Language Processing to develop machine translation algorithms in a start-up company. She considered that the greatest rewards of her work consisted in improving the services that people were already using. In this way, she originally had a perspective of the near future as enhancing existing services in machine translation and of the distant future as enriching human communication. She however noted that she and the other AI developers in her company had to shift their perspective on the near and distant future services that her company should offer, because users were not using current offerings in the way they had been anticipating. She explained that, at first, AI developers at her company were:

"Imagining the user personas [that would be] traveling and studying. Those were the two big personas. But, after we released the product, in reality, actually people use the product to... make friends all over the world, particularly, to make romantic friends. So that was some usage that we didn't expect." (Dasom interview)

AI developers at her company had adjusted their work according to changes in imagined users and types of uses. They had at first imagined who the users of the service would be and what they would do. Then, based upon actual users, they shifted their expectations of the distant future (from enabling communication to creating new potentials for human connection) and their work as well. They worked towards new future offerings that suited better their current users. Adjustments of what future AI developments should cover were ongoing.

Discussion, Implications, and Conclusion

Discussion

Recent developments in AI have prompted reflections regarding the ways in which work, management, and organizations will be affected. This chapter draws attention to how AI developers navigate sociotechnical imaginaries (Jasanoff and Kim 2009) that project competing visions of AI's societal role and consequences. It reveals how AI developers' consideration of AI's potential future consequences shapes their development work and professional identity. Developers' engagement with these imaginaries intersects with the social representations (Moscovici 1984) they hold about themselves and their work, representations that anchor their understanding of AI and their role in shaping its societal impact.

This chapter identified two ideal-typical ways in which AI developers relate to sociotechnical imaginaries of AI. These models reflect distinct temporal and ethical orientations. The tangential model aligns with a distant future imaginary, where AI developers emphasize long-term, speculative potentials and maintain a narrower social representation of themselves as detached scientists primarily responsible for technical development. The integral model, by contrast, aligns with near future imaginaries, in which developers adopt a more expansive representation of themselves as socially embedded practitioners mindful of AI's immediate societal consequences.

Boundary work (Gieryn 1983) also reflects this distinction. The tangential model reinforces a rigid boundary between AI developers' technical domain and societal concerns, delegating ethical or social considerations to external actors (Abbott 2014). In contrast, the integral model promotes a more porous boundary, extending the remit of AI development to include proactive engagement with issues such as fairness, transparency, and harm mitigation.

While these models represent stylized extremes, this chapter reveals that AI developers routinely move between them. Developers blend elements of both models as they negotiate organizational pressures, ethical considerations, and evolving sociotechnical imaginaries. Integrating the integral model into the tangential one involves expanding developers' social representations to include experiential knowledge and sensitivity to the social identities of users. Developers may thus shift from representing themselves solely as technical experts to viewing themselves as ethical agents or mediators who bear some responsibility for AI's societal impact (Shilton 2018).

Moreover, changing sociotechnical imaginaries and shifts in developers' social representations enable blending of the models. Developers may reimagine AI's future role as they observe real-world consequences and recalibrate expectations based on prior technological trajectories. This dynamic interplay between sociotechnical imaginaries and social representations influences how developers frame their work priorities, responsibilities, and ethical boundaries.

Implications

Implications for Scholarship on AI and the Future

This chapter brings a sociotechnical imaginaries' perspective to the role of AI in shaping societal futures (Jasanoff 2015). Prior research has positioned AI as a driver of societal transformation (Canals and Heukamp 2020), but this study instead highlights how developers' navigation of imaginaries and social representations influences AI's development. Rather than viewing AI as a technology that precipitates the future, this chapter shows that developers' sensemaking about AI's possible futures recursively shapes the present work of AI development.

Developers' active engagement with both near and distant sociotechnical imaginaries reveals how the imagined societal role of AI becomes embedded in development practices long before deployment. The ambivalence and contestation within the AI community stemming from competing imaginaries help explain AI's ambiguous societal impacts.

Additionally, this chapter demonstrates that developers' social representations are dynamic. As developers' understandings of AI's risks and potentials evolve, their professional identities and boundary work shift accordingly. By documenting how sociotechnical imaginaries and social representations change over time, this chapter shows that AI development practices are contingent and shaped by developers' reflexive engagement with their own visions of the future.

Implications for Scholarship Engaging With the Future

This chapter shows that practitioners such as AI developers engage in speculative work. AI developers do not passively follow technological trajectories. Instead, they actively shape and re-shape sociotechnical imaginaries in their everyday work, influencing what kinds of AI futures are made possible.

Further, by connecting boundary work, identity work, and social representations to engagement with sociotechnical imaginaries, this chapter deepens existing understandings of expertise within technology development communities. While prior scholarship emphasized homogeneity in epistemic cultures (Forsythe 1993; Knorr Cetina 1999), this chapter reveals that within a single community, distinct social representations of AI developers coexist and are shaped by differing relationships to near and distant imaginaries of AI's societal role.

This reframing also sheds new light on the literature on unanticipated consequences (Merton 1936). Developers' engagement with sociotechnical imaginaries illustrates that some consequences are not entirely unforeseen. Rather, potential risks (e.g., such as bias, surveillance, or social disruption) are anticipated, albeit unevenly, depending on which imaginaries and representations developers privilege. Thus, seeds of "unanticipated" consequences are often embedded in the contested imaginaries circulating within development communities.

Implications for Scholarship on Technology Development

This chapter contributes to scholarship on technology development by demonstrating how sociotechnical imaginaries can create internal tensions within developer communities. Developers' differing visions of AI's future generate competing notions of what is considered within or beyond the legitimate boundaries of their work (Abbott 2014; Gieryn 1983). While prior research has emphasized how external pressures shape boundary work, this chapter reveals how internal tensions linked to developers' positioning regarding near and distant imaginaries lead to contests over what constitutes legitimate expertise and responsibility.

Additionally, this chapter highlights how developers' social representations are often shaped by their demographics and social identities. This study reveals that developers from underrepresented backgrounds may be more attuned to the potential near future negative consequences of AI. Their lived experiences and social identities can serve as anchors for social representations that foreground harm reduction and ethical engagement, thus shaping how they interpret sociotechnical imaginaries.

Conclusion

AI holds the potential to reshape work, organizations, and societies. This chapter illuminates how AI developers imagine and navigate AI's future consequences, highlighting the interplay between sociotechnical imaginaries and social representations in shaping development practices. By revealing how developers blend and contest competing visions of AI's future, this chapter deepens our understanding of how AI is shaped as much by cultural and social dynamics as by technical imperatives. Addressing these tensions will be crucial as AI becomes more pervasive.

References

Abbott, A. 2014. The system of professions: An essay on the division of expert labor. University of Chicago press.

Augustine, G., S. Soderstrom, D. Milner and K. Weber. 2019. Constructing a distant future: imaginaries in geoengineering. Academy of Management Journal, 62(6): 1930–1960. https://doi.org/10.5465/amj.2018.0059

Bailey, D. E. and S. R. Barley. 2020. Beyond design and use: How scholars should study intelligent technologies. Information and Organization, 30(2): 100286. https://doi.org/https://doi.org/10.1016/j.infoandorg.2019.100286

Bareis, J. and C. Katzenbach. 2022. Talking AI into Being: The Narratives and Imaginaries of National AI Strategies and Their Performative Politics. Science, Technology, & Human Values, 47(5): 855–881. https://doi.org/10.1177/01622439211030007

Borup, M., N. Brown, K. Konrad and H. Van Lente. 2006. The sociology of expectations in science and technology. Technology Analysis & Strategic Management, 18(3–4): 285–298. https://doi.org/10.1080/09537320600777002

Brown, N. and M. Michael. 2003. A Sociology of Expectations: retrospecting prospects and prospecting retrospects. Technology Analysis & Strategic Management, 15(1): 3–18. https://doi.org/10.1080/0953732032000046024

Buolamwini, J. and T. Gebru. 2018. Gender Shades: Intersectional Accuracy Disparities in Commercial Gender Classification Proceedings of the 1st Conference on Fairness, Accountability and Transparency, Proceedings of Machine Learning Research. https://proceedings.mlr.press/v81/buolamwini18a.html

Canals, J. and F. Heukamp (Eds.). 2020. The Future of Management in an AI World. Springer.

Charmaz, K. 2006. Constructing grounded theory: A practical guide through qualitative analysis. Sage publications.

Crawford, K. 2016, June 25. Artificial intelligence's white guy problem. The New York Times.

Crawford, K., R. Dobbe, T. Dryer, G. Fried, B. Green, E. Kaziunas et al. 2019. AI now 2019 report. New York, NY: AI Now Institute.

Eubanks, V. 2018. Automating inequality: How high-tech tools profile, police, and punish the poor. St. Martin's Press.

Feldman, M. S. 1995. Strategies for interpreting qualitative data. Sage.

Forsythe, D. E. 1993. Engineering Knowledge: The construction of knowledge in Artificial Intelligence. Social Studies of Science, 23(3): 445–477. https://doi.org/10.1177/0306312793023003002

Gieryn, T. F. 1983. Boundary-Work and the Demarcation of Science from Non-Science: Strains and Interests in Professional Ideologies of Scientists. American Sociological Review, 48(6): 781–795. https://doi.org/10.2307/2095325

Griffin, T. A., B. P. Green and J. M. V. Welie. 2024. The ethical wisdom of AI developers. AI and Ethics. https://doi.org/10.1007/s43681-024-00458-x

Irani, L. 2015. Hackathons and the making of entrepreneurial citizenship. Science, Technology, & Human Values, 40(5): 799–824.

Jasanoff, S. 2015. Future imperfect: Science, technology, and the imaginations of modernity. In Dreamscapes of modernity: Sociotechnical imaginaries and the fabrication of power (pp. 1–33).
Jasanoff, S. and S.-H. Kim. 2009. Containing the Atom: Sociotechnical Imaginaries and Nuclear Power in the United States and South Korea. Minerva, 47(2): 119. https://doi.org/10.1007/s11024-009-9124-4
Kellogg, K., M. Valentine and A. Christin. 2020. Algorithms at work: The new contested terrains of control. Academy of Management Annals, 14(1): 366–410. https://doi.org/10.5465/annals.2018.0174
Knorr Cetina, K. 1999. Epistemic cultures : How the sciences make knowledge. Harvard University Press.
Locke, K., K. Golden-Biddle and M. S. Feldman. 2008. Perspective—Making Doubt Generative: Rethinking the Role of Doubt in the Research Process. Organization Science, 19(6): 907–918. https://doi.org/10.1287/orsc.1080.0398
Makridakis, S. 2017. The forthcoming Artificial Intelligence (AI) revolution: Its impact on society and firms. Futures, 90: 46–60.
Malmio, I. 2023. Ethics as an enabler and a constraint – Narratives on technology development and artificial intelligence in military affairs through the case of Project Maven. Technology in Society, 72, 102193. https://doi.org/https://doi.org/10.1016/j.techsoc.2022.102193
Marabelli, M., E. Vaast and J. L. Li. 2021. Preventing the digital scars of COVID-19. European Journal of Information Systems, 30(2): 176–192. https://doi.org/10.1080/0960085X.2020.1863752
Markova, I. 1996. Toward an epistemology of social representations. Journal for the Theory of Social behaviour, 2(2): 177–196.
Merton, R. K. 1936. The Unanticipated Consequences of Purposive Social Action. American Sociological Review, 1(6): 894–904. https://doi.org/10.2307/2084615
Miles, M. B. and A. M. Huberman. 1994. Qualitative data analysis—An expanded sourcebook (2nd ed.). Sage.
Moscovici, S. 1984. The phenomenom of social representations. pp. 3–69. *In:* Farr, R. M. and S. Moscovici (Eds.). Social representations. Cambridge University Press, Editions de la Maison des Sciences de l'Homme.
Moscovici, S. 2001. Why a theory of social representations? pp. 14–17. *In:* Deaux, K. and G. Philogene (Eds.). Representations of the Social. Blackwell Publishing.
Moss, E. and J. Metcalf. 2020. Ethics owners: A new model of organizational responsibility in data-driven technology companies.
Orr, W. and J. L. Davis. 2020. Attributions of ethical responsibility by Artificial Intelligence practitioners. Information, Communication & Society, 23(5): 719–735. https://doi.org/10.1080/1369118X.2020.1713842
Piorkowski, D., S. Park, A. Y. Wang, D. Wang, M. Muller and F. Portnoy. 2021. How AI developers overcome communication challenges in a multidisciplinary team: A case study. Proceedings of the ACM on Human-Computer Interaction, 5(CSCW1), 1–25.
Roux-Rosier, A., R. Azambuja, and G. Islam. 2018. Alternative visions: Permaculture as imaginaries of the Anthropocene. Organization, 25(4): 550–572. https://doi.org/10.1177/1350508418778647
Shilton, K. 2018. Values and ethics in human-computer interaction. Foundations and Trends® in Human–Computer Interaction, 12(2): 107–171.
Strauss, A. L. and J. Corbin. 1998. Basics of qualitative research—Techniques and procedures for developing grounded theory. Sage.
Teodorescu, M. H., L. Morse, Y. Awwad and G. C. Kane. 2021. Failures of fairness in automation require a deep understanding of human-ML augmentation. MIS quarterly, 45(3): 1483–1499.
Vaast, E. 2007. Danger is in the eye of the beholders: Social representations of Information Systems security in healthcare. Journal of Strategic Information Systems (forthcoming).
Vaast, E. and N. Levina. 2015. Speaking as one, but not speaking up: Dealing with new moral taint in an occupational online community. Information and Organization, 25(2): 73–98.
Wagner, W., G. Duveen, R. Farr, S. Jovchelovitch, F. Lorenzi-Cioldi, I. Markova. et al. 1999. Theory and method of social representations. Asian Journal Of Social Psychology, 2(1): 95–125.
Washington, A. L. and R. Kuo. 2020. Whose side are ethics codes on? Power, responsibility and the social good. Proceedings of the 2020 Conference on Fairness, Accountability, and Transparency.

CHAPTER 7

Social Representation of GenAI (Generative Artificial Intelligence) and ChatGPT in Educational Policy Discourse and Among Teachers

Jean-Marie Seca[1,*] and *Matthieu Seca*[2]

Introduction

If we are essentially preoccupied with the theory of social representations and its various methodological and epistemological orientations, we risk losing semantics and content concerning the institutional, sociological, political and techno-scientific interest of studying generative artificial intelligence (GenAI). Of course, in this paper, we will still try to identify which sociocognitive structures and attitudes emerge. We will therefore observe, through these structures and attitudes, the components of a social representation of GenAI, both in educational policy discourse and among teachers. To achieve this, we will explore the SRs of GenAI and of ChatGPT in education, analyzing them at two levels:

1. an analysis of institutional and public policy discourses on this technological phenomenon, combined with controversies over the value of integrating ChatGPT into universities, schools and colleges. This integration is deeply linked to its modes of regulation and institutional legitimacy. It is not easy to describe,

[1] University of Lorraine, Department of sociology/TETRAS Research Centre, 23, boulevard Albert 1er BP 60446 F-54001 NANCY CEDEX (France).
Email: jean-marie.seca@univ-lorraine.fr

[2] Université Paris-Saclay, IAE/ISM/Research Center : UVSQ/LAREQUOI, 23 boulevard Vauban 78180 GUYANCOURT (France).
Email: matthieuseca@outlook.com

DOI: 10.1201/9781003502753-11

as the most popular model of generative AI, ChatGPT-3.5, was only released in November 2022 and has many limitations, although the ChatGPT-4 version (and its successor ChatGPT-5, released at the end of August 2025) appear to be better and more profitable. In this paragraph, therefore, we shall focus on an analysis of the ideological and strategic representations of institutions.

2. a summary of surveys carried out in this field, mainly among teachers and students. In addition, we will present some elements of our empirical study of teachers in higher education (public and private universities) and secondary education (high schools, apprenticeship centers) on social representations of GenAI in general and of ChatGPT in particular. The population studied is a sample whose size varies according to the number of non-responses, from 268 to 176 teachers, including a very small group of students from teacher training colleges in France. Various specialists (Rouquette, Moscovici, Moliner) have regularly stressed the essential contribution of opinion leaders (journalists, for example) to the genesis of SR and social thought (Seca 2010). Teachers and future educators play this leading role in the management and dissemination of AI-related RS, as institutional influencers within their classrooms, unions and institutions.

In our analysis of the social representations of GenAIs, we will primarily adopt a sociological and metasystemic perspective. A social representation can be defined as a cognitive metasystem. It aims to verify, control, and select what is worth developing, applying, or retaining in everyday thought processes. This cognitive process between the metasystem and the expression of opinions, attitudes, and behaviors occurs to either logical rules or social imperatives, or both. This process often develop around controversial topics and sometimes include arguments close to childish or illogical reasoning (Doise 1990, p. 115–116).

Furthermore, to justify our approach to opinions on GenAI's in relation to institutions' position on the same subject, we would point out that Willem Doise has thoroughly demonstrated the importance of studying these sociocognitive sets'discourse and content of at various levels. In a well-known contribution, he emphasized the importance of analyzing the normative regulations that control, verify, and direct psychosocial and sociological phenomena by describing them at four levels: intra-individual processes (level 1), inter-individual and situational (level II), positional (level III), and ideological (level IV). Thus, the explanation of a phenomenon such as intelligence or social influence can be explained by referring to one or more of these four levels, depending on the research (Doise 1982). The same applies to the study of SR.

The triangulation or hybridization of multiple data sources—secondary, primary, institutional, written, oral, statistical, legal, and discursive—is the methodological rule rather than the exception. This chapter therefore describes the objectification of the GenAI object and more specifically of ChatGPT in educational settings. It is based on secondary data (articles, reports, institutional opinions) and second- and first-hand field surveys. This cross-referenced and complementary approach to data analysis was first described several decades ago by Moscovici and his successors (Moscovici 1976; Palmonari and Emiliani 2009).

The Tumultuous Integration of ChatGPT Into Educational Institutions: Policy Principles and Ethical Controversies Over Academic and School Integrity

Until the late 1990s, the study of technology-related social representations was a fledgling scientific field. A number of contributions explored beliefs and attitudes towards technological innovation (Flick 1994; Silem et al. 1987). From the 2000s onwards, there has been an increase in the number of this type of publication (Bertoldo et al. 2016; Boy and Chanvril 2010; Brondi et al. 2021; Carugati and Tomasetto 2004; Castro et al. 2020; Christidou et al. 2004; Gal and Berente 2008; Kalampalikis et al. 2013; Lemonnier 2010), but in a limited proportion compared to the interest that technological innovations arouse in management (Dwivedi et al. 2012a, 2012b) or in cognitive psychology and ergonomics (Pasquier 2012; Rosen et al. 2015). Although this is an essential sector for forming opinions, especially given the cultural and political effects of digitization and the platform economy (Durand 2023), studying RS in relation to this subject remains a relatively unknown field of research until the publication of this book. An analysis of the social representation of artificial intelligence is, however, in line with the approaches outlined above. AI is both a social phenomenon discussed in the media and between social actors, and an emerging scientific object in the eyes of the public and citizens. These two criteria validate the existence of a viable, explicit and credible object of social representation (Abric 2003; Flament and Rouquette 2003; Gaymard 2021; Moliner and Guimelli 2015; Moliner 1996; Moscovici 1976; Rouquette 2009).

What's more, despite the existence in France of regulations governing the status of teachers, the two institutional criteria of academic freedom, in higher education, and relative individual autonomy in the management of one's own class, in primary and secondary education, also probably explain the existence of relatively heterogeneous and divided attitudes towards generative AI. According to the surveys carried out (see second part of this chapter), there is a spectrum of decisions, viewpoints and discourses, ranging from caution to enthusiasm to critical, even alarmist, contestation, although some French higher education unions are attempting to produce information documents that aim to be enlightened (Christofol and Pacteau 2024).

ChatGPT, the generative conversational chatbot, is neither banned nor strongly endorsed, although in France some institutions, such as Science Po Paris, are banning it while forming working groups to regulate it.[1] Other institutions, such as the Université de Lorraine in eastern France and the Université Gustave Eiffel[2] in the

[1] Science Po Paris (2023). ChatGPT: Sciences Po fixe des règles et lance une réflexion sur l'IA dans l'enseignement supérieur, Sciencepo.fr, 09 février 2023: https://www.sciencespo.fr/fr/actualites/sciences-po-fixe-des-regles-claires-sur-lutilisation-de-chat-gpt-par-les-etudiants/
Science Po Paris (2023). Sciences Po interdit sans mention par l'étudiant l'utilisation de l'outil ChatGPT, Newsroom.sciencepo.fr, 27 janvier 2023: https://newsroom.sciencespo.fr/sciences-po-interdit-lutilisation-de-loutil-chatgpt

[2] Université Gustave Eiffel (2024). ChatGPT dans l'enseignement supérieur: "La vraie question n'est pas son interdiction mais son usage": https://www.univ-gustave-eiffel.fr/luniversite/pages-speciales/vue-detaillee/chatgpt-dans-lenseignement-superieur-la-vraie-question-nest-pas-son-interdiction-mais-son-usage

Paris region, are also trying to regulate this tool, whose use nobody can control, since it is initiated by individuals, confined to their personal computers, whether students or researchers. These ChatGPT practices are sometimes beneficial, as they may involve assisting a quadriplegic write up assignments or a cautious, exploratory and critical approach to ChatGPT output. They can also pose problems, such as the pre-writing of a text to be included in a dissertation, or outright cheating to obtain good marks for a file to be handed in as part of a directed assignment, without making the slightest effort to research, read and synthesize.

Public Policy Discourse on Artificial Intelligence and Education

It is important to describe the emergence of public policy discourses and pedagogical and scientific justifications for the use of GenAI, and mainly ChatGPT, in educational systems and particularly in higher education. Various critical and/or laudatory statements from the journalistic and academic literature can be found in journals such as *Management et datascience* (Moussavou 2023; Quinio and Bidan 2023; Tran 2021, 2023; Zgoulli 2023). However, there are many other academic articles describing the advantages and disadvantages of ChatGPT or GPT in education (Hua et al. 2024; Kiryakova and Angelova 2023; Sok and Heng 2023; Yang 2024).

The first use of artificial intelligence in education dates back to 1961. It involved a computerized system organized around pedagogical content using graphical interfaces. Experts at the University of Illinois devised an intelligent tutoring system (ITS) called PLATO (Programmed Logic for Automatic Teaching Operations) (Bitzer et al. 1961; Braunfeld and Fosdick 1962). ITS became increasingly popular in the 1970s, with the spread of personal computers. One example is TICCIT (Time-shared, Interactive Computer-Controlled Instructional Television) at the University of Pittsburgh. This software linked television and computers for individualized multimedia education, either at home or in a school setting (Stetten 1971).

All these approaches were structured around the linear or programmed teaching model theorized by Frederic Burrhus Skinner, for whom individualized tutoring was a crucial element in improving pedagogy (Skinner 1968). Another term used is "self-determined learning" (Baskara 2023; Ryan and Deci 2000). Overall, these early inventions were expert systems, organized around rules and cognitive tree structures. It was only with the spread of personal computing and the multiplication of networks linked to the Internet, followed by the emergence of Web 2.0 and very large datasets, that significant advances were made in the organization of ITS.

With Web 2.0 and the exploitation of Big Data, a first form of digitalized control over behavior has developed, through the use of algorithms for tracking user and consumer behavior by major communications and digital multinationals such as Microsoft, Google, Netflix and Amazon. However, without wishing to go into the details of the organizational structure of generative AIs, they could not function without the exploitation, or even plundering, of global data, whether copyrighted or not (Bertolucci 2023). So, for the time being, generative AIs, in order to achieve a minimum of efficiency, are heavy consumers of big data. This is a fact, not an opinion. The use of generative AIs, and ChatGPT in particular, in educational organizations therefore implies acceptance of this global predatory geopolitics of digital data (Miailhe 2018; Stacciarini and Gonçalves 2025).

The use of ChatGPT has been historically and socio-economically justified, even before its release, in international consensus and deliberations within global symposia and various AI reports, including from UNESCO (Miao et al. 2021; UNESCO 2019), from the OECD (Lorenz et al. 2023; OECD 2024) or emanating from other entities (France;[3] European Union on ethics and AI[4]). There is little documentation from ministerial educational authorities in France on ChatGPT.

A report by the Direction du Numérique (digital department) of the Department of National Education and Youth (France), dated January 2024: *Intelligence artificielle et éducation. Apports de la recherche et enjeux pour les politiques publiques* (French Ministry of Education and Youth 2024), more informative than reflexive, claims to establish a state of the art in this field. The authors of this report draw on the *Beijing Consensus*, debated at the UNESCO (United Nations Educational, Scientific and Cultural Organization) International Conference on AI and Education (May 16–18, 2019) (UNESCO 2019).

This biased, pro-AI integration view, dates back to 2019, when the GPT model was not being disseminated, and therefore few convincing technological and educational results were discernible. It was subsequently strengthened around strategic priorities and a master plan for educational AI: inclusive, equitable use; improved learning; increased skills for living in a world with AI; transparency and control of education-related data; defense of a "humanistic" conception of AI (sustainable development, "social well-being", control); and finally cross-sectoral AI management taking into account all stakeholders (Miao et al. 2021).

The "AI" object in institutional discourse is thus very favorably addressed and described. The term "pre-objectification of GenAIs" is one that could be used to describe this process. We then wonder about the scientific and technological references of the authors of this report, given the uncertainties and performance controversies that still exist today in AIs used in educational settings (Wach et al. 2023; Walczak and Cellary 2023; Yang 2024).

Familiarizing the Strange: AI as an Extension of Long-standing Practices and as an Enhancement of Learners' and Teachers' Cognitive Abilities

Finally, to give it an impression of déjà-vu, the integration of generative AI in educational establishments is brought closer to the old Open or Distance Education and Open Educational Resources devices. This approach is further developed in a

---

[3] French Department of Higher Education (2018). La stratégie IA, pour faire de la France un acteur majeur de l'intelligence artificielle, November 2018: https://www.enseignementsup-recherche.gouv.fr/fr/la-strategie-ia-pour-faire-de-la-france-un-acteur-majeur-de-l-intelligence-artificielle-49298

[4] European Commission (2019). *Ethics guidelines for trustworthy AI.* April 2019: https://digital-strategy.ec.europa.eu/en/library/ethics-guidelines-trustworthy-ai

French thematic group: a portfolio called "*GTnum #IA_EO*" (*The impact of Artificial Intelligence through Open Education*[5]). This process of assimilating the new with something older is quite familiar in any objectification of a representational object. It was described by Moscovici in his seminal research: the "*familiarization of the strange*" (Moscovici 1976).

Similarly, the growing influence of personalized tutoring theories aimed at improving learning, assessment, and teaching dynamics strongly reflects also this familiarization process of AI. Furthermore, the arguments for personalized tutoring and "human-machine complementarity" are associated with the promotion of new professional skills that claim to respect the rights and status of teachers. Based on this single idea of "human-machine" complementarity, teachers' work would focus more on providing intellectual, moral and methodological support, as well as assistance with knowledge transfer, human relations and "*high-level thinking*" (Miao et al. 2021).

We will not discuss here the transhumanist philosophy (Hottois 2017) or the longstanding prediction of a technological singularity (Vinge 1993), linked to this concept of "human-machine"complementarity. However, it must be acknowledged that digital prostheses, such as GenAIs, can help people with disabilities improve their education and socio-professional integration (El Morr et al. 2024). People with disabilities could well be really "augmented" by AI.

But, from another perspective, some experts see risks in AI, such as the diminishment of human abilities (Benasayag 2016; Bertolucci 2023). Others diagnose a potential collapse of generative AI. This collapse could be caused by their over-repeated self-feeding on purely synthetic content that they reproduce (Bertrand et al. 2023).

From a functionalist and improvement-oriented perspective, these UNESCO analysts suggest that the integration of AI into education systems also requires the creation of reliable databases for learning. Dysfunctional situations, such as violations of academic integrity and cheating by learners, are not even addressed as a theme to be dealt with, even though there are many quality publications on this subject (Bin-Nashwan et al. 2023; Karkoulian et al. 2024; McCarthy 2023; Sullivan et al. 2023; Susnjak 2022; Sweeney 2023; Ventayen 2023; Wach et al. 2023).

The obsession with improving both the educator and the person being educated is significant. In our view, it is, and will continue to be, a major marketing theme for the dissemination of GenAIs in education systems, and an argument for faciliting their acceptability. This theme enables us to reconcile the industrial and political goals of governments and major corporations, on the one hand, and the legitimate reservations or reticence of teaching teams, on the other. The proposed strategy

[5] French Ministry of Education and Youth. 2024. Intelligence artificielle et éducation. Apports de la recherche et enjeux pour les politiques publiques (p. 56 pages) [Carnet Hypothèses: éducation, numérique et recherche]. Paris: Ministère de l'Education nationale et de la Jeunesse/GTnum/ Direction du numérique pour l'éducation. Consulté à l'adresse Ministère de l'Education nationale et de la Jeunesse/GTnum/ Direction du numérique pour l'éducation website: https://edunumrech.hypotheses.org/files/2024/04/MEN_DNE_brochure_IA_2024_web_fr.pdf (2023). L'impact de l'IA à travers l'éducation ouverte: https://edunumrech.hypotheses.org/files/2023/09/GTnum_LS2N-IA_EO_portfolio_juin2023_vf.pdf.

seems to be to offer education professionals the prospect of moderate integration of these AIs, while preserving the intrinsic value of their profession (transmission, validation, evaluation). The same triumphant coloring of technocratic discourse in these fields can be observed when we peruse an OECD report in which the progress of generative AI is described as "brilliant" (Lorenz et al. 2023). It also suggests "seizing the tremendous opportunity" offered by generative AIs. But this posture is only justifiable if the generative AI tools can be backed up by real reliability. This is far from being the case to date (Floridi and Chiriatti 2020; Hicks et al. 2024). For the time being, there is indeed a utopia—or, more bluntly, an ideology—associated with GenAI development strategies.

AI and Respect for Learners' Rights?

Parallel reservations are expressed, in particular, by the publication and validation of the European Union's *AI Act*: a regulation on artificial intelligence, adopted on December 8, 2023, and published in the *EU's Official Journal* on June 13, 2024.[6] One of its major objectives is to regulate the strong and medium risks associated with artificial intelligence. The French Ministry of Education and Youth's analysis of ethical issues, while correct in terms of documentation but biased and partisan, is in the same vein, making references to the AFIA (French Association for Artificial Intelligence) or UNICEF, etc. The fundamental rights of learners in the face of these AI devices are described, with reference to the Council of Europe (Holmes et al. 2022). Reference is made to the dignity of learners, the rejection of a complete delegation of teaching, assessment and accreditation of taught content to an AI device. The autonomy of young people in relation to these technologies is a priority, so as not to inflict on them a "dictated" training path. Other rights are highlighted: rejection of discrimination, respect for confidentiality, data protection, transparency and "explicability" of AI (Floridi and Cowls 2019).

In the same spirit of industrial deployment, since March 2025, the French Ministry of Higher Education and Research[7] has appointed two project managers to launch an official national survey on AI:[8] François Taddei, President of the Learning Planet Institute, an international think-tank, and Frédéric Pascal, Director of the DataIA Institute, an applied AI research center at the University of Paris-Saclay. These two leaders of technology interest groups are first and foremost stakeholders in the development of AI. This practice of entrusting serious tasks to committed and

[6] The EU Artificial Intelligence Act. Up-to-date developments and analyses of the EU AI Act: https://artificialintelligenceact.eu/
See also: Regulation (eu) 2024/1689 of the European Parliament and of the Council of 13 June 2024 laying down harmonised rules on artificial intelligence and amending Regulations: https://eur-lex.europa.eu/legal-content/EN/TXT/HTML/?uri=OJ:L_202401689

[7] French Department of Higher Education and Research: Lancement de la mission sur l'intelligence artificielle dans les pratiques pédagogiques: https://www.enseignementsup-recherche.gouv.fr/fr/lancement-de-la-mission-sur-l-intelligence-artificielle-dans-les-pratiques-pedagogiques-98140

[8] Individual questionnaire to measure AI usage and expectations: French Ministry of Education and Youth: March 18, 2025: https://mission-ia-sup.fr/index.php?r=survey/indexandsid=935462andlang=fr

potentially biased actors is consistent with a pattern of government behaviour that has been in place in France for more than twenty years. Isn't it a kind of marketing action research that doesn't say its name?

Private or partisan experts are entrusted with a survey that could be carried out as a public service in an objective and critical manner by specialists in the social sciences and technology, employed as researchers in renowned French research centers. Everything is being done to influence professional bodies in higher education and research to conform to political decisions, so that the forced process of equipping education with artificial intelligence devices follows its course.

Some enlightened and unbiased researchers insist on classifying the use of chatbots and ChatGPT in education by evoking the existence of three paradigms (Adiguzel et al. 2023; Ouyang and Jiao 2021). The first describes a context where artificial intelligence has a predominant function and the learner becomes a mere recipient. The second defines the learner as a collaborator with the machine. The third refers to the learner as a manager of his or her own training, autonomous in the face of prosthetic devices. Which of these three paradigms will be chosen in French higher education and, more generally, in European education systems?

Some Available Empirical Data on Teachers' Attitudes and Social Representation Towards Genais

Several sources of data on attitudes toward GenAI were used. The first paragraph of this section describes the impact of management research in terms of SR study. Next, we will develop comparisons between primary data from our own survey of teachers and secondary statistics compiled by researchers who are closest to approaches on attitudes and SR research. Whenever possible, we will specify how these results relate to social representation theory.

Technology Adoption and Representational Universes

Many studies of ChatGPT integration are modeled on positivist and functionalist approaches, such as those concerning technology adoption models from cognitive and neobehaviorist social psychology[9] although they have mainly been applied and developed in management science. These models are influenced by the theory of reasoned action (TRA) (Ajzen and Fishbein 1980) or the theory of planned behavior (TPB) (Madden et al. 1992) by management approaches to information systems (Delone and McLean 2003). The best-known model, and one on which a large number of publications have been based, including applications involving the use of ChatGPT, is the TAM (Technology Acceptance Model).

The TAM was initially developed by Fred Davis (Davis 1989) and subsequently reformulated, mainly in collaboration with Viswanath Venkatesh (Venkatesh and

[9] There are other approaches to management, derived from social psychology, such as expectation confirmation theory (Oliver 1977).

Davis 1996, 2000; Venkatesh et al. 2003). These functionalist models are organized around the detection of factors (such as the perceived effectiveness of a technological tool or its ease of use) that determine the intention to change behavior. They are organized around the goal of reducing resistance to innovation. They presuppose that organizations, and society in general, should function properly by adopting innovative technologies.

The questions of "*why this technology*" and "*how and why it might actually be useful and effective*" are therefore rarely asked. Similarly, such paradigms never ask whether *resistance to change* is justified. These models are inspired by a three-dimensional theoretical vision of the quite "dated" attitude, even though it is not wrong in general terms: affective component/cognitive dimension/conative or behavioral intention. In this theorization, attitude is an intermediary variable (or stimulus) between the subject performing an act and its response, within the framework of the famous "stimulus-response" paradigm.

From the 1980s–1990s, however, theorists such as Anthony Praktanis and Anthony Greenwald proposed a more nuanced, cautious conception, historically linked to the initial sociological conceptualization of this notion (Dewey 1922; Mead 1934; Thomas and Znaniecki 1918) by describing not its function as a predisposition to action, which is always virtually possible, but rather its evaluative and descriptive form of social objects, and thus bringing it closer to the currents of social cognition and social representations (Pratkanis and Greenwald 1989).

Since 2023, a number of empirical research publications have been catalogued on the factors influencing ChatGPT adoption in educational settings, particularly in higher education (Almogren et al. 2024; Duong et al. 2023; Foroughi et al. 2024; Habibi et al. 2023; Lai et al. 2024; Park 2009; Strzelecki 2024; Yilmaz et al. 2023).

We won't be able to enumerate them all here. Indeed, given their prior conceptualizations, many of these empirical studies on the adoption of an innovation are not theoretically neutral. In these publications, the authors may pre-structure the way in which these common objects of representation, namely generative AI and ChatGPT, are considered. These conceptions and their epistemology contribute to generating a certain incremental representation of generative AI, with pro-contra or intermediate options, based on the goal of reducing resistance to change at all costs.

Cross-representations of Uses Between Learners and Teachers

The main findings of French quantitative surveys on the representation of generative AI (GENAI) highlight the three polarizations mentionned at the beginning of this chapter (enthusiasm, fear of threats, exploration of individualized, self-determined educational services), as well as contrasting positions according on the status of the social actors (students, teachers). According to Alain Goudey *et alii*, the main results of their survey of French management and administration students and professors reveal a striking contrast. This online questionnaire was administrated in the first quarter of 2024, and received responses from 668 students, 204 professors and 29 managers:

"*Our survey shows, first of all, that students use GenAIs far more frequently than their professors. In fact, 45% say they use them regularly or even very frequently. Only 7% never use it. Conversely, 44% of teachers never or very rarely use GenAIs, and only 6% say they use them very frequently. At first glance, then, using an GenAI seems to have become a reflex among students, but not at all among their teachers.*"

(Goudey et al. 2024, p. 21).

Each statutory group projects its own representations and practices onto the other. Teachers would tend to think that students only occasionally resort to GenAIs. Conversely, students consider that their teachers use GenAIs as much as they do. This is a fairy well-known phenomenon of asymmetry depending on the reference point for social comparison (Codol 1984) analyzed in particular in gender relations. But this cognitive bias seems to be more widespread than you might think.

The GenAI is becoming increasingly well known among educational audiences. In our own survey of teachers, carried out between October and December 2024, mainly in French higher education across a wide range of disciplines (with around 25% of respondents coming from secondary education), we found that an extreme minority: 1.12% (3 individuals out of 268) knew neither GenAI nor ChatGPT. What's more, some 32.71% of our sample of teachers had never used ChatGPT. We can therefore say that, in 2025, the proportion of non-users of GenAIs in general (44% for teachers, 7% for students in the FNEGE survey) or ChatGPT (32.7% for the teachers in our study) will be around the 35%–40% mark for the teaching staff.

Will this proportion of non-users decrease? We can assume so, as a growing share of the management of various institutions and ministries (survey on AI in higher education launched in March 2025 referenced in footnotes 7 and 8) are beginning to mobilize to find solutions for integrating GENAIs. The reaction, however, is typically slow and significantly exemplary of French institutional temporalities, almost three years after the launch of ChatGPT, in November 2022.

We found a small indication of this downward trend among non-users. A survey of 1,233 students at a Belgian university carried out by Marie Lobet and her colleagues in the first quarter of 2023 (four to five months after the release of the OpenIA model), indicated the existence of a minority use of ChatGPT: 6.3% of students, i.e. 78 out of 1,233, claimed to have used it more than 10 times; 69 out of 1,233 respondents, i.e. 5.6%, said they had used it between one and five times (Lobet et al. 2023 p. 5). This contrast in the proportions of students using this model, between the survey by Loubet *et alii* (February 2023) and the study by Goudey *et alii* (one year later, in March 2024), indicates that the proportion of non-users is likely to fall sharply in a single year.

LeSphinx (a polling firm) and Compilatio (an IT services company specializing in anti-plagiarism) carried out a survey of 1,242 teachers and 4,443 students, aGenAIn in France. The authors of this survey describe teachers' poor knowledge of learners' behaviors: "In fact, 88% of teachers believe that their students use AI at least occasionally, while 55% of students surveyed said they have used it irregularly. Teachers are also concerned about the impact of Artificial Intelligence on learning. Eight out of ten teachers (80%) believe that their students copy and paste AI-generated content into their homework. Whereas only 9% of students claim to do so."[10] In our own teacher survey (based on 176 responses), we found roughly the same proportions: 84.7% of our respondents thought that "learners will produce textual or mathematical content to write their assignments, presentations and exercises"; 84.1% also subscribed to the idea that ChatGPT would be used to "write chapters or paragraphs of dissertations or thesis".

Is GenAI a Threat to the Proper Development of Educational Practices?

Overall, only 9% of teachers think that GenAI would improve students' writing skills. Furthermore, there is a real fear that academic integrity will be compromised. And this calls into question the usual ways of assessing submitted work by the learners. On the other hand, half of the students surveyed by LeSphinx/Compilatio expect to use ChatGPT with the same frequency in the future (49% say they will not use it in the future). More than half the students questioned in this survey claim to use generative AI, and 34% of teachers say they use it at least occasionally.

The time lag between invention and its social dissemination remains an important factor to consider (Alter 2010). It would be interesting to look at the rate of non-users over the next five years to see the level of resistance to what might be considered "propaganda around AI". It seems to us that this resistance is fairly strong among teachers, even though by definition, these staff are always curious, open and attentive to innovation.

In fact, about a quarter of the respondents to our survey declined to comment when asked if GenAI tools like ChatGPT pose a threat to educational practices. Meanwhile, only 30.7% saw no risk or threat at all (see Table 1). The same question, asked on the basis of the item "ChatGPT is an advantage for the proper development of educational practices", gave 41.5% of responses "disagreeing" or "disagreeing somewhat", 31.2% "not expressing an opinion", and 27.3% "agreeing" or "agreeing completely" with this statement. We therefore have a strong base of opposition to this technology in our study sample, with a fraction of around a quarter of the sample in favor of integrating GENAIs into education.

Out of a large list of GenAIs (23 items), proposed in our October-December 2024 survey, ChatGPT is by far the best-known tool: 96.64% of teaching respondents say they are aware of its existence (259 out of 268). The best-known GenAI models, after ChatGPT, have a much lower awareness ranking: Google Bard aka Gemini (identified by 34%, or 91 out of 268 individuals), Midjourney (known by 32.46%, or 87 out of 268 individuals) and Dall-E2 or 3 (known by 28.9%, or 72 out of 268 individuals). The other models (Anthropic Claude AI 1 or 2; Character.AI; VortexAI; ChatPDF; ChatSonic; CivitAI; Hello History; Hugging Face; Meta LiamMA2; Microsoft Copilot or BingChat; Perplexity AI; PhiloGPT; Photoroom; Poe Chat; Quilibot IA; Runway Gen-2; Stable Diffusion; Youchat; other minority GenAI structures) receive a percentage of notoriety between 0.4% and 10.4%.

[10] Blog LeSphinx développement: Intelligence artificielle-enseignement-étude: 6 novembre 2023 (questionnaire administré du 21 juin au 15 août 2023): https://www.lesphinx-developpement.fr/blog/enseignement-et-ia-generative/. See also LeSphinx infographic on this survey: https://infos.lesphinx.eu/mc/Infographie_IA_LeSphinx.pdf.

Table 1: Teacher survey: Seca and Seca, October–December 2024.

Online percentages (headcount)	**Not at all a threat to the proper development of educational practices**	**Not a threat...**	**No opinion**	**Is a threat...**	**Is a threat to the development of educational practices**	**Total**
In my opinion, the use of AI (generative or conversational artificial intelligence), like ChatGPT...	10,23% (18)	20,45% (36)	23,30% (41)	24,43% (43)	21,59% (38)	176

We can make the hypothesis, to be verified later, that since ChatGPT is known by almost all the teachers surveyed, it represents an obvious socio-cognitive attractor and is therefore an element of the core SR of generative AI. However, the social definition of generative AI is not easy to establish. Marie Loubet *et alii* had noted that "the definition of AI is not clear to them, since students wrote the name of applications not using AI in the field '*I use other AI*'" (Lobet et al. 2023, p. 5).

This lack of clarity surrounding the definition of AI in general can be reinforced when we consider applications that are not always perceived as "AI." They are single-task, highly-finalized, such as spell-checking software on Word, or those for translation (Deepl.com, Google translate), image management (Google search Image) or board games such as online chess (Ches.com). All these applications are essentially driven by AI models. Some users who focus on writing a text, translating or playing chess may therefore not spontaneously make the connection to single-task AI models when asked about social or professional practice and AI.

ChatGPT owes its notoriety to its disruptive effect, starting in November 2022, and to its conversational-like functionality involving successive relaunches. These characteristics of radical change, rapid dissemination and perceived "dialogue with the machine" have influenced its marketing imprint. The general public soon thought, rightly or wrongly, that they could "chat" with this artificial intelligence. Through a combination of buzz, word-of-mouth information and announcements in traditional media and social networks, a snowball effect quickly generated an image of credibility and notoriety for the product.

Historically speaking, ChatGPT has seen the strongest growth in usage, with 13 million daily users and 100 million monthly users, two months after its launch in January 2023 (Macq *2024a*: 23). ChatGPT's traffic curve, drawn up by Exploding Topics Similar Web, from November 2022 to February 2025, shows a quasi-arithmetic progression. In March 2005, ChatGPT.com was visited around 5.2 billion times each month, with a total of 601.5 million unique visitors.[11] OpenIA has thus greatly contributed to the popularization of AI (in the broadest sense of the word) among consumers worldwide. Generative AI would go on to be debated throughout the world's societies.

[11] Duarte Fabio (2025). Number of ChatGPT Users (March 2025), Explodingtopics.com, March 13: https://explodingtopics.com/blog/chatgpt-users.

The Question of Academic Integrity and Whether ChatGPT Should be Banned or Integrated Into Teaching Practices

The effectiveness attributed to the GenAI is also highly differentiated according to academic social actors (teachers or students): in the LeSphinx/Compilatio survey of November 6, 2023, 76% of students believe that this tool would make them more effective. Conversely, 27% of teachers felt the same. However, consensual opinions are perceptible: around 60%–64% (teachers/students) of both types of academic stakeholder in this study are against the banning of GenAIs in higher education, although there are concerns about their impact: 81%–72% teachers/students). A consensus exists to qualify the acts of copying content from GenAIs to write assignments as cheating: 76%–65% (teachers/students). We have seen that 80% of teachers think that students cheat, compared to 9% of learners who say that students violate academic integrity.

On the other hand, in a context where GenAIs are institutionally authorized, the intra-institutional survey carried out by Pôle Léonard de Vinci (a private Parisian higher education group) reveals that 99% of students in their higher education schools would use these technologies, 92% of them frequently. We learn that 30% of the sample would use paying options (access to ChatGPT4) costing up to 20 euros per month (Massias 2024). Given that this private institution encourages the use of GenAIs, the students' overall view is positive (70%). In detail, 95% of students at these engineering colleges say that they are easy-to-use tools; 68% that they are reliable instruments; 65% that they increase their knowledge and productivity/performance; 62% that they save time in daily life and even increase their creativity (45%).

Only 25% of students believe that GenAIs replace them in doing their work. In its preamble to the survey report, the private Vinci éducation group claims to have included this empirical study and their liberal/permissive attitude authorizing all types of GenAI use, within the framework of France's national artificial intelligence policy, to want to illustrate the Recommendations of the report by the Commission de l'Intelligence Artificielle *IA: Notre Ambition pour la France*, published in March 2024 (Aghion and Bouverot 2024). Doubts may be raised about the possible cognitive and educational consequences of these assertions and orientations, even if they are draped in the formality of official recommendations.

This trend towards integrating GenAI into educational practices is specific to private higher education and management training in France. To a lesser extent, it affects faculties of Humanities and Social Sciences, Arts and Letters, and Law. Overall, the main ethical risks identified as characteristic of ChatGPT by 176 teachers in our own survey are as follows: "Many users seem to place too much trust in the content generated by ChatGPT" (75%); "ChatGPT encourages cheating by pupils or students in the completion of exercises and assignments" (82.4%). The risks "associated with confidentiality" (59.7%) and "dependence on digital technologies" (65.9%) are less strongly emphasized.

Educational risks are seen above all in the "weakening of users' critical abilities" due to a biased vision of the knowledge produced by ChatGPT (77.8%), the "greater difficulty in assessing learners" (63.1%), the "insufficient regulation of ChatGPT by

French educational and ministerial institutions" (63.1%) and the "genesis of content from overly heterogeneous databases mixing journalism and science" (60.8%). Overall, the teachers questioned in our empirical study see few positive consequences in the educational uses of ChatGPT.

Conclusion: Comments on the Various Results

Of course, all of the observations made in this chapter must be confirmed by more systematic surveys on GenAIs adoption in different regions of the world. The goal is to improve our understanding of how this social representation is formed among various professional, cultural, social, geographical, and national groups. For example, we have seen that the use of GenAIs seem more legitimate in engineering schools and in management or science programs. Why is this? Under what circumstances are these views generated among both teachers and learners? At EMLV, where the intra-institutional study was conducted (Massias 2024), the official educational policy supports the government's national ambitions for AI industry development. In this case, it is the political orientation of the institution that generates educational support for GenAIs.

In line with Doise's approach to analyzing levels of explanation (Doise, 1982), we noticed that opinions about an SR object can be reinforced by the perceived legitimacy of GenAIs practices. These patterns of alteration and transformation linked to a metasystem are commonly observed. The ideological level, linked to political and financial interests, also strongly influences the practices expressed and attitudes. Teachers also strongly resist GenAIs, either through embarrassed silence (selecting the "no opinion" option on the five-point Likert scale) or by rejecting ChatGPT outright. Around 50 to 55% of teachers reject ChatGPT. Basically, only about a third of the surveyed teachers seem to have a favorable or fairly favorable opinion of this tool.

Another element of analysis to highlight is the slow response of official national institutions such as ministries and higher and secondary education institutions. We had already noticed a tendency toward minimal genetic change in social representation, when it exists and has acquired a certain stability. If we take the social representation of democracy, for example, its thêmata are described as universal and as ontology of humanity even some differences between old and new democracies are observed (Marková 2001). Perhaps social and professional resistance to technological imposition is one reason many academic institutions react slowly. Therefore, the adoption and SR of AI would take place at two levels of reality, both of which must be assessed: institutional and intra-individual.

Describing the legal, sociological, cultural and pedagogical interpretations and implications of the spread of ChatGPT and of national or local state policies is a complex task, given the long history of this innovation and the heterogeneous postures of educational institutions and social actors. Indeed, AI, in the broadest sense of the term, generates many financial, economic, military, political and philosophical fantasies, linked to the quest for state power and economic growth. Because of these financial, political, strategic and ideological investments, it is difficult to distinguish between, on the one hand, rational or even reasonable activities leading to pragmatic and appropriate decisions (Bertolucci 2023; Haton et al. 2023) and, on the other hand, risky or hazardous anticipations.

Studying the social representation of this technology also implies taking into account the financial interests of its producers and designers, in the sense given to them by contemporary economists (Guellec 2017) or other specialized economists (Perroux 1972; Schumpeter 1934). The ideological pressure associated with these capitalist interests cannot be ignored by social representation theorists. The infatuation with this technology may therefore also be the result of marketing influence and the need to find a return on very heavy investments that translate into billions of dollars or euros on the part of companies and governments. Last but not least, we can also observe the resurgence of Frankensteinian imaginations, where the machine becomes a threat, or when self-interested expectations of social, cultural and political control of populations emerge, as well as hopes for educational care and the enhancement of human capabilities.

Acknowledgements

Having promised complete anonymity to the French educational institutions and teachers' federations that agreed to distribute our questionnaire at the end of 2024, we will not name them. However, we would like to express our sincere thanks for their help and for distributing our questionnaire.

References

Abric, J.-C. (Éd.). 2003. Méthodes d'étude des représentations sociales. Ramonville-Saint-Agne: Érès.

Adiguzel, T., M. H. Kaya and F. K. Cansu. 2023. Revolutionizing education with AI : Exploring the transformative potential of ChatGPT. Contemporary Educational Technology, 15(3): ep429. https://doi.org/10.30935/cedtech/13152

Aghion, P. and A. Bouverot. 2024. IA : Notre Ambition pour la France. Paris: Commision de l'Intelligence Artificielle / République Française. Consulté à l'adresse Commision de l'Intelligence Artificielle/ République Française website: https://www.economie.gouv.fr/files/files/directions_services/cge/commission-IA.pdf?v=1710339902.

Ajzen, I. and M. Fishbein. 1980. Understanding attitudes and predicting social behavior. Englewoof Cliffs: Prentice Hall.

Almogren, A. S., W. M. Al-Rahmi and N. A. Dahri. 2024. Exploring factors influencing the acceptance of ChatGPT in higher education : A smart education perspective. Heliyon, 10(11): e31887. https://doi.org/10.1016/j.heliyon.2024.e31887

Alter, N. 2010. L'Innovation ordinaire (PUF). Paris. Consulté à l'adresse https://doi.org/10.3917/puf.alter.2010.01

Baskara, F. R. 2023. The Promises and Pitfalls of Using Chat GPT for Self-Determined Learning in Higher Education : An Argumentative Review. Prosiding Seminar Nasional Fakultas Tarbiyah Dan Ilmu Keguruan IAIM Sinjai, 2: 95–101. https://doi.org/10.47435/sentikjar.v2i0.1825

Benasayag, M. 2016. Cerveau augmenté, homme diminué. Paris: La Découverte.

Bertoldo, R., C. Mays, M. Poumadère, N. Schneider and C. Svendsen. 2016. Great deeds or great risks? Scientists' social representations of nanotechnology. Journal of Risk Research, 19(6): 760–779. https://doi.org/10.1080/13669877.2015.1042503

Bertolucci, M. 2023. L'Homme diminué par l'IA. Paris: Hermann.

Bertrand, Q., J. Bose, A. Duplessis, M. Jiralerspong and G. Gidel. 2023. On the Stability of Iterative Retraining of Generative Models on their own Data. Présenté à The Twelfth International Conference on Learning Representations. Consulté à l'adresse https://openreview.net/forum?id=JORAfH2xFd

Bin-Nashwan, S. A., M. Sadallah and M. Bouteraa. 2023. Use of ChatGPT in academia : Academic integrity hangs in the balance. Technology in Society, 75: 102370. https://doi.org/10.1016/j.techsoc.2023.102370

Bitzer, D., P. Braunfeld and W. Lichtenberger. 1961. PLATO : An Automatic Teaching Device. IRE Transactions on Education, 4(4): 157–161. https://doi.org/10.1109/TE.1961.4322215

Boy, D. and F. Chanvril. 2010. Les représentations sociales des technologies du vivant en Europe. Sociologie et sociétés, 42(2): 17–42. https://doi.org/10.7202/045354ar

Braunfeld, P. G. and L. D. Fosdick. 1962. The Use of an Automatic Computer System in Teachin. IRE Transactions on Education, E-5(3/4): 156–167. https://doi.org/10.1109/TE.1962.4322279

Brondi, S., M. Pivetti, S. Di Battista and M. Sarrica. 2021. What do we expect from robots? Social representations, attitudes and evaluations of robots in daily life. Technology in Society, 66: 101663. https://doi.org/10.1016/j.techsoc.2021.101663

Carugati, F. and C. Tomasetto. 2004. Le corps enseignant face aux technologies de l'information et de la communication : Un défi incontournable. Revue des sciences de l'éducation, 28(2): 305–324. https://doi.org/10.7202/007356ar

Castro, A., M. M. Vitali, A. B. S. Bousfield and B. V. Camargo. 2020. Social representations of the internet for the elderly. Journal of Human Growth and Development, 30(2): 227–240. https://doi.org/10.7322/jhgd.v30.10369

Christidou, V., K. Dimopoulos and V. Koulaidis. 2004. Constructing social representations of science and technology : The role of metaphors in the press and the popular scientific magazines. Public Understanding of Science, 13(4): 347–362. https://doi.org/10.1177/0963662504044108

Christofol, H. and C. Pacteau. (Éds.). 2024. Intelligence artificielle : Révolution ou outil ? > SNCS-FSU. La Vie de la Recherche scientifique / FSU-SNESUP, (437). Consulté à l'adresse https://sncs.fr/portfolio/intelligence-artificielle-revolution-ou-outil/

Codol, J.-P. 1984. La perception de la similitude interpersonnelle : Influence de l'appartenance catégorielle et du point de référence de la comparaison. L'Année psychologique, 84(1): 43–56. https://doi.org/10.3406/psy.1984.29000

Davis, F. 1989. Perceived Usefulness, Perceived Ease of Use, and User Acceptance of Information Technology. MIS Quarterly, 13(3): 319–340. https://doi.org/10.2307/249008

Delone, W. H. and E. R. McLean. 2003. The DeLone and McLean Model of Information Systems Success : A Ten-Year Update. Journal of Management Information Systems, 19(4): 9–30. https://doi.org/10.1080/07421222.2003.11045748

Dewey, J. 1922. Human Nature and Conduct : An Introduction to Social Psychology. New York: Henry Holt and Company.

Doise, W. 1982. L'Explication en psychologie sociale. Paris: PUF.

Doise, W. 1990. Les représentations sociales. pp. 111–174. *In:* Ghiglione, R., C. Bonnet and J.-F. Richard (Éds.). Traité de psychologie cognitive, tome 3 : Cognition, representations et communication. Paris: Dunod.

Duong, C. D., T. N. Vu and T. V. N. Ngo. 2023. Applying a modified technology acceptance model to explain higher education students' usage of ChatGPT : A serial multiple mediation model with knowledge sharing as a moderator. The International Journal of Management Education, 21(3): 100883. https://doi.org/10.1016/j.ijme.2023.100883

Durand, C. 2023. Techno-féodalisme : Critique de l'économie numérique. Paris: La Découverte.

Dwivedi, Y. K., M. Wade and S. Schneberger (Éds.). (2012a). Information Systems Theory. Explaining and Predicting Our Digital Society, vol. 1. New York, Dordrecht, Heidelberg, London: Springer. Consulté à l'adresse https://doi.org/10.1007/978-1-4419-6108-2

Dwivedi, Y. K., M. Wade and S. Schneberger (Éds.). (2012b). Information Systems Theory : Explaining and Predicting our Digital Society, vol. 2. New York, Dordrecht, Heidelberg, London: Springer. Consulté à l'adresse https://doi.org/10.1007/978-1-4419-9707-4

Eke, D. O. 2023. ChatGPT and the rise of generative AI : Threat to academic integrity? Journal of Responsible Technology, 13: 100060. https://doi.org/10.1016/j.jrt.2023.100060

El Morr, C., B. Kundi, F. Mobeen, S. Taleghani, Y. El-Lahib and Gorman, R. 2024. AI and disability : A systematic scoping review. Health Informatics Journal, 30(3): 14604582241285743. https://doi.org/10.1177/14604582241285743

Flament, C. and M.-L. Rouquette. 2003. Anatomie des idées ordinaires : Comment étudier les représentations sociales. Paris: Armand Colin.

Flick, U. 1994. Social representations and the social construction of everyday knowledge: Theoretical and methodological queries. Social Science Information, 33(2): 179–197. https://doi.org/10.1177/053901894033002003

Floridi, L. and M. Chiriatti. 2020. GPT-3: Its Nature, Scope, Limits, and Consequences. Minds and Machines, (30): 681–694. https://doi-org/10.1007/s11023-020-09548-1

Floridi, L. and J. Cowls. 2019. A Unified Framework of Five Principles for AI in Society. Harvard Data Science Review, 1(1). https://doi.org/10.1162/99608f92.8cd550d1

Foroughi, B. et al. 2024. Determinants of Intention to Use ChatGPT for Educational Purposes: Findings from PLS-SEM and fsQCA. International Journal of Human–Computer Interaction, 40(17): 4501–4520. https://doi.org/10.1080/10447318.2023.2226495

French Ministry of Education and Youth. 2024. Intelligence artificielle et éducation. Apports de la recherche et enjeux pour les politiques publiques (p. 56 pages) [Carnet Hypothèses: éducation, numérique et recherche]. Paris: Ministère de l'Education nationale et de la Jeunesse/GTnum/ Direction du numérique pour l'éducation. Consulté à l'adresse Ministère de l'Education nationale et de la Jeunesse/GTnum/ Direction du numérique pour l'éducation website: https://edunumrech.hypotheses.org/files/2024/04/MEN_DNE_brochure_IA_2024_web_fr.pdf

Gal, U. and N. Berente. 2008. A social representations perspective on information systems implementation : Rethinking the concept of "frames". Information Technology and People, 21(2): 133–154. https://doi.org/10.1108/09593840810881051

Gaymard, S. 2021. Les fondements des représentations sociales. Sources, théories et pratiques (Paris). Dunod: Dunod. Consulté à l'adresse https://shs.cairn.info/les-fondements-des-representations-sociales--9782100822195

Goudey, A., P. Loupiac and B. Quinio. 2024. Regards croisés sur les IA Génératives dans l'Enseignement Supérieur en Gestion—Panorama des pratiques et perspectives [FNEGE] [FNGE]. Paris- Nanterre-Mont-Saint-Aignan: FNEGE - Neoma Business School - TBS Education - Université Paris-Nanterre. Consulté à l'adresse FNEGE - Neoma Business School - TBS Education - Université Paris-Nanterre website: https://www.calameo.com/read/00193017163b774979682

Guellec, D. 2017. Économie de l'innovation. La Découverte. https://doi.org/10.3917/dec.guell.2017.01

Habibi, A., M. Muhaimin, B. K. Danibao, Y. G. Wibowo, S. Wahyuni and A. Octavia. 2023. ChatGPT in higher education learning : Acceptance and use. Computers and Education: Artificial Intelligence, 5, 100190. https://doi.org/10.1016/j.caeai.2023.100190

Haton, J.-P., E. Haton and M.-C. Haton. 2023. Intelligences artificielles : De la théorie à la pratique : Modèles, applications et enjeux des IA. Paris: Dunod.

Hicks, M. T., J. Humphries and J. Slater. 2024. ChatGPT is bullshit. Ethics and Information Technology, 26(2): 38. https://doi.org/10.1007/s10676-024-09775-5

Holmes, W., J. Persson, I.-A. Chounta, B. Wasson and V. Dimitrova. 2022. Artificial Intelligence and Education : A Critical View Through the Lens of Human Rights, Democracy and the Rule of Law (1st ed). Namur: Council of Europe.

Hottois, G. 2017. Philosophie et idéologies trans/posthumanistes. Paris: Vrin.

Hua, S., S. Jin and S. Jiang. 2024. The Limitations and Ethical Considerations of ChatGPT. Data Intelligence, 6(1): 201–239. https://doi.org/10.1162/dint_a_00243

Kalampalikis, N., M. W. Bauer and T. Apostolidis. 2013. Science, technology and society: The social representations approach. Revue internationale de psychologie sociale, 26(3): 5–9. Consulté à l'adresse https://www.cairn.info/revue-internationale-de-psychologie-sociale-2013-3-page-5.htm

Karkoulian, S., N. Sayegh and N. Sayegh. 2024. ChatGPT Unveiled : Understanding Perceptions of Academic Integrity in Higher Education - A Qualitative Approach. Journal of Academic Ethics. https://doi.org/10.1007/s10805-024-09543-6

Kiryakova, G. and N. Angelova. 2023. ChatGPT—A Challenging Tool for the University Professors in Their Teaching Practice. Education Sciences, 13(10): 1056. https://doi.org/10.3390/educsci13101056

Lai, C. Y., K. Y. Cheung, C. S. Chan and K. K. Law. 2024. Integrating the adapted UTAUT model with moral obligation, trust and perceived risk to predict ChatGPT adoption for assessment support : A survey with students. Computers and Education: Artificial Intelligence, 6, 100246. https://doi.org/10.1016/j.caeai.2024.100246

Lemonnier, P. 2010. L'Étude des systèmes techniques. Techniques and Culture. Revue semestrielle d'anthropologie des techniques, (54–55): 46–67. https://doi.org/10.4000/tc.4989
Lobet, M., A. Honet and V. Wathelet. 2023. ChatGPT, l'usage par les étudiants de première année à l'université ChatGPT. Consulté à l'adresse https://shs.hal.science/halshs-04190226
Lorenz, P., K. Perset and J. Berryhill. 2023. Initial policy considerations for generative artificial intelligence [No. 1, OECD Publishing]. Paris: OECD Artificial Intelligence Papers. Consulté à l'adresse OECD Artificial Intelligence Papers website: https://doi.org/10.1787/fae2d1e6-en
Macq, B. (Éd.). 2024. Face aux défis de l'intelligence artificielle générative. Tome 1 (Académie Royale de Belgique). Bruxelles.
Madden, T. J., P. S. Ellen and I. Ajzen. 1992. A comparison of the theory of planned behavior and the theory of reasoned action. Personality and Social Psychology Bulletin, 18(1): 3–9. https://doi.org/10.1177/0146167292181001
Marková, I. 2001. La démocratie comme thème de psychologie sociale. Bulletin de psychologie, 54(456): 601–609. https://doi.org/10.3406/bupsy.2001.15079
Massias, J. 2024. L'impact des IA génératives sur les étudiants (sondage intra-institutionnel réalisé en février-mars 2024, auprès de 1830 élèves d'écoles supérieurs du groupe privé Vinci Education). Courbevoie: De Vinci éducation-RM Conseil and Talan. Consulté à l'adresse De Vinci éducation-RM Conseil and Talan website: https://www.devinci.fr/pdf/etude-impact-des-IA-generatives-sur-les-etudiants.pdf
McCarthy, C. 2023. ChatGPT use could change views on academic misconduct. Dean and Provost, 24(10): 1–4. https://doi.org/10.1002/dap.31202
Mead, G. H. 1934. Mind, Self and Society from the Standpoint of a Social Behaviorist. Chicago: University of Chicago Press.
Miailhe, N. 2018. Géopolitique de l'Intelligence artificielle : Le retour des empiresand#160;? Politique étrangère, (3): 105–117. https://doi.org/10.3917/pe.183.0105
Miao, F., W. Holmes, H. Ronghuai and Z. Hui. 2021. IA et éducation : Guide pour les décideurs politiques. Paris: UNESCO: Bibliothèque numérique. Consulté à l'adresse UNESCO: Bibliothèque numérique website: https://unesdoc.unesco.org/ark:/48223/pf0000380006
Moliner, P. and C. Guimelli. 2015. Les représentations sociales. Grenoble: PUG. https://doi.org/10.3917/pug.guime.2015.01
Moliner, P. 1996. Images et représentations sociales : De la théorie des représentations à l'étude des images sociales / Pascal Moliner. Grenoble: PUG.
Moscovici, S. 1976. La psychanalyse, son image et son public. Paris: PUF.
Moussavou, J. 2023. ChatGPT dans la rédaction scientifique: Des perspectives prometteuses, des préoccupations légitimes. Management and Datascience. https://doi.org/10.36863/mds.a.23913
OECD. 2024. Assessing potential future artificial intelligence risks, benefits and policy imperatives (N° No. 27, OECD Publishing). Paris: OECD Artificial Intelligence Papers. Consulté à l'adresse OECD Artificial Intelligence Papers website: https://doi.org/10.1787/3f4e3dfb-en
Oliver, R. 1977. Effect of expectation and disconfirmation on postexposure product evaluations : An alternative interpretation. Journal of Applied Psychology, 62(4): 480–486. (1978-13415-001). https://doi.org/10.1037/0021-9010.62.4.480
Ouyang, F. and P. Jiao. 2021. Artificial intelligence in education : The three paradigms. Computers and Education: Artificial Intelligence, 2: 100020. https://doi.org/10.1016/j.caeai.2021.100020
Palmonari, A. and F. Emiliani. 2009. Paradigmi delle rappresentazioni sociali. Bologna: Il Mulino.
Park, S. Y. 2009. An analysis of the technology acceptance model in understanding university students' behavioral intention to use e-learning. Journal of Educational Technology and Society, 12(3): 150–162. Consulté à l'adresse https://www.jstor.org/stable/jeductechsoci.12.3.150
Pasquier, H. M.-L. 2012. Définir l'acceptabilité sociale dans les modèles d'usage : Vers l'introduction de la valeur sociale dans la prédiction du comportement d'utilisation (Thèse pour le doctorat en psychologie, Université de Rennes-2). Université de Rennes-2, Rennes. Consulté à l'adresse https://theses.hal.science/tel-00840220
Perroux, F. 1972. Le profit et la part de l'innovation. Économie appliquée, 25(4): 493–505. https://doi.org/10.3406/ecoap.1972.2926

Pratkanis, A. R. and A. G. Greenwald. 1989. A sociocognitive model of attitude structure and function. *In:* Berkowitz, L. (Éd.). Advances in Experimental Social Psychology (Vol. 22, p. 245–285). New York: Academic Press. https://doi.org/10.1016/S0065-2601(08)60310-X

Quinio, B. and M. Bidan. 2023. ChatGPT : Un robot conversationnel peut-il enseigner? Management and Datascience, 7(1). https://doi.org/10.36863/mds.a.22060

Rosen, L. D., N. Cheever and M. Carrier (Éds.). 2015. The Wiley Handbook of Psychology, Technology, and Society. Malden, Oxford, Chichester: John Wiley and Sons, Ltd.

Rouquette, M.-L. 2009. La pensée sociale : Perspectives fondamentales et recherches appliquées/sous la direction [et avec une introduction] de Michel-Louis Rouquette. Toulouse: Erès.

Ryan, R. M. and E. L. Deci. 2000. Self-determination theory and the facilitation of intrinsic motivation, social development, and well-being. American Psychologist, 55(1): 68–78. https://doi.org/10.1037/0003-066X.55.1.68

Schumpeter, J. 1934. The theory of economic development : An inquiry into profits, capital, credit, interest, and the business cycle. New Brunswick/London: Transaction Publishers.

Seca, J.-M. 2010. Les Représentations sociales. Paris: Armand Colin.

Selcer, P. 2015. Fabricating Unity : The FAO-UNESCO Soil Map of the World. Historical social research-Historische sozialforschung, 40(2): 174–201. (WOS:000352520400009).

Silem, A., J.-M. Albertini, P. Vergès and J.-B. Grize. 1987. Salariés face aux nouvelles technologies. Vers une approche sociologique des représentations sociales (CNRS éditions). Paris.

Skinner, F. B. 1968. The Technology of Teaching. New York: Meredith Corporation.

Sok, S. and K. Heng. 2023. ChatGPT for Education and Research: A Review of Benefits and Risks [SSRN Scholarly Paper]. Rochester, NY. https://doi.org/10.2139/ssrn.4378735

Stacciarini, J. H. S. and R. J. de A. F. Gonçalves. 2025. Data Centers, Critical Minerals, Energy, and Geopolitics: The Foundations of Artificial Intelligence. OSF. https://doi.org/10.31235/osf.io/2zvkt_v1

Stetten, K. J. 1971. TICCIT: A delivery system designed for mass utilization. Présenté à EASCON Conference, Waschington D.C. Waschington D.C.: The Mitre Corporation.

Strzelecki, A. 2024. Students' Acceptance of ChatGPT in Higher Education : An Extended Unified Theory of Acceptance and Use of Technology. Innovative Higher Education, 49(2): 223–245. https://doi.org/10.1007/s10755-023-09686-1

Sullivan, M., A. Kelly and P. McLaughlan. 2023. ChatGPT in higher education : Considerations for academic integrity and student learning. Journal of Applied Learning and Teaching, 6(1): 31–40. https://doi.org/10.37074/jalt.2023.6.1.17

Susnjak, T. 2022. ChatGPT: The End of Online Exam Integrity? arXiv. https://doi.org/10.48550/arXiv.2212.09292

Sweeney, S. 2023. Who wrote this? Essay mills and assessment – Considerations regarding contract cheating and AI in higher education. The International Journal of Management Education, 21(2): 100818. https://doi.org/10.1016/j.ijme.2023.100818

Thomas, W. I. and F. Znaniecki. 1918. The Polish peasant in Europe and in America : Monograph of an immigrant group. Boston: Richard G. Badger: The Gorham Press.

Tran, S. 2021. Vers la fin des cours en présentiel et des campus? Management and Datascience, 5(3). https://doi.org/10.36863/mds.a.15837

Tran, S. 2023. ChatGPT va-t-il disrupter l'enseignement supérieur? Management and Datascience, 7(2). https://doi.org/10.36863/mds.a.23569

UNESCO. 2019. Beijing Consensus on Artificial Intelligence and Education [International Conference on Artificial Intelligence and Education 'Planning education in the AI era: Lead the leap]. Paris: UNESCO: Bibliothèque numérique. Consulté à l'adresse UNESCO: Bibliothèque numérique website: https://unesdoc.unesco.org/ark:/48223/pf0000368303

Venkatesh, V. and F. D. Davis. 1996. A Model of the Antecedents of Perceived Ease of Use : Development and Test. Decision Sciences, 27(3): 451–481. https://doi.org/10.1111/j.1540-5915.1996.tb00860.x

Venkatesh, V. and F. D. Davis. 2000. A Theoretical Extension of the Technology Acceptance Model : Four Longitudinal Field Studies. Management Science, 46(2): 186–204. Consulté à l'adresse https://www.jstor.org/stable/2634758

Venkatesh, V., M. G. Morris, G. B. Davis and F. D. Davis. 2003. User Acceptance of Information Technology : Toward a Unified View. MIS Quarterly, 27(3): 425–478. https://doi.org/10.2307/30036540

Ventayen, R. J. M. 2023. ChatGPT by OpenAI: Students' Viewpoint on Cheating using Artificial Intelligence-Based Application [SSRN Scholarly Paper]. Rochester, NY: Social Science Research Network. https://doi.org/10.2139/ssrn.4361548

Vinge, V. 1993. The coming technological singularity: How to survive in the post-human era. Whole Earth Review, (Winter).

Wach, K., C. D. Duong, J. Ejdys, R. Kazlauskaitė, P. Korzynski, G. Mazurek. et al. 2023. The dark side of generative artificial intelligence : A critical analysis of controversies and risks of ChatGPT. Entrepreneurial Business and Economics Review, 11(2): 7–30. https://doi.org/10.15678/EBER.2023.110201

Walczak, K. and W. Cellary. 2023. Challenges for higher education in the era of widespread access to generative AI. Economics and Business Review, 9(2). https://doi.org/10.18559/ebr.2023.2.743

Yang, H. 2024. Towards Responsible Use: Student Perspectives on ChatGPT in Higher Education. European Conference on E-Learning, 23(1): 415–422. https://doi.org/10.34190/ecel.23.1.2790

Yilmaz, H., S. Maxutov, A. Baitekov and N. Balta. 2023. Student Attitudes towards Chat GPT: A Technology Acceptance Model Survey. International Educational Review, 1(1): 57–83. https://doi.org/10.58693/ier.114

Zgoulli, S. 2023. Systèmes d'IA Générative à l'université : 20 propositions à débattre. Management and Datascience, 7(2). https://doi.org/10.36863/mds.a.23466

Health, Law and Ethics of AI

CHAPTER 8

Between Copy and Distortion
Social Representations of Synthetic Data and Digital Twins in the Shifting Epistemic Landscape of Primary Care Research[a]

Eduard Moreno-Gabriel,[1,*] *Núria Vallès-Peris*[2] and
Brenda Biaani León-Gómez[1]

Introduction

In ancient Egypt, the royal placenta was worshipped and considered the Pharao's twin. In fact, everyone was supposed to possess a spiritual twin, the Ka or astral body, through whom it was supposed to be possible to perform multiple operations (Rachewiltz et al. 1976). Although Ka is one of the most significant categories among Ancient Egyptians' notions about human nature, more than a hundred years of research have led to different interpretations, often contradictory. For some, Ka is a double of the human body, a material copy of a person, but produced from a different kind of substance, affording distinct possibilities. It is also interpreted as an image, similar to a genius or defender, a protector of a person during his or her lifetime. Others find in Ka the vitality and life force inherent of the living creatures, as well as it is also considered as the individuality of a person, its character (Lekov 2005).

[1] Primary Care Research Institute Jordi Gol (IDIAPJGol), CAP La Llàntia, 08303, Mataró, Spain. Email: bleongo.mn.ics@gencat.cat

[2] Spanish National Research Council (CSIC), Campus de la UAB, 08193, Cerdanyola del Vallès, Spain. Email: nuria.valles@iiia.csic.es

* Corresponding author: emoreno@idiapjgol.info

[a] This project received a research grant from the Spanish Research Agency, Ministry of Science, Innovation and Universities (Spain), awarded on the call for Projects for Knowledge Generation, with reference PID2024-160387NA-I00.

DOI: 10.1201/9781003502753-13

Despite its apparent remoteness, the figure of *Ka*, along with the diverse interpretations surrounding it, serves as a potent metaphor for understanding the emerging social representations of Artificial Intelligence (AI) in primary healthcare. More specifically, parallels can be drawn between the ancient Egyptian concept of the Ka, which represented the human soul or "life force", and the way synthetic data (SD) and digital twins (DT) are currently being objectified and anchored into the values and common understanding of healthcare researchers, managers, and practitioners.

In brief, just as the human soul was objectified as a placenta or a twin, data (including SD) is seen as the "lifeblood of modern healthcare" (Giuffrè and Shung 2023). Similarly, much like *Ka*, digital twins are integrated in different areas of health care research, education, data management, and health IT development (Gonzales et al. 2023) as realistic representations of physical entities, yet they transcend the limitations imposed by such material reality (Korenhof et al. 2021).

Indeed, the swift 'popularisation' of DT and SD creates fertile ground for the development of diverse conceptions, particularly within expert and grey literature (cf. Jacobsen 2025). Primarily, these efforts aim to 'naturalize' inherently 'artificial' data, enabling individuals who work with them to comprehend and master this uncertain world (Räty and Snellman 1992). However, there have been few systematic attempts to map the social representations that permeate the daily lives of those actively engaging with SD and DT. Therefore, in this chapter, we undertake a theoretical reflection based on a case study to examine the emerging social representations of SD and DT, as well as the blind spots and controversies they conceal or provoke.

To do so, we draw on Zulato's (2023, 2021) approach, integrating the dialogical perspective of social representations theory (SRT) with the notion of liminal hotspots (Greco and Stenner 2017). In doing so, we aim to capture both the products and processes of representing SD and DT (Castro and Santos 2020). More precisely, we begin by identifying how SD is often objectified as a distorted copy. Conceiving this image as a liminal hotspot highlights how the interdependent poles of the themata at the core of emerging representations of SD (e.g., privacy concerns/public utility) are mobilized in the process of making sense of this emergent phenomenon (Marková 2015). Furthermore, Science and Technology Studies (STS) sheds light on how the contradictory nature of emerging representations of SD and DT is often anchored in the context of healthcare research. Finally, we discuss emerging ethical and democratic concerns that are often acknowledged but also overlooked and minimized within this dialogical dynamic, reflecting a transitional moment in health and biomedical epistemologies.

Synthetic Data and Digital Twins in Primary Care

SD and DT are often used interchangeably in healthcare applications, although their meanings are unclear and not concise (Cabrera et al. 2025; Popa et al. 2021). SD encompasses a spectrum of datasets, that range from partially to fully synthetic. While partially SD incorporates real-world data, fully SD are generated *de novo* (Giuffrè and Shung 2023). SD is helpful for DT, which is a "living" replica of a physical system (human or non-human). DT combines various emerging technologies such as AI, the Internet of Things, big data, and robotics, resulting in a renewed artefact representation of the Cartesian idea of the "automata", possibly the first known idea of twin-like creatures (Popa et al. 2021). The DT is a model that is continually adapted to changes based on the collected digital data that can forecast the future of the corresponding physical counterpart (Liu et al. 2018).

DT and SD emerge within a broader shift in medicine, from treating diseases to predicting and preventing them, as part of what some scholars call the '4P approach', which aims for predictive, personalized, preventive, and participatory models (Younesi and Hofmann-Apitius 2013). In general, DT and SD refer to artificially generated datasets that replicate the statistical properties of real-world data without containing any identifiable patient information. These data are becoming a promising and frequently used tool for health research, particularly in contexts where legal and ethical constraints hinder access to real clinical data. The benefits associated with the use of SD include the ability to improve access to large datasets, foster international collaboration, and accelerate the development of new medical technologies (Mendes et al. 2025).

In their potential applications in primary care, SD and DT as virtual replicas of physical systems or processes could be used to predict individual patient's behaviour or public health, assuming that society is a system that can be replicated. DT enables the real-time simulation and prediction of human systems, whether at the individual or group level. Although the use of DT in healthcare is still in its early stages, it offers the potential to develop personalized patient models that support more effective treatment planning and better health outcomes. One area where DT are increasingly being applied is hospital management, where it helps build realistic models to test various scenarios, such as changes in patient flow, staff training levels, or equipment availability. In addition, the DT of individual patients enable tracking of their predicted clinical pathways. These personalized DT allow for the testing of different treatment strategies, helping to predict outcomes and optimise care while potentially lowering overall healthcare expenses (Giuffrè and Shung 2023).

AI technologies involved in the generation of these datasets range from deep learning structures such as Generative Adversarial Networks (GANs) and Variational Auto-encoders (VAEs), to agent-based econometric models or stochastic differential equations simulating physical or economic systems. These methods enable the creation of realistic data that preserve the clinical correlations necessary for the development and validation of algorithms in the healthcare field. In both primary and hospital care settings, SD and DT can be used to train predictive models without compromising patient privacy, thereby facilitating biomedical research and the development of clinical decision-support systems (Yoon et al. 2020).

Case Study

Spain is emerging as a key player in healthcare innovation through the use of SD projects. Various research projects and applications have been developed in recent years. SYNTHIA, led by IIS La Fe in Valencia, generates synthetic multimodal patient data for AI-driven, personalized medicine. During COVID-19 pandemic, the Barcelona Supercomputing Center used synthetic clinical reports for disease modelling. Madrid's SERMAS DxGPT project uses GPT-4 for diagnosing rare diseases. BIGAN Aragón provides infrastructure for analyzing anonymized health data. The European SEARCH project, involving Tecnalia and Trinity College Dublin, is deveoloping AI diagnostic tools using federated learning and SD. Hospital Sant Joan de Déu applies automated systems in paediatric care and leads the Únicas network for rare diseases. This trend reflects European dynamics, where the pandemic accelerated SD adoption, highlighting the need for privacy-preserving data sharing in healthcare research and the fostering of cross-border collaborations.

To examine the emerging configurations and controversies of SD and DT in healthcare, we chose the case of the "Information System for the Development of Research in Primary Care" (SIDIAP). SIDIAP is a team of researchers and data scientists from the Catalan Health Service (Spain) responsible for developing and maintaining an information platform of excellence with data from the health system and offering research support services in order to generate knowledge and disseminate results to improve healthcare practice and people's health ("SIDIAP: Què és el SIDIAP?" 2023). Basically, this service centralizes, maintains and guards a database gathering electronic health records from up to 80% of the Catalan population who attend primary healthcare centres.

Recently, a small group of SIDIAP workers undertook the project of developing what they call "pseudosynthetic data", drawing from the original SIDIAP database. In 2024, as a member of the research institute where SIDIAP is integrated (IDIAP Jordi Gol), the first author of this chapter attended a workshop where this group presented this project to potential users (researchers and primary care physicians). Drawing on the accessibility granted by this shared membership, this author openly shared with them his ongoing and rather critical interest in exploring challenges and possibilities in AI adoption in healthcare, which was shared with the two co-authors of this chapter (León-Gómez et al. 2023). This facilitated access to grey literature and internal documents disseminating the project, and eased the interviewing of the two leading researchers of this group. To maximize variation and gain a broader understanding of how SD is currently being implemented and conceived in the Catalan healthcare context, we drew on purposive sampling to identify key informants who could complement the views of these experts. As a result, we conducted three more interviews with representatives from key stakeholder groups: a primary care physician conducting research with SD, a governmental digital health officer, and a researcher using SD in a different healthcare setting (i.e., hospital) (see Table 1).

To explore the perceptions, experiences, and challenges related to the use of DT and SD in primary care and public health, a semi-structured interview guide was developed, featuring open-ended questions addressing the following topics:

- Perceptions of the use of DT and SD in healthcare.
- Barriers and facilitators of implementation.
- Ethical and social implications of using SD and DT in healthcare.
- Potential applications in primary healthcare.

Interviews had an average duration of 60 minutes, were conducted both online and offline between March and June 2025. All interviews were recorded with prior

Table 1: Overview of interview participants' roles.

Participant	Role/sector
EE01	Data Scientist – researcher (SIDIAP - primary care)
EE02	General Practitioner – researcher (primary care)
EE03	Head of Hospital Department – researcher (public hospital)
EE04	Data Scientist – researcher (SIDIAP - primary care)
EE05	Head of Catalan Government body guiding digital transformation in healthcare (Government)

informed consent granted from participants, even though the IDIAP Jordi Gol IRB waived the project from going through ethical review given its exploratory nature and non-sensitive data generation. Interviews were conducted in Catalan, transcribed verbatim, and subsequently coded following the precepts of a reflexive thematic analysis approach (Braun and Clarke 2019). This process included:

- Exploratory reading of the transcripts.
- Initial coding of the themes.
- The codes were grouped into themes informed by the SRT and STS.
- Reviewing and naming themes.

The analysis was conducted iteratively, initially by the two first authors independently, and then triangulating themes and interpretations between all three authors. This process was conducted using Atlas.ti 25. The excerpts selected for illustrating codes were translated from Catalan to English by the authors, prioritizing keeping the original expressions over grammatically correct English.

Findings and Discussion

As a result of the analytic process outlined above, we identified some key mechanisms involved in the ongoing development of social representations of SD in the Catalan healthcare research context (see Table 2 for a summary of the most relevant).

Reconciling Re-presentation and Manipulation: Objectifying SD as a "Distorted Copy"

In recent decades, an increasingly rich body of literature has explored the imaginaries associated with science and technology from multiple perspectives (McNeil et al. 2017). Among these contributions, Jasanoff and Kim (2013) stand out as particularly influential. According to them, sociotechnical imaginaries are 'collectively held and performed visions of desirable futures (or of resistance against the undesirable), that are also animated by shared understandings of forms of social life and social order attainable through, and supportive of, advances in science and technology' (Jasanoff and Kim 2013:19). STS research demonstrates that imaginaries are deeply embedded in the practices and institutional arrangements of technoscience, guiding the paths of innovation and technological development (Fujimura 2003). Moreover, these

Table 2: Key Mechanisms in the Development of Social Representation of synthetic data in Catalan Healthcare Research.

SRT Mechanism	**Theme**	**Quote(s)/metaphor**
Thematas	private/public social/technical	*The trade-off between, for example, privacy and utility is a challenge. But that's a technical issue*
Objectification	Distorted copy	*The synthetic data would be the document that comes out of the printer with stripes or blurred because there is not enough ink. Therefore, it is a copy of the original document but with distortions, because the original dataset has been manipulated*
Anchoring	SD as a pattern shift and problem-solver in data configurations	*So, yes, we do have a problem with data access, and since that issue is still quite blocked for now, okay, we have explored the possibility of, instead of giving out data, helping with the development of synthetic data.*
Hegemonic Representation	AI and SD as unbiased and asepctic technical issue	*Artificial intelligence does not have biases, nor does it intend to harm anyone.*
Polemic Representation	SD as an unavoidable contradiction: protecting data privacy by using synthetic data generated from unprotected data.	*We are back to the double-edged sword, right? Personally, I think certain things need to be protected, but at the same time, someone else is doing it—and I am using it. I mean, the U.S. is doing it, they are doing it over there without restrictions, and we are using it here.*

imaginaries extend beyond individual research initiatives; they carry an implicit understanding of the social world (Wynne 2007). Rather than unfolding in a linear fashion, this process involves negotiation and contestation between competing imaginaries and those concerning worldviews, the artefacts themselves, and the relationships interwoven with them (Vallès-Peris and Domènech 2020).

Throughout our interviews, we prompted participants to explain SD to an outsider or lay people, with the main aim of exploring these social imaginaries surrounding its use in healthcare. They often struggled to offer a clear commonsensical description, mainly because they acknowledged the novelty mentioned above ("I am not an expert, because it is something relatively new for me" EE02), and quickly turned to unpacking their potential uses and capacities to address some organizational challenges within health research, which we will discuss below.

This struggle to define SD and DT is reflected in the expert literature, where there is no unified definition or meaning. In many cases, even the stakeholders responsible for working with these technologies or developing them lack a clear and unified understanding of what they mean (Cabrera et al. 2025). They function as umbrella terms used to categorise various practices and methods that can bring together the interests of various stakeholders, including development companies, commercial enterprises, and research groups (Popa et al. 2021). As noted in the interviews, SD is conceptualised as a technological and semiotic cocktail, containing combinations of already existing technologies and social representations.

Then [10 years ago], we did not call it artificial intelligence, and what has ended up happening, as always, is that they have changed the names. What did we call them at that time? Before, it was Big Data, then it became advanced statistics, then it went on to talk about the digital twin, and in the end we are now talking about generative intelligence and synthetic data (EE03)

Incompatible accounts and concerns about synthetic datasets in primary care research are articulated in relation to diverse expectations, responsibilities and epistemic commitments. This multiplicity tends to increase ambivalence and multiple perspectives or interpretative repertoires in STS terms (Gilbert and Mulkay 1984). For instance, SD and DT are frequently constructed as representations or copies of so-called real-world data. However, these are sometimes presented as fabricated data that prevent identification. Nevertheless, as we see in the following excerpt, the metaphor of the "distorted copy" seems to open some space between these contradictory views. Moreover, this affords some materiality to the rather abstract notion of SD while preserving the ambiguity (Jodelet 1986).

Look, synthetic data... (...) to give a graphic example: Imagine you have a printer at home and its ink is running out. When a printer is running out of ink, or the ink has completely run out, you can print, but the document will not be printed well because it will be blurred or striped or so. The printer also has a scanner. Therefore, if you used this printer to scan or print a document that you have as an "original", the synthetic data would be the document that comes out of the printer with stripes or blurred because there is not enough ink. Therefore, it is a copy of the original document but with distortions because the original dataset has been manipulated. (EE04)

This quotation introduces some critical and controversial aspects of the socio-technical process of constructing SD, such as resorting to manipulation and insufficient resources. However, what interests us here is how this attempt to turn the unfamiliar into the familiar captures a themata (same/different), or rather a paradox, at the core of SD: they are deemed to be identical to real data (which, in turn, we assume are faithful copies of external reality) and, simultaneously, SD is considered an alteration of real data, bearing no identifiable resemblance to the real data.

At first glance, turning an abstract concept into something inherently contradictory can seem a bit unhelpful and at odds with the functionality often attributed to social representations: to transform the abstract, technical concepts of science into comprehensible, familiar, and safe ones in the consensual universe (Räty and Snellman 1992). However, as Billig (1988) stressed in his efforts to contribute to and extend the focus on social representations, it is precisely the contradictory themes within social representations which allow their anchoring and circulation within an "arguing" (rather than "thinking") society.

This emerging image of SD is not just inserted in a set of rather static networks of values and beliefs. Instead, it is anchored in our everyday discourses and therefore becomes familiar, manageable, and useful precisely because it can be embedded within a so-called argumentative context that allows for certain topics to be discussed and thought of in order to solve specific practical problems. Interviewees' discourse is context-dependent; therefore, their descriptions of social life and opinions are potentially variable. The social world is a multiple reality in which different ways of explaining the patterns of action and beliefs co-exist (Mol 2002).

SD and DT as a Liminal Hotspot: Anchoring Them into Technical Practices to Escape the Paradox

In this context, experts and those gradually becoming familiar with SD and DT have a rather unified and manageable image of these concepts and, at the same time, a multitude of different repertoires at their disposal. Arguably, this double-sided representation serves as an attempt to reconcile an agonistic field or controversy (Latour 1992), precluding some debates and criticisms they face.

For example, critics can highlight how SD and DT rely on large volumes of high-quality data to produce accurate copies that function well. This dependence on continuous data flow raises significant ethical concerns. It involves decisions about what counts as valuable data, how it is collected, who is included or excluded, and what assumptions are embedded in the technologies used (Rullo 2024), especially when re-identification is still possible (Teller 2021).

A key concern is the implications of bypassing consent protocols, which weakens the obligation to engage affected communities (Cabrera et al. 2025). Several studies highlight how private companies that often play a central role in developing SD potentially gain access to sensitive data that could be used for purposes unrelated to users' well-being (Ferlito et al. 2024). Especially when embedded in a healthcare ecosystem with a growing presence of private actors in public provision, as is the case in Catalonia, these concerns are particularly relevant. Catalonia has a mixed healthcare system with complex partnerships and interactions between the public and private healthcare sectors. Public-private collaboration ranges from managing and increasing tenders and agreements to creating and diversifying a healthcare market involving both public and private (profit and non-profit) companies (Vallès-Peris and Pareto 2025). This model is characterised by minimal accountability and a gradual transfer of decision-making power to the providers themselves (Benach et al. 2017), a scenario that facilitates the emergence of concerns about the risk of data leaks involving patient information, which could be exploited for illicit purposes (Iqbal et al. 2022) and repurposed for non-clinical goals, such as risk scoring, behaviour tracking, or even insurance pricing (Braun 2021).

Against this background, DT and SD can be presented as a different kind of data (i.e., a "distortion") that makes most of these critiques regarding data property, patient involvement, and data leaks rather irrelevant and stresses their potential to reduce bureaucratic burdens and ease restrictions on the use of private data.

Therefore, the paradoxical objectification of SD and DT as a distorted copy facilitates the unfolding of a rhetorical scenario where various interpretations and critiques of SD and DT (mainly, those of SD as copy and SD as distortion) co-exist and dialogue between them. This type of dilemma can also be seen as an example of what psychosocial literature describes as liminal hotspots (Greco and Stenner 2017; Zulato et al. 2023). Even though it is a concept mainly and originally used to describe experiences of being caught in a transition between positions, its proponents argue that it can be useful to examine dynamic "phenomena whose features are not adequately grasped by the norms of intelligibility associated with established forms-of-process" (Greco and Stenner 2017:152). Here, the established forms-of-process or social representations that collide (and merge) are those that view "data as copy" and, on the other hand, "data as distortion".

Greco and Stenner (2017) distinguish four key features of liminal hotsports: paradox, paralysis, polarization, and pattern shift. We have already pointed out the paradoxical nature of SD and DT. In our data, we also found instances of a certain paralysis at the institutional level:

We are very interested in making progress on this issue, synthetic data, but we have not made any further progress. We don't know more, we haven't done anything... right? (EE05)

However, our interviews also revealed strategies to avoid getting stuck in the controversies around privacy or in the foundational paradox of SD and DT (namely, that we can use them because they are distorted and anonymous copies but cannot rely on them because they are altered data). For example, as this researcher puts it, the paradox or trade-off between privacy (derived from "SD as copy") and utility (SD as distortion) becomes manageable if it is anchored in the more pragmatic and technical aspects of research:

The trade-off between, for example, privacy and utility is a challenge. But that's a technical issue. (EE04)

This is consistent with highlighting and praising SD generation techniques that somehow escape the copy/distortion paradox and portray SD as a complement to overcome the limitations of so-called real data, mainly scarcity, by way of so-called data augmentation:

I believe that, right now, the most important and potentially transformative aspect is augmentation. Because in the end, for example, we need to be clear that all rare diseases are, theoretically, among those that should benefit the most from data augmentation. If you have 20 patients worldwide with a particular disease, it is extremely difficult to draw conclusions. And I'm not just thinking in terms of imaging, but also in terms of synthetic data and data augmentation—it makes a lot of sense, especially when it comes to the design of drugs, medications, proteins... (EE03)

In summary, stakeholders tend to polarise between epistemic surveillance and techno-optimism. While critics raise ethical concerns regarding data access and validity that may be slowing some developments, those most directly involved in this kind of research tend to anchor SD within the specific healthcare context leaning towards different forms of techno-optimism (as in the quotation above: "it is a technical issue") that, in turn, can become hegemonic and imply a pattern shift in healthcare research.

As noted, SD generation is often represented in ways that discursively reproduce the bifurcation between technological development and psychosocial dynamics. The dominant depictions of SD and DT tend to anchor these artefacts within the technical side of this coin, therefore precluding socio-political discussions (Moore 2005).

This gives way to a sort of technological fetishism that represents AI as a neutral solution, devoid of biases, therefore hiding power relationships and socioeconomic determinants of health (Sharon 2018).

I mean, we all have our biases. Artificial intelligence doesn't have biases, nor does it intend to harm anyone. The one who has biases and might cause harm is the person who trained it. If I train a model today, here, I will bring my own biases, because I will be thinking about the local population, I am a man, so maybe I will think about what affects me as a man, as a male... (...) And as someone who is in their forties... (...) And maybe I am a worker, and maybe I won't have as much... Or I've just had a child and maybe I am more sensitive to certain things. In the end, how I generate the dataset (the database I will later use to train a model) is key. If someone has trained it in a certain way... obviously, most models are trained in the United States. (...) Because we have to keep in mind that in Europe things are much more regulated. (...) We'd be getting into geopolitical issues now, but it's clear that both China and the United States are the major producers of both text and image models, because they don't have restrictions. (EE01)

Despite the techno-imprinting and techno-optimist views governing SD and DT research (Rullo 2024), critics argue that these may reduce individuals to data, overlooking subjective aspects and embedding normative assumptions in healthcare processes (De Boer et al. 2022). In this context, the critics of the metaphor of "twins" implied in the DT term resonate, as DT do not correspond to exact and neutral representations of what they are meant to depict. This is not only an issue of distortion. DT are shaped by selective data collection, modelling assumptions, and normative intentions (Korenhof et al. 2021). This assumption is misleading in healthcare contexts, where human bodies, identities, and lived experiences cannot be fully captured by data. In this sense, SD and DT are presented as objective representations that aim to eliminate the subjective component inherent in healthcare (Popa et al. 2021).

In fact, from a techno-fetishist perspective, this situated experience (often safeguarded by "ethical" procedures and committees) is acused of somehow blocking innovation in the healthcare sector ("just with the data we have, we could do much more"). Thus, as this interviewee put it, SD and DT are not only portrayed as an effective alternative for addressing privacy challenges at minimal cost, but also as reframing how we relate to data altogether. Here, we envisage a shift from seeing data as external (and difficult to gain access to, "a problem with data access") to something that could be "developed" from within. This pattern shift allows the problematization of "data access" to be undone.

"Data—we usually don't give them out because we have a problem, which we'll talk about now. Okay, so, they are not easily accessible. We often have them (...) but they are not easy to access, especially for third parties, right? (...) So, we are working on guidelines for how to carry out these developments [synthetic data] so that they are not only ethical, but also secure, and, well, aligned with the agreements we are entering into regarding how these developments should be done. That is where we are, okay. You are asking me about synthetic data. So how do we see it—or at least

how do I see it, right? My opinion is that we don't have a data problem. Just with the data we already have, we could do much more than we are currently doing. So, yes, we do have a problem with data access, and since that issue is still quite blocked for now, okay, we have explored the possibility of, instead of giving out data, helping with the development of synthetic data, okay." (EE05)

SD are envisioned not only as a context-specific solution to data access problems but are also deemed to hold the potential to replace real data altogether ("instead of giving out data", developing data). This optimistic view fantasizes about a "research without (annoying and being annoyed by) participants", withdrawing any associated risks, such as questions of informed consent or public engagement in research.

I: And do you see any risks? Assuming they manage to develop this synthetic data and all, which would have those advantages, what drawbacks do you see?

P: Nope. None, because it solves all the problems, all the drawbacks that real data has. I mean, with real data, I could identify you and harm you. Not with this, no way.

You see what I mean? That's the point. (...) I do not see any risks, only benefits—as long as we use synthetic data for the common good. Of course, if we start using it to make personalized drones to kill people, then obviously it's a risk. (EE05)

Towards the end of this excerpt, we see how the dichotomy (neutral)technology/human(biased use) is at play to locate the risks in the latter. Hence, by inserting SD and DT within the modern socio-technical dichotomy, stakeholders have a "double face" at their disposal that allows them to manage and safeguard the acceptance of technological developments based on their deemed neutrality. However, this hegemonic depiction of SD and DT as pure technologies displaces problematic issues such as thecurse of recursion(Shanley et al. 2024),potential biases, or the role of "noise" within the data.

From Hotspots to Blindspots: Curse of Recursion, Dataification, and Context and Complexity as Noise

The 'curse of recursion' in the context of SD and in health research refers to a specific issue that arises when SD are used to train models, and then those models (or new models trained on SD) are used to generate further data, as the technician below describes this issue. Over time, this recursive process can lead to data degradation, bias amplification, and loss of realism.

P: "The United States has few restrictions; we have many more."

I: "Is that a good thing?"

P: "We are back to the double-edged sword, right? Personally, I think certain things need to be protected, but at the same time, someone else is doing it—and I'm using it. I mean, the U.S. is doing it, they're doing it over there without restrictions, and we're using it here. So, the European Union is putting restrictions in place—which I agree with—but in the end, these models come back to us..." (EE01).

Additionally, this insertion of SD within the socio-technical dichotomy reinforces the inevitability rhetoric, manifested in phrasings such as "in the end, these models come back to us". This narrative, which frames technological advancement as unstoppable and necessary, functions as both justification and a mechanism for social adaptation. Within this framework, critique does not disappear but is subordinated to the logic of inevitability: the question is no longer whether we should proceed but how we should manage the consequences. In this way, the narrative of inevitability not only shapes technical decisions but also conditions the possibilities for truly inclusive participation in the design and deployment of these technologies in primary care and, more broadly, in health data government, particularly within vulnerable populations (Sekalala and Chatikobo 2024).

Another prominent criticism of current practices related to the 4P approach embedded in healthcare digitalisation, which also seems applicable to SD, is that they threaten to turn citizens and patients into mere sets of data points, such that the person represented is reduced to the model used to represent them (De Boer et al. 2022). SD does not merely reflect or replicate bodies or social collectives; these developments are already loaded with normative ideas embedded in the processes of data collection, analysis, and interpretation (Korenhof et al. 2023). Additionally, they produce specific body and social images based on data, algorithmic inferences, and modelling decisions (Korenhof et al. 2021) that may influence how individuals and social groups are perceived, treated, and even how they perceive themselves (Guerrero Quiñones and Puzio 2025). While SD or DTs are often used to protect privacy by removing individuals from datasets, some argue that meaningful engagement with both data subjects and researchers is essential to ensure responsible and inclusive data practices (Susser et al. 2024).

Furthermore, when SD are trained on non-diverse datasets, they tend to misrepresent marginalised or minorised groups or fail to include them, reinforcing existing disparities in healthcare (Ferlito et al. 2024; Iqbal et al. 2022; Popa et al. 2021). This underrepresentation may lead to unreliable diagnoses, exclusion from treatment, and distorted assumptions about certain populations. Consequently, many of the biases already present in the healthcare system may be further amplified (Giuffrè and Shung 2023). Regardless of whether individuals understand, control, or consent to their DT or data use to construct synthetic datasets, the representations generated risk becoming illegitimate or harmful (Braun 2021), the underrepresentation of data could have adverse effects on minority or vulnerable groups.

Additionally, the use of opaque AI technologies to generate or analyze health data can lead to misrepresentations of the patient, as the underlying processes may overlook contextual nuances, reinforce biases, or produce outputs that the patient, clinician or healthcare researchers cannot easily interpret or address (Giuffrè and Shung 2023). In this context, from the dichotomic perspective outlined above, "real" data, data from vulnerable minority groups, and situated experiences are at risk of being interpreted as potential "noise" that disrupts "clean" research.

P: "This idea of building models with less data? I think it makes sense. (...) In fact, one of the big gurus of artificial intelligence or machine learning talks about this—that it is more important to have a dataset that is clean and as good as possible than to spend hours building a super-complex model. So yes, it does make sense, and I actually think we are heading in that direction, because the more data you feed into a model, the more noise you can also introduce—variables that do not contribute anything. And that's exactly where we need to go."

I: "But what is noise in a dataset?"

P: "Variables that, even though they're unrelated, can end up affecting or influencing the prediction, for example." (EE01)

Again, Stenner and Greco's (Greco and Stenner 2017; Stenner 2017) process-oriented perspective is helpful in making sense of the broader and shifting epistemological dynamics at play. Following the work of Michel Serres, these authors conceive "noise" as a symptom of transition or, in our present case, how the "old, real" data is becoming some sort of nuisance for the better, emerging technology-driven medicine of the 4P approach.

The Birth of the "Data Gaze": Building Trust Towards a Simulation-Based Medicine

Hence, such is the primacy and autonomy assigned to SD that these are also deemed to hold the potential of enacting a "pattern shift" in the entire biomedical perspective: from evidence-based medicine to what could be called 'hipothesis based or probabilistic medicine'. In other words, the possibility of developing trustworthy SD means, to some extent, bringing the future to the present and treating it with the same status as the evidence from the past, thereby developing treatments and interventions based on predictions.

Another important aspect is that synthetic data—although this is still quite preliminary—could also end up helping to generate entirely new models of disease. What does that mean? Basically, we already have retrospective data on the symptoms and signs of certain diseases, and we know that when there is a combination of, say, five symptoms and three signs, we classify it as Disease A. But we could potentially generate new diseases by altering the proportion of symptoms and signs. Does that mean generating a new pathogen? No. I am talking about theoretically generating a new disease. And if you can generate a new disease, you can also generate a treatment. Take the flu virus, for example. Instead of waiting for last year's flu to develop this year's vaccines—which we already know will not target this year's virus, but last year's—we could be generating thousands of variations of the flu virus and potentially creating specific vaccines for each of those variations. That is one of the most interesting aspects of synthetic data: it allows you to generate hypotheses and, therefore, to anticipate rather than just look back retrospectively. (EE03)

In a context dominated by techno-optimism and people and their singularities seen as noise, the benefits of SD and DT for health research vastly outweigh the (unseen) risks. Paraphrasing the participant speaking above, although this is a pre-liminal stage, we might be at the verge of an epistemological shift within which "generating thousands of variations" (of data and models) are at the core of health research, contributing to stop seeing scarcity, reliability, and bias of data as problems.

As with any emerging mode of knowing willing to become hegemonic, to lay bare and realise the potentials of this new paradigm, an entire disciplinary effort is required, directed towards building trust and skills to understand and evaluate datasets (not patients).

Then there's adoption. Not just our own adoption, I mean more generally. In the end, the challenges related to how research groups adopt these data are closely tied to the broader adoption of artificial intelligence. One of the key issues is the trust in these synthetic datasets. So, it is necessary to promote a culture of trust in these data and be very open and transparent about the methods used to generate them so that researchers understand how they were created and whether they can actually be useful. For example, as mentioned earlier, evaluating datasets across different dimensions—if multiple datasets are generated and evaluated, and the results of those evaluations are made available, then depending on the dataset and the application, they will have different values. In this way, researchers can make informed decisions about which dataset to use and in which context. So, adoption and trust are general challenges in this field, and it's a matter of transparency, education, and promoting a culture of trust. (EE04)

Among the most promising rhetoric-technical strategies to promote such a "culture of trust", is the so-called "explainability". From a process-oriented standpoint, explainability can be seen as a sort of intermediary concept-practice that could work to bridge the transition between the "old" and "new" paradigms. Although it is described as a "set of algorithms", following the logic of the emerging paradigm (where simulation is at the core of knowledge construction practices), explainability is depicted as a "safety mechanism" by metaphorically appealing to an underlying substance (i.e., physiology), reminiscent of the "old" positivistic reliance on hard evidence.

The most important risk is the unreliability of the models—that is, generating models that are not real, that do not actually simulate reality, but are so far removed from it that they lack any meaningful explanation. That is why there are mechanisms, such as the concept of explainability, which is another set of algorithms focused on identifying whether the synthetic result being produced meets the same physiological explainability criteria as when you look at a real image. You see a finding, and that finding has an underlying physiological explanation. So, the fact that we can now observe that same kind of explanation in synthetic images is a safety mechanism. (EE03)

Thus, explainability emerges as a discursive effort to present and account for the so-called "black box" of AI in acceptable neopositivistic terms. However, the idea of explainability implicitly assumes that the data can be "wrong". Despite this, based on our participants' discourses, this does not seem to lead to a "turn back to the patient" to cross-check data representation with singularities, local contexts, or vulnerable social conditions. In contrast, the social representations that emerged during the interviews point towards an emerging "clinical gaze" or what could be called a "data gaze" (c.f. Beer 2019), more concerned with interpreting and explaining data than with looking at and making sense of symptoms or evidence.

Well, the loss of the clinical eye is real. I think that is true. Now, is this a problem? Maybe it is not. That is, when I was an undergraduate student and we listened to the cardiac blows, the cardiac soples, they told you, this is a mesocystolic blow with mitral focus, with a reinforcement of the third tone. And today, there is not a single student who knows this. And that's bad, but they take an ecocard and they tell you: "I do not mind if it's a mesocystic systemic buff or else. With the echocard I see that the mitral valve closes 3 millimetres less than it should close". Do you know what I mean? That is, the clinical eye will be lost with things that may not have great interest or great impact on patients. All of this means, then, that at the level of medical education we must know how to play very well with the great advantages that all this can bring us and keep the things that are still important. The clinical eye will be lost, for sure, but it does not have to be bad always. Sometimes, it will be bad, and sometimes it will not. And that is where we will also have to make decisions as we see this loss of clinical eye and the repercussions it has. That is what I said about the auxiliary. We are now seeing that it is not so important. In the beginning, everyone said that if you do not know if a blow is more assistoric or not. And not now. And this will also require a period of learning, and of making mistakes and polishing them. (EE04)

Hence, as the concluding chapter of this volume highlights (Negura 2026), we see how the ways in which AI is being socially represented in healthcare settings pose epistemic challenges and potential pattern shifts for biomedical disciplines and practices.

Conclusion

We begin this chapter on SD and DT in primary healthcare research by evoking the idea of Ka, as a kind of "double" or vital copy of humans that co-exists with its original in a relationship of tension and interdependence. This metaphor underscores the inherent ambiguity of SD and DT, situated between originality and distortion, and is therefore embedded in a series of related dichotomies and tensions. These tensions reveal that the challenges surrounding SD and DT are not only about accuracy, explicability, or security but also about redefining what counts as valid knowledge in healthcare.

Although SD and DT have a long history in certain circles, in recent years, these concepts have expanded beyond specialised and technological fields to resonate in conversations linked to more mundane and significant aspects for many, such as primary healthcare researchers and managers. Specifically, in the Spanish context, SD and DT are beginning to be perceived as solutions with enormous potential to boost research in multiple areas, from knowledge of rare diseases to public health, and organisational and professional developments. When it comes to circulating and "popularizing" the concepts of SD and DT, those who work with them most directly find it difficult to provide clear frameworks for understanding them. As crystallised in the metaphor of the "distorted copy" that we find in interviews with representatives of these groups, the very notion of SD and DT is inherently contradictory or paradoxical.

As we approach this emerging representation as a process of configuration, it can be argued that we are facing what Greco and Stenner (2017) have called a "liminal hotspot". That is, a scenario in which the development of this type of data is threatened by the ethical aspects of working with copies and, at the same time, is stalled by mistrust of distorted data. This results in a threshold moment that demands new frameworks of meaning.

Nevertheless, if we look closely, we see how some social representations of SD and DT bring along a kind of cultural change or a pattern shift (in Greco and Stenner's terms) in the way science relates to evidence and data. Pointing out "safety mechanisms" and strategies to build trust in this type of data is symptomatic of the turning point in health research epistemologies, moving from evidence-based medicine (concerned with copies or snaphsots of reality) to a so-called "data gaze" that builds knowledge-production practices drawing on simulation (e.g., predictions, probabilities, and hypotheses about the future).

Arguably, the study of social representations of SD and DT does more than simply point towards how a technical innovation such as SD and DT is objectified and anchored into existing common sense within primary healthcare research. Additionally, this study reveals how mobilizing a set of social representations of SD and DT (and AI more broadly) is, in turn, reorganising the primary healthcare research domain. Moreover, rhetorical devices such as the discourse of inevitability or the importance of datafication are slowly reconfiguring our relationships with participants and, ultimately, patients and vulnerable groups.

As with discussions around the status of Ka for Ancient Egyptians' and modern-day Egyptologists, we find ourselves negotiating the status we give to SD and DT as a way of discussing what it means to be human in our technoscientific contexts.

Acknowledgments

This project received a research grant from the Spanish Research Agency, Ministry of Science, Innovation and Universities (Spain), awarded on the call for Projects for Knowledge Generation, with reference PID2024-160387NA-I00.

We would like to thank Víctor Bermejo for reviewing the first draft of this chapter. We are particularly thankful to the editor, an anonymous reviewer, and Professor George Gaskell for their valuable feedback on the subsequent versions of the chapter.

References

Beer, D. 2019. The Data Gaze: Capitalism, Power and Perception. SAGE Publications Ltd, London.

Benach, J., J. M. Pericàs, A. Llop and A. Martínez. 2017. Les polítiques de salut : orígens, contradiccions i reptes. Canvi d'època i polítiques públiques a Catalunya, 114–123.

Billig, M., S. Condor, D. Edwards, M. Gane, D. Middleton and A. Radley. 1988. Ideological dilemmas: A social psychology of everyday thinking. Sage London.

Braun, M. 2021. Represent me: please! Towards an ethics of digital twins in medicine. J. Med. Ethics, 47: 394–400.

Braun, V. and V. Clarke. 2019. Reflecting on reflexive thematic analysis. Qual. Res. Sport. Exerc. Heal.

Cabrera, L.Y., J. Wagner, S. Gerke and D. Susser. 2025. Tempered enthusiasm by interviewed experts for synthetic data and ELSI checklists for AI in medicine. AI Ethics, 5: 3241–3254.

Castro, P. and T. R. Santos. 2020. Dialogues with the absent Other: Using reported speech and the vocabulary of citizenship for contesting ecological laws and institutions. Discourse Soc., 31: 249–267.

de Boer, B., C. Strasser and S. Mulder. 2022. Imagining digital twins in healthcare: Designing for values as designing for technical milieus. Prometheus, 38.

Ferlito, B., M. De Proost and S. Segers. 2024. Navigating the Landscape of Digital Twins in Medicine: A Relational Bioethical Inquiry. Asian Bioeth. Rev., 16: 471–481.

Fujimura, J. H. 2003. Future imaginaries: Genome scientists as sociocultural entrepreneurs. pp. 176–199. *In:* Goodman, D. H., D. Health and S. M. Lindee (Eds.). Genetic Nature/Culture: Anthropology and Science beyond the Two-Culture Divide. University of California Press, Los Angeles.

Gilbert, G. N. and M. J. Mulkay. 1984. Opening Pandora's box : a sociological analysis of scientists' discourse. Cambridge University Press.

Giuffrè, M. and D. L. Shung. 2023. Harnessing the power of synthetic data in healthcare: innovation, application, and privacy. npj Digit. Med. 6: 186.

Gonzales, A., G. Guruswamy and S. R. Smith. 2023. Synthetic data in health care: A narrative review. PLOS Digit. Heal. 2: e0000082.

Greco, M. and P. Stenner. 2017. From paradox to pattern shift: Conceptualising liminal hotspots and their affective dynamics. Theory Psychol. 27: 147–166.

Guerrero Quiñones, J. L. and A. Puzio. 2025. Digital twins for trans people in healthcare: queer, phenomenological and bioethical considerations. J. Med. Ethics 0, jme-2024-110403.

IDIAP 2023. SIDIAP: Què és el SIDIAP? [WWW Document]. URL https://www.sidiap.org/index.php/es/ (accessed 7.25.25).

Iqbal, J. D., M. Krauthammer and N. Biller-Andorno. 2022. The Use and Ethics of Digital Twins in Medicine. J. Law, Med. Ethics, 50: 583–596.

Jacobsen, B. N. 2025. Machine Learning, Synthetic Data, and the Politics of Difference. Theory, Cult. Soc. 42: 41–57.

Jasanoff, S. and S. H. Kim. 2013. Sociotechnical Imaginaries and National Energy Policies. Sci. Cult. (Lond). 22: 189–196.

Jodelet, D. 1986. La representación social: fenómenos, concepto y teoría. Psicol. Soc. Vol. 2: 1986 (Psicología Soc. II), ISBN 84-7509-343-4, 494 págs. 469–494.

Korenhof, P., V. Blok and S. Kloppenburg. 2021. Steering Representations—Towards a Critical Understanding of Digital Twins. Philos. Technol., 34: 1751–1773.

Korenhof, P., E. Giesbers and J. Sanderse. 2023. Contextualizing realism: An analysis of acts of seeing and recording in Digital Twin datafication. Big Data Soc., 10: 20539517231155060.

Latour, B. 1992. Ciencia en acción: cómo seguir a los científicos e ingenieros a través de la sociedad. Labor, Barcelona.

Lekov, T. 2005. Ancient Egyptian notion of ka according to the Pyramid Texts. J. Egyptol. Stud. 2: 11–37.

León-Gómez, B. B., E. Moreno-Gabriel, L. A. Carrasco-Ribelles, C. V. Fors and L. Liutsko. 2023. Retos y desafíos de la inteligencia artificial en la investigación en salud. Gac. Sanit., 37.

Liu, Z., N. Meyendorf and N. Mrad. 2018. The role of data fusion in predictive maintenance using digital twin. In: AIP Conference Proceedings. American Institute of Physics Inc., p. 48.

Marková, I. 2015. On Thematic Concepts and Methodological (Epistemological) Themata. Pap. Soc. Represent, 24: 4.1–4.31.

McNeil, M., M. Arribas-Ayllon, J. Haran, A. Mackenzie and R. Tutton. 2017. Conceptualizing Imaginaries of Science, in: The Handbook of Science and Technology Studies. MIT Press, Cambridge, Mass, pp. 435–463.

Mendes, J. M., A. Barbar and M. Refaie. 2025. Synthetic data generation: a privacy-preserving approach to accelerate rare disease research. Front. Digit. Heal. 7: 1563991.

Mol, A. 2002. The body multiple: ontology in medical practice. Duke University Press, London.

Moore, K. 2005. Politics of Nature: How to Bring the Sciences into Democracy. Contemp. Sociol. A J. Rev., 34: 168–169.

Negura, L. 2026. AI at the crossroads of representation: object, agent, and social imaginary.*In*: Negura, L. (ed.). Beyond Algorithms: Social Representations and the Public Understanding of AI, Science Publishers, CRC Press/Taylor & Francis Group.

Popa, E. O., M. van Hilten, E. Oosterkamp and M.-J. Bogaardt. 2021. The use of digital twins in healthcare: socio-ethical benefits and socio-ethical risks. Life Sci. Soc. Policy, 17: 6.

Rachewiltz, B. De, P. Parisi and V. Castellani. 1976. Twins in Myth. Acta Genet. Med. Gemellol. (Roma), 25: 17–19.

Räty, H. and Snellman, L. 1992. Making the unfamiliar familiar: Some notes on the criticism of the theory of social representation. Ongoing Prod. Soc. Represent.

Rullo, L. 2024. Paths of Digital twins in the public sector. A systematic review of the social sciences literature. Riv. di Digit. Polit., 631–653.

Sekalala, S. and T. Chatikobo. 2024. Colonialism in the new digital health agenda. BMJ Glob. Heal., 9: e014131.

Shanley, D., J. Hogenboom, F. Lysen, L. Wee, A. Lobo Gomes, A. Dekker et al. 2024. Getting real about synthetic data ethics: Are AI ethics principles a good starting point for synthetic data ethics? EMBO Rep., 25: 2152–2155.

Sharon, T. 2018. When digital health meets digital capitalism, how many common goods are at stake? Big Data Soc., 5.

Stenner, P. 2017. Liminality and Experience. Palgrave Macmillan UK, London.

Susser, D., D. S. Schiff, S. Gerke, L. Y. Cabrera, I. G. Cohen, M. Doerr. et al. 2024. Synthetic Health Data: Real Ethical Promise and Peril. Hastings Cent. Rep., 54: 8–13.

Teller, M. 2021. Legal aspects related to digital twin. Philos. Trans. R. Soc. A Math. Phys. Eng. Sci., 379: 20210023.

Vallès-Peris, N. and M. Domènech. 2020. Roboticists' Imaginaries of Robots for Care: The Radical Imaginary as a Tool for an Ethical Discussion. Eng. Stud., 12: 157–176.

Vallès-Peris, N. and J. Pareto. 2025. Artificial intelligence as a mode of ordering. Automated-decision making in primary care. Information, Commun. Soc., 28: 2015–2033.

Wynne, B. 2007. Public Participation in Science and Technology: Performing and Obscuring a Political–Conceptual Category Mistake. East Asian Sci. Technol. Soc. an Int. J., 1: 99–110.

Yoon, J., L. N. Drumright and M. Van Der Schaar. 2020. Anonymization through data synthesis using generative adversarial networks (ADS-GAN). IEEE J. Biomed. Heal. Informatics, 24: 2378–2388.

Younesi, E., and M. Hofmann-Apitius. 2013. From integrative disease modeling to predictive, preventive, personalized and participatory (P4) medicine. EPMA J.

Zulato, E., L. Montali and M.W. Bauer. 2021. Understanding a liminal condition: Comparing emerging representations of the "vegetative state." Eur. J. Soc. Psychol., 51: 936–950.

Zulato, E., L. Montali and P. Castro. 2023. Regulating liminality: Making sense of the vegetative state and defining the limits of end-of-life action. Br. J. Soc. Psychol., 62: 1733–1752.

CHAPTER 9

Ethics and Governance in AI Integration for Law Enforcement

Societal Aspects and Public Perceptions

Yasmine Ezzeddine

Introduction

The rapid evolution of artificially intelligent technologies has ushered in a new era for law enforcement, one marked by both promising innovations and profound ethical, governance, and societal challenges. As AI systems increasingly inform decisions ranging from predictive policing to real-time surveillance, the stakes for ensuring that these tools are employed justly and transparently have never been higher. This chapter builds upon the foundational insights explored in other chapters—from the intricate social representations of AI to its cultural and educational transformations—and offers a critical synthesis of how these emerging technologies are reshaping law enforcement practices. We believe that by examining the dual imperatives of ethical governance and public legitimacy in AI integration, we are shedding the light on a discussion that is not only timely but essential in an era of accelerated technological change (Jobin et al. 2019).

Recent years have witnessed a significant shift in the operational landscape of policing. Traditional methods are giving way to data-driven approaches that leverage machine learning algorithms to predict criminal activity, allocate resources, and even anticipate behavioural patterns. However, as these sophisticated systems become more embedded in the fabric of law enforcement, questions concerning accountability, transparency, and fairness have come to the forefront. Contemporary debates in academic and policy circles underscore the need to critically assess

Sheffield Hallam University, BTE, Sheffield, UK.
Email: zzedine@shu.ac.uk, ORCID: 0000-0002-2810-2231
LinkedIn: Yasmine-Ezzeddine

DOI: 10.1201/9781003502753-14

whether AI tools are enhancing public safety without compromising individual rights or exacerbating existing societal inequalities (IEEE Global Initiative 2019).

Central to this examination is an exploration of the ethical imperatives that should guide AI deployment in policing. The principles of transparency, accountability, and fairness are pivotal. Transparency demands that the methodologies and decision-making processes underlying AI systems are accessible not only to technical experts but also to the broader public, ensuring that the operations of law enforcement agencies remain open to scrutiny. Moreover, the debate surrounding AI policing is not solely about its ethical deployment but also about who controls these technologies. Recent research explores the implications of AI accessibility, arguing that centralized control by law enforcement agencies may reinforce power imbalances, while broader access could pose security risks (Ezzeddine and Bayerl 2024). The findings highlight the need for governance structures that balance innovation, accountability, and ethical constraints.

Accountability, meanwhile, involves establishing clear lines of responsibility for both the design and the outcomes of AI-driven actions. Fairness is equally critical; it requires that these technologies do not reinforce systemic biases or marginalize vulnerable communities—a concern that is particularly pressing given the documented instances of algorithmic discrimination in other domains (Crawford 2021).

Underpinning this discussion is Serge Moscovici's theory of social representations, which provides a robust framework for understanding how societies interpret and integrate new technologies. Moscovici's perspective helps illuminate the dynamic interplay between collective perceptions and technological innovations. As communities interpret AI's role in law enforcement, their beliefs, cultural narratives, and historical experiences collectively shape the legitimacy and acceptance of these systems. This theoretical lens is particularly useful in unpacking the complexities of public sentiment, as it reveals not only how trust is constructed but also how fears and misconceptions can undermine technological adoption. By applying this framework, the chapter aims to unravel the intricate ways in which societal perceptions both influence and are influenced by the integration of AI into policing practices (Garvie et al. 2018).

Ongoing discussions have laid a comprehensive foundation by exploring AI's broader societal, cultural, and political dimensions. For instance, different contributions examined how AI is represented in popular culture (Moreira 2026, in this volume), the media's role in shaping public discourse (Gatt, Borg and Lauri 2026, in this volume), and the transformative impact of technology on education (Séca and Séca 2026, in this volume) and work (Vultur 2026, in this volume). These discussions collectively point to a multifaceted landscape where AI is both celebrated for its potential and critiqued for its risks. Against this backdrop, the current chapter synthesizes these perspectives by focusing on a critical arena—law enforcement—where the stakes are inherently high. The integration of AI in policing is not merely a technical upgrade; it represents a fundamental shift in how society administers justice, manages public safety, and upholds civil liberties (Richardson et al. 2019).

Furthermore, by examining existing governance frameworks, policy challenges, and the nuances of public perceptions, this chapter offers a comprehensive roadmap for the ethical integration of AI in law enforcement. In doing so, it highlights the importance of adaptive regulatory mechanisms that can keep pace with technological advancements while ensuring that ethical standards are rigorously maintained. The need for such frameworks is underscored by recent legislative efforts and empirical studies that call for more stringent oversight of AI systems, particularly those with the potential to significantly affect human lives (Tutt 2019).

Another crucial aspect addressed in this introduction is the role of community engagement in fostering trust and legitimacy. The legitimacy of AI in law enforcement hinges not only on robust technical and regulatory safeguards but also on the active involvement of the public. Engaging with community stakeholders, through public consultations, transparency initiatives, and participatory oversight, can help demystify AI technologies and address concerns regarding surveillance and discrimination. This approach ensures that the deployment of AI is not perceived as a top-down imposition but as a collaboratively managed tool that serves the interests of all citizens (Crawford 2021).

This chapter also draws on varied insights to propose a holistic vision for the future of AI in law enforcement. It challenges policymakers, technologists, and community leaders alike to consider not only how AI can enhance operational efficiency but also how it can be harnessed to promote fairness, accountability, and democratic participation. By advocating for a balanced approach that marries technological innovation with ethical governance, the chapter underscores the broader societal imperative to ensure that AI serves as a force for justice rather than as a catalyst for division and inequality.

In summary, the integration of AI into law enforcement is emblematic of a broader societal transformation—one that is replete with both opportunities and challenges. This chapter sets the stage for a detailed exploration of the ethical, governance, and public perception issues that arise from this integration. It draws on recent scholarship and emerging policy debates to offer a forward-looking perspective that is grounded in robust theoretical frameworks and empirical evidence. That said, this chapter also charts a course for future research and practice where the ultimate goal is to ensure that the deployment of AI in law enforcement is guided by ethical imperatives and democratic values (Jobin et al. 2019; IEEE Global Initiative 2019).

Ethical Principles and AI Governance

The integration of artificial intelligence into law enforcement has intensified ethical debates concerning fundamental democratic principles, particularly in balancing public security with individual freedoms. As AI-driven surveillance expands its reach, concerns over accountability, transparency, and fairness become increasingly urgent. Unlike traditional policing tools, AI operates through complex, often opaque decision-making processes, making ethical oversight particularly challenging. Addressing these concerns requires a governance framework that embeds ethical principles into both the design and deployment of AI technologies, ensuring that their use does not compromise civil liberties or disproportionately impact marginalized communities. This section delineates the core ethical principles underpinning AI integration in policing and examines how contemporary governance frameworks can ensure that these technologies serve society equitably and transparently.

Ethical Foundations in AI Deployment

At the heart of responsible AI deployment are three interrelated ethical principles: transparency, accountability, and fairness. These principles guide not only the development of AI algorithms but also the policies that govern their application in policing.

Transparency, Accountability, and Fairness

Transparency is the cornerstone of ethical AI use. In law enforcement, where decisions can have profound personal and societal impacts, it is imperative that the inner workings of AI systems are understandable and accessible. This openness enables both technical experts and the general public to scrutinize algorithmic processes. By demystifying the "black-box" nature of many AI models, transparency fosters trust and facilitates independent oversight. Recent scholarship emphasizes that accessible documentation, clear decision logs, and public disclosure of algorithmic methodologies are essential for maintaining legitimacy in the eyes of stakeholders (IEEE Global Initiative for Ethical Considerations in Artificial Intelligence and Autonomous Systems 2019).

Indeed, recent research has underscored the critical role of governance frameworks in mitigating AI-related risks in law enforcement. Ezzeddine et al. (2022) highlight that public concerns around AI policing largely stem from a lack of transparency and the absence of clear safeguards. Their findings suggests that institutional accountability mechanisms, such as independent audits and public reporting, are necessary to maintain public trust and ensure the ethical deployment of AI technologies in policing.

This brings us to the second most important ethical pillar which is Accountability which is intrinsically linked to transparency. When AI systems influence policing, establishing clear lines of responsibility is crucial. Decision-making processes must be designed so that accountability is embedded at every stage, from data acquisition and algorithm development to implementation and review. Clear accountability frameworks ensure that when errors or biases emerge, they can be traced back and rectified. For instance, independent audits and oversight committees can provide external validation and hold developers and law enforcement agencies answerable for the outcomes of their AI systems (Jobin et al. 2019).

Fairness, as an ethical pillar, requires that AI systems operate without perpetuating historical biases or reinforcing systemic inequalities. In the context of law enforcement, fairness means ensuring that AI tools do not disproportionately target or disadvantage certain demographic groups. Recent research underscores the potential for algorithmic bias, noting that data used for training often reflect societal prejudices. Addressing these concerns involves deploying bias detection protocols and ensuring that the algorithms are trained on diverse and representative datasets. Fairness in AI is not merely about equal treatment; it is about rectifying imbalances that could lead to discriminatory practices (Eubanks 2018). Together, these principles form a robust framework that supports ethical AI integration. However, translating these ideals into operational practices requires both technical and institutional measures.

Technical and Institutional Measures

Technical measures play a critical role in operationalizing ethical principles in AI systems. One key strategy is the implementation of regular algorithmic audits. These audits involve comprehensive testing and evaluation of AI systems to detect biases, errors, or unintended behaviours. For example, algorithmic impact assessments can be conducted prior to deployment to simulate potential outcomes and identify vulnerabilities. Such proactive measures are essential in a high-stakes field like law enforcement, where the consequences of biased or opaque algorithms can be far-reaching.

As for bias detection protocols, these are capable of further enhancing the reliability of AI systems. In fact, these protocols use statistical methods and machine learning techniques to identify disparities in how different groups are affected by algorithmic decisions. Once detected, these biases can be mitigated through algorithmic adjustments, improved data curation, or revised decision criteria. The development of these technical solutions has been a focal point in recent studies, which advocate for ongoing research and innovation in algorithmic fairness (Tutt 2019).

Institutional measures complement technical strategies by establishing frameworks for ethical oversight. Law enforcement agencies must incorporate ethical guidelines into their operational protocols. One effective approach is the creation of independent oversight bodies tasked with monitoring AI implementations. These bodies, often comprising multidisciplinary experts, can provide unbiased reviews and ensure that ethical standards are upheld. Additionally, transparent reporting mechanisms are vital. By making audit results and performance evaluations publicly available, agencies can reinforce public trust and demonstrate their commitment to ethical practices. Also, robust training programs for law enforcement personnel are essential. Officers and administrators must understand the ethical implications of AI tools, including their potential biases and limitations.

International Guidelines and Standards

As AI technologies proliferate, international organizations have and continue to develop guidelines to steer ethical AI integration. These guidelines not only offer best practices but also create a common language for discussing ethical issues across jurisdictions.

Overview of Key Guidelines

Recent international guidelines on AI ethics have emerged as authoritative references for governments, technologists, and law enforcement agencies. The IEEE Global Initiative for IEEE Global Initiative for Ethical Considerations in Artificial Intelligence and Autonomous Systems (2019) emphasizes that ethical design should be a fundamental consideration throughout the entire lifecycle of AI systems.

Similarly, the European Commission's Ethics Guidelines for Trustworthy AI (2021) provide a comprehensive framework that stresses the importance of human-centric AI. These guidelines call for AI systems that are lawful, ethical, and robust, insisting that AI technologies must be designed to augment human capabilities without compromising individual rights. Both sets of guidelines converge on the importance of designing systems that are not only technically proficient but also socially and ethically responsible. The significance of these guidelines is underscored by a growing body of empirical research. For instance, studies have shown that when law enforcement agencies align their practices with international ethical standards, there is a measurable increase in public trust and acceptance (Jobin et al. 2019). This alignment is critical because ethical lapses in AI deployment can lead to public backlash and legal challenges, ultimately undermining the effectiveness of policing efforts.

Overall, these international guidelines serve as a benchmark for developing national policies. They provide a common set of expectations that can be integrated into local regulatory frameworks, ensuring consistency across different jurisdictions.

Integration into Law Enforcement Practices

Operationalizing international guidelines within law enforcement necessitates a multifaceted approach that adapts these principles to local contexts. One of the primary challenges in this integration is the dynamic nature of AI technologies, which requires regulatory frameworks to be both robust and adaptable. To address this challenge, several recent case studies have explored the implementation of ethical AI within policing agencies, offering valuable insights into best practices.

A notable example is the deployment of facial recognition systems in metropolitan police forces. In jurisdictions where ethical guidelines have been rigorously applied, agencies have instituted independent review panels that assess the technology's impact on privacy and civil liberties before it is put into operational use. These panels not only review technical specifications but also consult with community representatives to ensure that public concerns are adequately addressed. Such initiatives have demonstrated that when ethical guidelines are effectively operationalized, the resulting systems are more likely to be accepted by the public and less prone to legal disputes (Garvie et al. 2018).

Another area where international guidelines have been influential is in the use of predictive policing algorithms. Recent implementations have emphasized the need for continuous monitoring and reassessment of algorithmic performance. By integrating real-time data and feedback loops, law enforcement agencies can quickly identify and rectify any emergent biases. Moreover, collaborative efforts between policymakers, technologists, and academic researchers have led to the development of standardized evaluation metrics that ensure these systems adhere to ethical benchmarks. These metrics include measures of fairness, accuracy, and transparency, and they provide a concrete basis for accountability (Richardson et al. 2019).

The integration process also involves revising existing legal and institutional structures to accommodate new technological realities. For instance, updating data protection laws and creating new legal frameworks specifically tailored to AI applications in policing are critical steps in this evolution. Recent legislative reforms in specific jurisdictions have focused on enhancing data privacy and implementing accountability mechanisms for AI-related decisions. Notably, the European Union's AI Act (2024) introduces a risk-based approach that imposes strict requirements on high-risk systems such as facial recognition and predictive policing used by law enforcement agencies (European Commission 2024). In the United States, some states like California have enacted laws such as the California Consumer Privacy Act (CCPA) and implemented bans on facial recognition in policing (Garvie 2019), although there is no unified federal framework. These regulatory efforts frequently align with international human rights frameworks such as the Council of Europe's Convention 108+ and the UN's Guiding Principles on Business and Human Rights, which underscore accountability, proportionality, and the protection of civil liberties in technological governance.

Nevertheless, the role of cross-sector partnerships cannot be overstated. Collaboration between government agencies, academic institutions, and private technology companies has proven essential in translating ethical guidelines into practice. These partnerships foster an environment of shared expertise and responsibility, enabling the development of AI systems that are both innovative and ethically sound. For example, The Algorithmic Justice League, in collaboration with academic institutions and civil rights groups, has developed auditing tools aimed at identifying algorithmic discrimination in facial recognition systems (Buolamwini 2020). Similarly, the Partnership on AI, a multi-stakeholder initiative involving public agencies and civil society organizations, has produced frameworks for responsible AI auditing that are now being consulted by law enforcement technology providers in both the U.S. and Europe (Partnership on AI 2022).

While the operationalization of international guidelines into law enforcement practices represents a significant advancement, it is not without challenges. One persistent issue is the rapid pace of technological change, which can outstrip the ability of regulatory bodies to adapt. In response, several forward-looking initiatives have emerged to create adaptive regulatory frameworks, such as the UK Centre for Data Ethics and Innovation's (CDEI) "AI Assurance" framework, which outlines guidelines for auditing and certifying AI systems used in high-stakes public contexts, including policing (CDEI 2022). Similarly, the OECD's AI Principles Implementation Toolkit promotes national-level adaptation and continuous regulatory learning across member states (OECD 2021).

Furthermore, effective integration requires that all stakeholders, ranging from policymakers to the general public, are engaged in a dialogue about the ethical implications of AI in law enforcement. Recent public consultations have shown that direct community engagement increases public trust in AI technologies. For instance, the Ada Lovelace Institute's Citizens' Biometrics Council (2021) in the UK engaged citizens in deliberating on the ethical use of facial recognition and biometric data, revealing nuanced concerns about proportionality, bias, and consent. In contrast, the lack of public participation in New York's deployment of predictive policing led to widespread backlash and eventual moratoriums on algorithmic tools (Ferguson 2021).

In summary, the ethical foundations and governance of AI in law enforcement are interdependent domains that require a delicate balance between innovation and accountability. By embedding principles of transparency, accountability, and fairness into both technical and institutional measures, and by drawing on international guidelines to shape national practices, law enforcement agencies can navigate the complexities of AI integration. This holistic approach not only aims to improve operational outcomes but also aligns with growing evidence that carefully regulated AI systems can reduce false positives in facial recognition and optimize resource allocation in community patrol models (Babuta and Oswald 2020).

As we look to the future, it is clear that the continued evolution of AI will demand equally dynamic and responsive governance frameworks. The insights drawn from recent empirical studies and international guidelines provide a valuable roadmap for this journey. It is incumbent upon policymakers, technologists, and community stakeholders to work collaboratively in creating regulatory ecosystems that are not only robust but also flexible enough to adapt to the transformative potential of AI.

Public Perceptions and Social Representations

Public attitudes toward AI-driven policing are shaped by a combination of trust in law enforcement institutions, historical experiences with surveillance, and broader societal narratives around technology. While AI is often framed as a neutral tool for enhancing security, its implementation within policing is deeply influenced by societal biases, raising concerns about fairness and legitimacy. How AI is represented and discussed within public discourse significantly affects perceptions of its risks and benefits. Understanding these perceptions is critical, as public trust—or lack thereof—can determine the success or failure of AI adoption in law enforcement. Drawing from Moscovici's theory of social representations, this section examines how AI in law enforcement is framed within public discourse, the fears and expectations that arise, and the extent to which these representations influence policy and institutional responses.

The Role of Social Representations in AI Perceptions

Moscovici's (1984) theory of social representations provides a useful lens through which to analyse the public's attitudes toward AI in law enforcement. Social representations are the shared beliefs, symbols, and narratives that shape how individuals and groups interpret novel phenomena. In the context of AI policing, these representations are not formed in a vacuum; rather, they emerge from public discourse, media portrayals, historical experiences with policing, and broader societal concerns about technology and surveillance (Crawford 2021).

Public perception of AI policing is shaped by a tension between two dominant narratives: one of technological optimism and another of technological scepticism. The optimistic perspective sees AI as a neutral, objective tool that enhances law enforcement capabilities, reducing crime through predictive analytics and real-time surveillance. This view aligns with the broader discourse of "smart cities" and digital governance, in which AI is depicted as a means of optimizing public services (Jobin et al. 2019). Conversely, the sceptical narrative portrays AI as an opaque, biased, and potentially oppressive force that exacerbates systemic discrimination, enables mass surveillance, and erodes civil liberties (Benjamin 2019). These conflicting narratives illustrate the complexity of social representations surrounding AI policing and underscore the need for deeper public engagement and transparent governance.

Media and Cultural Influences on Public Perceptions

It has become apparent that media portrayals of AI play a fundamental role in shaping public perceptions, often oscillating between utopian and dystopian narratives (see Negura et al. 2026, in this volume). Hollywood films, television series, and news reports frequently depict AI as either an infallible force for good or an uncontrollable threat. Popular culture, for instance, as Moreira shows in this volume, has long associated AI with dystopian futures where automated systems override human control, leading to scenarios of excessive surveillance, false accusations, or pre-emptive policing based on flawed data (Eubanks 2018). These portrayals contribute to a collective anxiety about AI governance and fuel public concerns about accountability and ethical oversight. Empirical studies indicate that media narratives influence public trust in AI-driven policing. Research by Broussard (2020) found that negative media coverage of facial recognition technologies directly correlated with declining public trust in AI surveillance. Similarly, Ezzeddine et al. (2023) examined public perceptions of AI in policing and found that concerns over fairness, transparency, and privacy significantly shape attitudes toward AI adoption in law enforcement. Their study highlights that public apprehension often stems from a lack of clear accountability mechanisms and fears of biased decision-making rather than just viral incidents of wrongful arrests.

At the same time, governments and tech corporations have attempted to shape public perceptions through strategic messaging that emphasizes AI's role in enhancing security and efficiency. Public service announcements, policy briefs, and industry white papers often highlight AI's potential to make policing more "data-driven" and "objective" (Eubanks 2018). However, these efforts are often met with public scrutiny, particularly when AI deployments result in controversial outcomes.

Trust, Legitimacy, and Public Engagement

Trust and legitimacy are central to the acceptance of AI in policing. Unlike conventional technologies, AI-driven tools operate through complex, often opaque algorithms that can be difficult for the public to scrutinize. This mysterious "black box" nature of AI creates a transparency gap, which, in turn, fosters public mistrust as represented by Pasquale in as early as 2015. For AI policing technologies to gain legitimacy, law enforcement agencies must adopt robust transparency measures, including explainability frameworks that make AI decision-making processes more accessible to non-experts.

The extent to which AI policing is accepted by the public depends on how individuals perceive the trade-off between security and privacy. Ezzeddine and Bayerl (2023) found that while some citizens acknowledge AI's potential to enhance safety, others express concerns over mass surveillance and the erosion of civil liberties. Their research shows that trust in law enforcement institutions significantly affects these perceptions, with communities that have historically experienced over-policing exhibiting greater scepticism toward AI surveillance tools. The concern is that AI systems, trained on historical crime data, may perpetuate rather than mitigate racial and socio-economic biases (Richardson et al. 2019). This has led to calls for participatory AI governance models, wherein communities play an active role in shaping AI policies, oversight mechanisms, and implementation strategies.

In fact, recent initiatives have attempted to integrate public perspectives into AI governance. In the UK, initiatives such as privacy walks conducted by the Open Rights Group (ORG) and citizen juries hosted by the Ada Lovelace Institute have been used to solicit public input on surveillance technologies, including facial recognition in public spaces (Ada Lovelace Institute 2021; Veale and Borgesius 2021). These participatory experiments revealed critical concerns around racial profiling, lack of consent, and insufficient oversight.

Algorithmic Bias and Societal Inequalities

Concerns about algorithmic bias remain one of the most significant barriers to public acceptance of AI in law enforcement. Numerous studies have demonstrated that AI policing tools, particularly facial recognition systems and predictive analytics, exhibit racial and gender biases due to flawed training data and structural inequalities (Buolamwini and Gebru 2018). These biases manifest in discriminatory policing practices, disproportionately targeting racial minorities and low-income communities. A landmark study by Obermeyer et al. (2019) found that widely used predictive algorithms in the U.S. criminal justice system systematically disadvantaged Black defendants, reinforcing existing disparities in incarceration rates. Similar concerns have emerged in the UK regarding the deployment of the Harm Assessment Risk Tool (HART) by Durham Constabulary, which has been criticized for reinforcing racial and socioeconomic biases through its reliance on historical crime data (Babuta and Oswald 2021). These concerns mirror broader critiques of algorithmic opacity and systemic reinforcement of pre-existing inequalities.

Such findings have led to increased advocacy for binding legal mechanisms, such as the EU AI Act's mandatory conformity assessments and transparency obligations for high-risk AI systems. In addition, scholars like Veale and Borgesius (2021) have emphasized the importance of data governance and accountability standards that go beyond technical fixes, suggesting legal interventions such as data audits, public registers of AI use, and expanded rights to explanation.

Hence, public awareness of algorithmic bias has grown significantly, catalyzed by activist movements, investigative journalism, and policy debates. A notable example is the Algorithmic Justice League, founded by Joy Buolamwini, which campaigns against racial and gender bias in AI systems and has produced empirical evidence of facial recognition inaccuracies affecting marginalized communities (Buolamwini 2020). The organization's extensive work has amplified global calls for algorithmic transparency, fairness-aware machine learning, and inclusive AI design practices.

Ethical Considerations and the Future of Public Perceptions

As AI technologies continue to evolve, ethical considerations must remain at the forefront of policy discussions. While the principle of proportionality is a foundational concept, the binding nature of procedural safeguards under the European Convention on Human Rights (ECHR) and legal obligations under the AI Act offer more robust guidance for law enforcement than voluntary ethics statements. For example, Articles 5 and 10 of the AI Act require specific documentation, justification, and human oversight for AI applications used in law enforcement contexts (European Commission 2024).

It is imperative to acknowledge that public perceptions of AI in policing are not static; they evolve in response to real-world implementations, policy changes, and societal debates. Conversely, when AI applications lead to high-profile errors or ethical lapses, public concerns intensify. The future of AI policing, therefore, hinges on the ability of policymakers, technologists, and law enforcement agencies to cultivate a governance framework that is both transparent and responsive to public concerns. Ultimately, social representations of AI in policing will continue to be shaped by cultural narratives, empirical experiences, and participatory governance. By recognizing the public as active stakeholders in AI policymaking, rather than passive recipients of technological change, societies can foster a more inclusive and ethical approach to AI integration in law enforcement.

That said, public perceptions of AI policing are complex, fluid, and deeply intertwined with social, historical, and political contexts. As demonstrated in Ezzeddine et al. (2023), social representations of AI policing are shaped by media portrayals, previous experiences with surveillance, and demographic-specific concerns, including vulnerability to profiling and false identification. For instance, in their qualitative study of citizen views on AI surveillance in the UK, participants expressed significant concern that AI tools would amplify systemic inequalities, leading to withdrawal from digital spaces and public life. These insights reflect the interplay of affective and cognitive anchors central to Social Representation Theory, highlighting how AI is embedded in collective meaning-making and not merely evaluated on its technical merits.

The pathway to public acceptance lies not merely in technical improvements but in the adoption of governance models that prioritize transparency, fairness, and democratic engagement. By critically examining how AI policing is perceived and represented, this chapter underscores the necessity of aligning technological advancements with societal values. As scholars such as Ferguson (2021) and Yeung and Ulbricht (2022) argue, the integration of AI into law enforcement is not only a technical challenge but also a question of democratic legitimacy, public trust, and institutional accountability. The growing literature on AI legitimacy emphasizes that successful deployment requires not just legal compliance but also societal endorsement, achieved through transparent governance, independent oversight, and participatory design.

Governance Frameworks and Policy Challenges

Effective governance of AI in law enforcement requires a dynamic and responsive framework that reconciles innovation with fundamental rights. While regulatory discussions have gained momentum in recent years, existing frameworks often struggle to keep pace with rapid technological advancements and emerging ethical dilemmas. The challenge lies not only in drafting comprehensive policies but also in ensuring their enforceability across legal systems with divergent commitments to transparency, accountability, and rights protection. For example, while the EU AI Act introduces mandatory risk categorization, U.S. jurisdictions rely on voluntary guidelines or state-specific statutes. At the international level, instruments such as the Council of Europe's Framework Convention on AI and UNESCO's Recommendation on AI Ethics (2021) attempt to harmonize human rights protections across borders, though with varying degrees of implementation and legal effect.

The Need For AI Governance in Policing

The governance of artificial intelligence in policing has become an urgent policy concern as law enforcement agencies worldwide increasingly adopt AI-driven surveillance, facial recognition, and predictive analytics. While these technologies offer the promise of enhanced crime prevention and operational efficiency, they simultaneously introduce profound ethical, legal, and social dilemmas.

These concerns underscore the pressing need for robust governance frameworks that can safeguard fundamental rights while ensuring that AI applications in law enforcement adhere to principles of fairness, accountability, and proportionality. However, existing regulatory mechanisms remain fragmented and inconsistent across jurisdictions. For example, the European Union has recently enacted the EU AI Act (2024), which establishes risk-based classifications and places stringent requirements on high-risk AI systems, including those used in law enforcement. Meanwhile, in the United States, regulation is piecemeal, with states such as California and Illinois implementing facial recognition restrictions, while federal guidelines remain largely voluntary. This regulatory patchwork presents significant challenges for policymakers attempting to reconcile innovation, public safety, and civil liberties within coherent legal frameworks (European Commission 2024; Mozilla Foundation 2023).

Regulatory Gaps and Legal-ethical Dilemmas

A major challenge in AI governance is the lack of a unified international framework that establishes clear legal boundaries for AI deployment in law enforcement. Unlike other regulated domains, such as biomedical research or environmental protection, AI technologies have been developed and deployed at a rapid pace, often outstripping the capacity of legal institutions to keep up. National policies vary widely, with some countries adopting stringent oversight mechanisms while others operate with minimal regulatory constraints.

The European Union has made significant strides in AI regulation through the AI Act, which seeks to classify AI applications based on their risk levels and impose strict requirements on high-risk uses such as facial recognition and predictive policing (EU AI Act 2024). In contrast, the United States lacks a comprehensive federal AI governance framework. Instead, states such as Massachusetts and Illinois have implemented their own laws restricting biometric surveillance, including facial recognition, while federal guidance remains non-binding (ACLU 2022). This decentralization contrasts with the European Union's coordinated approach under the AI Act, further complicating cross-border regulatory alignment, particularly when AI policing tools are developed by multinational corporations subject to differing national obligations.

Key Governance Challenges: Facial Recognition and Predictive Policing

One of the most contentious areas of AI governance in policing is the use of facial recognition technology (FRT). While law enforcement agencies argue that FRT enhances security by identifying suspects and preventing criminal activity, numerous studies have revealed that these systems are prone to racial and gender biases, disproportionately misidentifying individuals from minority communities (Buolamwini and Gebru 2018). In response to growing public outcry, several jurisdictions have implemented bans or moratoriums on FRT use in policing. Cities such as San Francisco, Boston, and Portland have prohibited law enforcement from deploying facial recognition, citing concerns over privacy infringement and racial discrimination (Garvie 2021). However, governments such as China, Russia, and India have continued to expand their reliance on facial recognition technology (FRT), often framing its deployment within narratives of national security and counterterrorism (Access Now 2023). For instance, in India, FRT has been increasingly integrated into public surveillance infrastructure, while legal safeguards remain limited, prompting concerns about overreach and discrimination (Human Rights Watch 2023).

Beyond facial recognition, predictive policing algorithms present another significant governance challenge. These AI-driven systems analyse historical crime data to forecast where crimes are likely to occur, enabling law enforcement to allocate resources more efficiently. While proponents argue that predictive policing enhances crime prevention strategies, critics highlight its potential to reinforce systemic biases within the criminal justice system (Ferguson 2021). Since these algorithms are trained on existing crime data—data that itself may be shaped by biased policing practices—the risk of perpetuating discriminatory law enforcement patterns remains high (Richardson et al. 2019). Governance mechanisms must therefore address issues of data provenance, algorithmic transparency, and bias mitigation to prevent predictive policing from exacerbating existing inequalities. Some jurisdictions have responded by introducing algorithmic impact assessments (AIAs). In Canada, for example, the Treasury Board Secretariat mandates Algorithmic Impact Assessments for all federal agencies deploying automated decision systems, including in security contexts (Government of Canada 2022). Similarly, the City of Amsterdam has trialled AIAs for policing algorithms to promote transparency and democratic oversight.

However, the effectiveness of such assessments depends on their enforcement mechanisms and the willingness of policymakers to hold law enforcement accountable for AI-related harms.

Transparency, Accountability, and the Civil Liberties Debate

Transparency and accountability mechanisms remain foundational to AI governance, yet their implementation in policing has been inconsistent at best. One of the primary obstacles is the lack of explainability in many AI systems, also commonly referred to as the "black box" problem When AI algorithms are used in surveillance or predictive policing, law enforcement officers and the public often lack insight into how these decisions are reached. While Zuboff (2019) critiques opaque data practices primarily within corporate environments, similar concerns of algorithmic opacity apply to AI in policing, particularly where predictive tools influence operational decisions without transparent justification (Oswald 2022).

This opacity undermines public trust and raises serious concerns about due process, particularly where AI-generated risk scores or surveillance outputs are used in decisions about arrests or bail without meaningful explanation or legal recourse. In the U.S., concerns over procedural fairness have prompted calls for legal safeguards such as the Algorithmic Accountability Act and the Public Oversight of Surveillance Technology (POST) Acts (Crawford and Schultz 2021). Governance frameworks must therefore prioritize explainability and interpretability in AI policing technologies, ensuring that decision-making processes can be audited and challenged when necessary.

Another pressing policy challenge is the tension between AI governance and civil liberties, particularly concerning privacy rights. The widespread deployment of AI-driven surveillance tools raises significant concerns about mass surveillance and the erosion of individual freedoms. In democratic societies, the principle of proportionality requires that any infringement on civil liberties must be justified by a compelling public interest and subject to rigorous oversight (Jobin et al. 2019). However, AI policing technologies often operate in a legal gray area, with little judicial or legislative scrutiny. The expansion of real-time surveillance networks, automated license plate readers, and social media monitoring tools has led to growing fears of an encroaching surveillance state, where AI-driven law enforcement operates with minimal accountability (Benjamin 2019).

It appears that one of the underexplored aspects of AI-driven policing is its influence on digital behaviours. Ezzeddine and Bayerl (2023) argue that the mere awareness of AI surveillance alters online interactions, leading to self-censorship and shifts in digital engagement patterns. Their findings suggest that governance policies must account not only for direct AI interventions but also for their psychological and behavioural consequences on the public.

Pathways to Ethical AI Governance: Oversight and International Cooperation

Despite these challenges, there are emerging governance models that offer promising frameworks for regulating AI in policing. One approach is the implementation of independent oversight bodies tasked with monitoring AI use in law enforcement and ensuring compliance with ethical and legal standards. The establishment of AI ethics committees, composed of legal scholars, technologists, civil rights advocates, and community representatives, has been proposed as a means of fostering multi-stakeholder governance (Ferguson 2020). These committees could provide ongoing evaluations of AI policing tools, recommend policy adjustments, and facilitate public engagement in AI governance. Some jurisdictions have already begun experimenting with such oversight mechanisms. In Canada, the Office of the Privacy Commissioner (OPC) has conducted thorough investigations into law enforcement use of AI, including facial recognition technologies. Its 2021 report on the RCMP's use of Clearview AI found serious breaches of the Privacy Act and called for stronger oversight mechanisms and legislative reform (Office of the Privacy Commissioner of Canada 2021).

International cooperation is another critical component of effective AI governance. As AI policing technologies transcend national borders, there is a growing need for global regulatory coordination to prevent regulatory arbitrage and ensure consistent standards of ethical AI deployment. Organizations such as the OECD, the Council of Europe, and the UN have recently advanced discussions on international AI governance. The 2023 AI Safety Summit in the UK convened global policymakers to address cross-border challenges in AI deployment, including its use in security and law enforcement. These multilateral efforts increasingly emphasize harmonization around human rights, transparency, and democratic oversight (UK Government 2023). However, the success of these initiatives depends on the willingness of national governments to align their regulatory approaches and commit to upholding shared ethical principles.

In summary, the governance of AI in policing presents a complex and evolving challenge, requiring a multi-faceted approach that balances innovation with accountability, security with civil liberties, and efficiency with transparency. The path forward lies in the development of comprehensive, enforceable governance structures that prioritize human rights, public trust, and democratic oversight. By fostering international collaboration, strengthening transparency mechanisms, and institutionalizing public engagement in AI policymaking, societies can ensure that AI serves as a tool for justice rather than an instrument of unchecked power.

Final Reflections

The governance of artificial intelligence in policing is at a critical juncture, with ongoing debates over its ethical, societal, and regulatory dimensions shaping the future of law enforcement. The preceding discussions have underscored the complex interplay between AI capabilities, public perceptions, and the legal mechanisms designed to ensure responsible deployment. As AI technologies continue to evolve, their integration into policing must be guided by frameworks that prioritize human rights, transparency, and accountability. The governance frameworks explored throughout reveal both progress and persistent challenges in regulating AI in law enforcement, highlighting the need for a balanced approach that neither stifles innovation nor compromises fundamental freedoms. As this chapter draws to a close, it is essential to reflect on the key insights presented and consider their implications for the future of AI-driven policing.

One of the most pressing concerns is the potential for AI systems to reinforce existing biases, leading to discriminatory policing practices that disproportionately affect marginalized communities. Despite growing awareness of algorithmic bias, many AI policing tools still operate in ways that lack transparency and fairness (Buolamwini and Gebru 2018; Richardson et al. 2019). While some jurisdictions, including the Netherlands, Canada, and the UK, have piloted bias mitigation strategies such as algorithmic audits, algorithm registers, and public consultations, these initiatives remain inconsistently applied and often lack legal enforceability. For instance, the City of New York's Local Law 144 mandates audits for automated employment decision tools, though similar mandates for law enforcement remain rare (AI Now Institute 2023). The persistence of these ethical challenges highlights the urgent need for governance structures that can effectively identify and rectify AI-induced injustices before they become entrenched in law enforcement practices.

Beyond the technical and ethical concerns, this chapter has examined the broader societal implications of AI policing. Public perceptions of AI-driven surveillance and predictive analytics are shaped by multiple factors, including media portrayals, historical experiences with law enforcement, and the perceived legitimacy of technological interventions. The theory of social representations has been particularly useful in understanding how different communities interpret and respond to AI policing tools. Nevertheless, public attitudes toward AI policing remain deeply contextual and influenced by socio-political factors along with local experiences with urban surveillance shape varying degrees of trust in AI-driven policing (Ezzeddine et al. 2025). Their findings reaffirm the necessity of context-sensitive AI governance frameworks that prioritize transparency and community engagement to ensure ethical AI integration in law enforcement.

While some view AI as a means of enhancing security and reducing crime, others perceive it as a mechanism of control that erodes civil liberties and exacerbates existing inequalities (Ezzeddine and Bayerl 2023). This divergence in public opinion suggests that AI governance cannot be solely a top-down regulatory process; rather, it must involve meaningful community engagement to ensure that AI systems align with the needs and values of those most affected by their deployment.

Not to forget that, the governance frameworks discussed in this chapter reveal a fragmented and evolving regulatory landscape. While international bodies such as the European Union have taken proactive steps toward AI regulation through measures like the AI Act, other regions have yet to establish comprehensive governance structures. The lack of harmonized policies across jurisdictions has resulted in inconsistencies in AI oversight, enabling some law enforcement agencies to operate with minimal accountability. This regulatory gap is particularly concerning given the transnational nature of AI development, where technologies created in one country can be deployed in another without clear legal safeguards. As AI policing tools become increasingly integrated into global security networks, the need for international cooperation in AI governance will become more pronounced (Cath 2018).

These findings have several implications for future research, policy development, and law enforcement practices. First, there is a pressing need for interdisciplinary research that bridges the gap between AI ethics, criminology, and public policy. While scholars in computer science have made strides in developing fairer and more explainable AI systems, their work must be complemented by research in the social sciences to understand the real-world impacts of AI policing. Further studies should examine how AI systems influence law enforcement decision-making, particularly in high-stakes situations such as criminal investigations, surveillance operations, and sentencing recommendations. Additionally, research on the lived experiences of communities subjected to AI surveillance can provide valuable insights into the social consequences of these technologies, informing more equitable governance approaches.

From a policy perspective, governments and regulatory bodies must take a proactive role in establishing enforceable standards for AI policing. This includes mandating transparency in AI decision-making, requiring bias audits before deploying AI systems, and implementing mechanisms for public accountability. Policymakers should also consider the role of independent oversight bodies in monitoring AI use in law enforcement, ensuring that AI-driven policing tools are subject to democratic scrutiny.

For law enforcement agencies, the responsible adoption of AI requires a shift toward more participatory governance models. AI policing should not be implemented as a purely technological solution but as part of a broader strategy that integrates community engagement, human rights protections, and oversight mechanisms. Law enforcement officers must receive adequate training on the limitations and ethical considerations of AI systems to avoid over-reliance on algorithmic decision-making. Additionally, fostering partnerships with civil society organizations, technologists, and academic researchers can help law enforcement agencies develop AI tools that are both effective and ethically sound.

Nevertheless, it is important to reflect on the broader societal imperative to ensure that AI in law enforcement remains aligned with democratic values and human rights principles. AI policing is not merely a technical advancement; it is a societal transformation that will shape the future of justice, security, and civil liberties. If left unchecked, AI-driven law enforcement risks exacerbating existing inequalities and eroding fundamental freedoms. However, if governed responsibly, AI has the potential to enhance fairness, accountability, and efficiency in policing. The challenge lies in ensuring that AI is designed and deployed in ways that uphold the principles of justice and public trust.

By fostering a culture of ethical AI development and enforcing robust regulatory safeguards, societies can work toward an AI-driven law enforcement system that is not only efficient but also just, transparent, and accountable. Ultimately, the success of AI governance in policing will be measured not by the sophistication of its algorithms but by its capacity to protect and uphold the fundamental rights of all individuals. Ensuring that AI remains a force for justice rather than oppression is not merely a policy challenge; it is a moral imperative that will shape the future of democracy itself.

References

Access Now. 2023. Facial Recognition Global Update: Authoritarian Expansion. https://www.accessnow.org

ACLU. 2022. U.S. State Facial Recognition Laws and Trends. https://www.aclu.org

Ada Lovelace Institute. 2021. Citizens' Biometrics Council: Final Report. Ada Lovelace Institute, London, UK. Available at: https://www.adalovelaceinstitute.org/our-work/citizens-biometrics-council/

Ada Lovelace Institute. 2021. Who Cares What the Public Think? – Public Attitudes and the Need for Meaningful Engagement with AI (briefing). Ada Lovelace Institute, London, UK. Available at: https://www.adalovelaceinstitute.org/wp-content/uploads/2022/05/Who-Cares-What-the-Public-Think_April-2022_v4.pdf

AI Now Institute. 2023. Tracking Global Algorithmic Auditing Policies. https://ainowinstitute.org

Babuta, A. and M. Oswald. 2020. Data analytics and algorithms in policing in England and Wales: Towards a new policy framework. Royal United Services Institute (RUSI)/University of Winchester study. RUSI, London, UK. Available at: https://rusi.org/explore-our-research/publications/occasional-papers/data-analytics-and-algorithms-policing-england-and-wales

Babuta, A. and M. Oswald. 2021. Machine learning, predictive algorithms and the policing of future crimes: Governance and oversight. *In:* McDaniel, J. L. M. and K. G. Pease. (Eds.). Predictive Policing and Artificial Intelligence. Taylor and Francis.

Benjamin, R. 2019. Race after technology: Abolitionist tools for the new Jim code. Policing, surveillance, and the digital divide. Policing, 43(1): 65–79.

Broussard, M. 2020. Artificial unintelligence: How computers misunderstand the world. MIT Press.

Buolamwini, J. 2020. The Algorithmic Justice League: Unmasking Bias in Artificial Intelligence. AJL Reports. https://www.ajl.org

Buolamwini, J. and T. Gebru. 2018. Gender shades: Intersectional accuracy disparities in commercial gender classification. Proceedings of the 1st Conference on Fairness, Accountability, and Transparency, 77–91. https://doi.org/10.1145/3287560.3287593

Cath, C. 2018. Governing artificial intelligence: Ethical, legal and technical opportunities and challenges. Philosophical Transactions of the Royal Society A: Mathematical, Physical and Engineering Sciences, 376(2133): 20180080.

Centre for Data Ethics and Innovation (CDEI). 2021. The Roadmap to an Effective AI Assurance Ecosystem. CDEI, London, UK. Available at: https://assets.publishing.service.gov.uk/government/uploads/system/uploads/attachment_data/file/1037195/The_roadmap_to_an_effective_AI_assurance_ecosystem.pdf

Centre for Data Ethics and Innovation (CDEI). 2022. Industry Temperature Check—Barriers and Enablers to AI Assurance. CDEI, London, UK. Available at: https://assets.publishing.service.gov.uk/media/638f3af78fa8f569f7745ab5/Industry_Temperature_Check_-_Barriers_and_Enablers_to_AI_Assurance.pdf

Crawford, K. 2021. Atlas of AI: Power, politics, and the planetary costs of artificial intelligence. Yale University Press.

Crawford, K. and J. Schultz. 2021. Transparency, Accountability, and Due Process in Algorithmic Justice. Harvard Law Review Forum, 134: 56–72.

Eubanks, V. 2018. Automating inequality: How high-tech tools profile, police, and punish the poor. St. Martin's Press.

European Commission/European Parliament and Council. 2021. Proposal for a Regulation of the European Parliament and of the Council laying down harmonised rules on artificial intelligence (Artificial Intelligence Act), COM(2021) 206 final. European Commission, Brussels, Belgium. Available at: https://eur-lex.europa.eu/legal-content/EN/TXT/?uri=CELEX:52021PC0206

European Commission. 2021. Proposal for a regulation laying down harmonized rules on artificial intelligence (Artificial Intelligence Act). https://ec.europa.eu/info/law/better-regulation/have-your-say/initiatives/12355-Artificial-Intelligence-Act_en

European Commission. 2024. AI Act: Regulation of Artificial Intelligence in the EU. https://digital-strategy.ec.europa.eu

European Digital Rights (EDRi). 2021. Facial Recognition and Biometric Mass Surveillance—Document Pool. EDRi, Brussels, Belgium. Available at: https://edri.org/our-work/facial-recognition-document-pool/

European Union. 2024. Artificial Intelligence Act (final text). European Parliament and Council (Regulation on Artificial Intelligence—adopted). Official Journal of the European Union, Brussels, Belgium.

Ezzeddine, Y. and P. S. Bayerl. 2023. Under AI watch: Understanding online behaviours under supposed AI surveillance. British Society of Criminology 2023 Conference Proceedings, 22. https://www.britsoccrim.org/wp-content/uploads/2024/01/BSC-Online-Journal-2023.pdf

Ezzeddine, Y. and P. S. Bayerl. 2024. Should everyone have access to AI? Perspectives on ownership of AI tools for security. In International Conference on AI Research ICAIR24 Conference Proceedings. Expected publication: December 2024.

Ezzeddine, Y., P. S. Bayerl and H. Gibson. 2022. Citizen perspectives on necessary safeguards for the use of AI by law enforcement agencies. arXiv.org. https://doi.org/10.48550/arXiv.2306.01786

Ezzeddine, Y., P. S. Bayerl and H. Gibson. 2023. Safety, privacy, or both: Evaluating citizens' perspectives around artificial intelligence use by police forces. Policing and Society, 33(7): 861–876. https://doi.org/10.1080/10439463.2023.2211813

Ezzeddine, Y., P. S. Bayerl and J. A. Rodriguez. 2025. Unveiling public sentiments towards AI-driven urban surveillance: A case study from Sheffield. In International Workshop on AI and Surveillance in Policing and Law and Order: Opportunities, Threats, Perspectives, and Cases. Routledge.

Ferguson, A. G. 2020. The rise of big data policing: Surveillance, race, and the future of law enforcement. NYU Press.

Ferguson, A. G. 2021. Surveillance and the Tyrant Test. Georgetown Law Journal, 110:205–290. Georgetown University Law Center, Washington, DC, USA. Available at: https://www.law.georgetown.edu/georgetown-law-journal/wp-content/uploads/sites/26/2022/02/Ferguson_Surveillance-and-the-Tyrant-Test.pdf

Garvie, C. 2019. Garbage In, Garbage Out: Face Recognition on Flawed Data. Georgetown Law Center on Privacy and Technology, Washington, DC, USA. Available at: https://www.flawedf.ai/

Garvie, C. 2021. Facial recognition technology: A primer. Georgetown Law Center on Privacy and Technology. Available at: https://www.nacdl.org/getattachment/5e6ea7e9-634a-45e5-866c-7172e201d155/face-rec-primer.pdf

Garvie, C. 2021. Garbage in, garbage out? Face recognition on flawed data. Georgetown Law Center on Privacy and Technology. Accessed through https://www.flawedfacedata.com/

Garvie, C., A. Bedoya and J. Frankle. 2018. The perpetual line-up: Unregulated police face recognition in America. Georgetown Law, Center on Privacy and Technology. https://www.perpetuallineup.org

Gatt, G., J. Borg and M. A. Lauri. 2026. The political impact of AI social representations. *In:* Negura, L. (Ed.). Beyond Algorithms: Social Representations and the Public Understanding of AI, Science Publishers, CRC Press/ Taylor and Francis Group.

Government of Canada. 2022. Algorithmic Impact Assessment Tool. https://www.canada.ca/en/government/system/digital-government/digital-government-innovations/responsible-use-ai.html.

Human Rights Watch. 2023. India's AI Surveillance: Legal Gaps and Civil Liberties. https://www.hrw.org

IEEE Global Initiative for Ethical Considerations in Artificial Intelligence and Autonomous Systems. 2019. Ethically aligned design: A vision for prioritizing human well-being with autonomous and intelligent systems. IEEE. https://standards.ieee.org/industry-connections/ec/autonomous-systems.html

IEEE Global Initiative for Ethical Considerations in Artificial Intelligence and Autonomous Systems. 2019. Ethically Aligned Design: A Vision for Prioritizing Human Well-being with Autonomous and Intelligent Systems (2nd ed.). IEEE, Piscataway, NJ, USA. Available at: https://standards.ieee.org/wp-content/uploads/import/documents/other/ead_v2.pdf

IEEE Global Initiative. 2019. Ethically aligned design: A vision for prioritizing human well-being with autonomous and intelligent systems. IEEE Standards Association. https://standards.ieee.org/industry-connections/ec/autonomous-systems.html

Jobin, A., M. Ienca and E. Vayena. 2019. The global landscape of AI ethics guidelines. Nature Machine Intelligence, 1(9): 389–399. https://doi.org/10.1038/s42256-019-0088-2

Moreira, L. 2026. AI in Utopian and Science Fiction Literature: From Fictional Accounts to Common Sense. *In:* Negura, L. (Eds.). Beyond Algorithms: Social Representations and the Public Understanding of AI. Taylor and Francis.

Moscovici, S. 1984. The phenomenon of social representations. pp. 3–69. *In:* Farr, R. and S. Moscovici (Eds.). Social Representations. Cambridge University Press.

Moscovici, S. 2000. Social representations: Explorations in social psychology. Polity Press.

Mozilla Foundation. 2023. State-Level AI Regulations in the US: An Overview. https://foundation.mozilla.org

Obermeyer, Z., B. Powers, C. Vogeli and S. Mullainathan. 2019. Dissecting racial bias in an algorithm used to manage the health of populations. Science, 366(6464): 447–453. https://doi.org/10.1126/science.aax2342

Office of the Privacy Commissioner of Canada. 2021. Investigation into the RCMP's Use of Facial Recognition Technology. https://www.priv.gc.ca

Organisation for Economic Co-operation and Development (OECD). 2021. State of Implementation of the OECD AI Principles: Insights from National AI Policies (OECD Digital Economy Papers). OECD Publishing, Paris, France. Available at: https://www.oecd.org/going-digital/ai/principles/

Oswald, M. 2022. Transparency and Explainability in AI Policing Tools. AI and Society, 37(4): 847–863.

Partnership on AI. 2022. PAI Annual Report 2022. Partnership on AI, San Francisco, CA, USA. Available at: https://partnershiponai.org/annual-report-2022/

Richardson, R., J. M. Schultz and K. Crawford. 2019. Dirty data, bad predictions: How civil rights violations impact police data, predictive policing systems, and justice. New York University Law Review, 94: 192–233.

Séca, J.-M. and M. Séca. 2026. Social Representation of GenAI (Generative Artificial Intelligence) and ChatGPT in educational policy discourse and among teachers. *In:* Negura, L. (Ed.). Beyond Algorithms: Social Representations and the Public Understanding of AI, Science Publishers, CRC Press/Taylor and Francis Group.

Tutt, A. 2019. An FDA for algorithms. Administrative Law Review, 69(1): 83–123. https://doi.org/10.2139/ssrn.2747994

UK Government. 2023. AI Safety Summit: Summary and Policy Commitments. https://www.gov.uk

United Nations Educational, Scientific and Cultural Organization (UNESCO). 2021. Recommendation on the Ethics of Artificial Intelligence. UNESCO, Paris, France. Available at: https://www.unesco.org/en/aiethics

Veale, M. and F. Z. Borgesius. 2021. Demystifying the Draft EU Artificial Intelligence Act. Computer Law Review International, 22(4): 97–112. De Gruyter.

Veale, M. and F. Zuiderveen Borgesius. 2021. Demystifying the Draft EU Artificial Intelligence Act—Analysing the Good, the Bad, and the Unclear Elements of the Proposed Approach. Computer Law Review International, 22(4): 97–112. Published by Verlag Dr. Otto Schmidt, Cologne, Germany. Available at: https://ssrn.com/abstract=3896852

Vultur, M. 2026. Artificial intelligence and world of work: social representations of AI in the context of job displacement and transformations. *In:* Negura, L. (ed.). Beyond Algorithms: Social Representations and the Public Understanding of AI, Science Publishers, CRC Press/Taylor and Francis Group.

Yeung, K. and I. Ulbricht. 2022. Algorithmic regulation: A maturing concept for investigating regulation of, and through, algorithms. Regulation and Governance, 16(1): 3–22. Wiley/Regulation and Governance, Oxford, UK. https://doi.org/10.1111/rego.12437

Zuboff, S. 2019. The age of surveillance capitalism: The fight for a human future at the new frontier of power. PublicAffairs.

Part II

Represented by AI—Content Generation and Social Effects

Epistemics & Simulation: AI as a Quasi-Agent

CHAPTER 10

Social Representations and Communication in the Era of Artificial Intelligence

Machine Learning as Subject, Other or Object of the Semiotic Triangle or Crossing All?

Annamaria Silvana de Rosa

Introduction

Responding to the recent explosive diffusion of AI applications through Machine Learning (ML)—almost seven decades after the birth of AI in 1956—this chapter offers a meta-theoretical reflection on where ML fits within the "semiotic triangle," considered from the multi-perspective of the 'modelling approach' and the supra-disciplinary paradigm of Social Representations and Communication (de Rosa 2014, 2024).

In today's digital era, Generative-AI (Gen-AI) systems have amplified public exposure to AI. At the same time, debates about Gen-AI and the prospects of Artificial General Intelligence (AGI) extend what, in the previous decade, became the era of data-driven knowledge—*big data* (Boyd and Crawford 2012; Crawford 2013; Crawford and Gray 2014)—and the *data deluge* (Anderson 2008) within the so-called *society of algorithms* (Knorr Cetina 2013; MacKenzie 2019). This continually developing digital society has recently been reframed within an updated analysis of

Founding Director of the EU-approved, European/International Joint Ph.D in Social Representations and Communication (1996–2020), relaunched in 2022 as Joint International Doctorate in Social Representations, Culture and Communication (2022-ongoing), Sapienza University of Rome. (http://http://www.europhd.net)

Postal address: via Giovanni Giolitti, 375 – 00185 Rome, Italy.

Email: annamaria.derosa@fondazione.uniroma1.it

DOI: 10.1201/9781003502753-17

the 'network society' (Castells 2024; de Rosa 2026a), with implications ranging from knowledge production to policy (Castells and Cardoso 2005).

After a brief contextual background on the current controversies around AI, we introduce our dialogues with ChatGPT-4o. Key geopolitical, ethical, educational, economic, environmental, and technological issues surrounding AI (de Rosa 2026b) were evoked as echoes of public debates in a free-association task with the chatbot. The central question of ML's role in the semiotic triangle is then reconsidered in light of textual and graphical outcomes from a second dialogue conducted nearly a year later, which enabled further meta-reflections. Finally, the conclusions present some prospective research agendas for social representation theory in the era of Gen-AI.

Main Controversies by Multiple Agents of the Discourse about AI

Like any powerful human creation, AI can be seen either as a risk or as a challenging opportunity for progress (see, for example, Sok and Heng 2023; Sloane 2024). On one hand, from a geopolitical perspective AI can fuel destructive competition between nations; on the other, it can serve as a tool for shared scientific and technological advancement if guided by ethical collaboration among individuals, companies, and governments.

In the context of the popularization of Gen-AI technologies such as ChatGPT, the general public is often less aware of their far-reaching implications—geopolitical, ethical, environmental, financial, labor-related, and educational. These topics remain subject to heated, polemical social representations among various stakeholders (de Rosa 2026b). Some authors have even called AI a "super-controversy," encompassing multiple debates (Marres et al. 2024). Such polemical representations are found not only among laypeople exposed to media discourse but also among experts, often carrying significant geopolitical implications (see among others: Munk et al. 2024; Spigarelli et al. 2024; Ortiz 2025).

The danger of ignoring these implications, and failing to recognize the limits of what has been termed "synthetic dis-information" (the hallucinations produced by algorithms drawing on incomplete or unreliable sources), is that society may come to view the technology as a neutral sign of progress, seamlessly integrated into daily life. This has happened before with the adoption of mobile phones, social media, and other digital innovations that transformed habits and customs—often beyond users' awareness of their impacts.

As researchers of the expert–lay knowledge interface, situated in specific cultural contexts yet connected globally, we must remain vigilant about the new processes driven by successive digital revolutions—from the age of the printing press to social media and microblogs, and now to Gen-AI with ML. Beyond stereotypes and prejudices for or against AI, we need openness to its processes in order to guide its development ethically and meet societal needs, rather than yield to polarized representations.

Machine Learning's Role in the Semiotic Triangle: Meta-reflections from the Multi-perspective 'Modelling Approach' to the Supra-disciplinary Paradigm of Social Representations and Communication

As scholars inspired by the supra-disciplinary paradigm of social representations—born from the open-mindedness of its founder Serge Moscovici (1961/1976, 1984, 1988)—and by his interrelated theories nourished by "cross-disciplinary nomadism" (de Rosa 2003, 2022), we should be particularly attentive to the role of Machine Learning applications in the knowledge–communication–action dynamic between science and common sense in the era of algorithms. We are not exempt from the need to critically monitor and guide AI's development—even in our own academic and research contexts—moving beyond simplistic "good vs. evil" depictions that have made AI both a crucial social issue and the focus of polemical social representations.

Technological innovations often evoke primal narratives, dreams, myths, and emotions oscillating between hope and fear (Cave and Dihal 2019; Liveley and Thomas 2020; Colamedici and Arcagni 2024). Historically, technological revolutions introduce unfamiliar innovations that people initially resist despite potential benefits. Over time, through communication and social practices, these innovations become familiar, and previously *polemical* representations can shift into *emancipated* and eventually *hegemonic* ones (when group-specific viewpoints fade away). Advanced technologies like ML evoke a paradox: a *humanized machine* that learns and "intelligently" imitates versus a *dehumanized human* whose unique capacity for thinking and decision-making is threatened. Fears about the "artificial" encroaching on the human underline why ML's extraordinary impact remains a crucial social issue, still marked by polemical representations (as illustrated in this volume by Negura et al. 2026). Yet understanding the process by which society makes sense of ML is necessary to overcome an "us vs. them" competition between ML and humans—remembering that humans are ML's creators and should remain able to guide its applications responsibly.

This chapter raises an overarching question for scholars within and beyond the social representations paradigm: *How and where does Gen-AI fit into the semiotic triangle?* Considering four options, we ask:

(a) Is ML simply an Object of social representations? If so, of what kind: polemical, emancipated, or hegemonic? How do communities from different socio-cultural contexts—characterized by slow or rapid technological development—represent it differently?

(b) Is ML an Other (Alter)? Given Gen-AI's capacity to interact with humans by producing words, images, and even videos, one might treat it as a quasi-interlocutor. In this case, do humans perceive this Other as a threatening enemy that destroys human roles, or as a supportive partner that takes over dangerous or tedious tasks (from hazardous industrial operations to complex medical procedures, or even time-consuming archival research)?

(c) Could ML take on the role of a Subject, effectively replacing humans even in creative and cognitive domains (writer, filmmaker, artist, decision-maker), thus producing knowledge and social representations on its own? Alternatively, could it be an "artificial subject" that merely co-produces information, misinformation, and disinformation based on datasets provided by humans and institutions in power?
(d) Finally, should ML be considered a new transversal entity, crossing all three vertices of the semiotic triangle—an advanced, interactive, multichannel medium that simultaneously partakes in subject, other, and object roles?

By exploring these options from the simplest (a) to the most sophisticated (d), the chapter illustrates the consequences each would have for research, including methodological choices and epistemological implications.

Methodological Insights for Dialogue with the ChatGPT-4o, along with Preliminary Comments on the Dynamics of the Response Set

As a deontological premise, we acknowledge that this qualitative study was our first experience engaging with ChatGPT-4o. The dialogue was an interesting communicative experiment between a chatbot and a first-time user—though a user with over four decades of research expertise in social representations and communication, and a strong interest in the academic, policy, and popular literature on AI. Two dialogues with Microsoft's Bing ChatGPT-4o Copilot were conducted: the first on May 1, 2024, and the second on March 13, 2025. A simplified version of the associative network technique, previously published in French, English, Brazilian Portuguese, and Arabic (de Rosa, 1995, 2003/2025) was employed for its projective value.

This technique allows respondents free rein to produce associated words, while the researcher captures not only the content but also the *order of elicitation* (*perceived importance*) of each association and an indication of its *evaluative polarity* (positive, neutral, or negative). In its full form, the associative network can include multiple stimulus words presented in sequence to build a richer semantic field and can integrate *indexes of neutrality and stereotyping* in the evoked representations. In this case, we used a series of stimulus words to probe the semantic network around AI, expecting each representation to be contextualized by others and to reveal complex configurations of meaning.

In our first exploratory study—conducted with an AI-based robot (UNESCO IESALC n.d.) rather than human participants—we posed a series of projective questions to a "Generative Pre-trained Transformer" (OpenAI 2024, May 1st The basic task was: "Build a network of associations to the keyword '.........' in order of importance, adding to each a sign (positive, negative, or neutral) according to you." The stimulus words—presented here only from Q.1 to Q.5 rather than the full Q.1–Q.10—covered key concepts: Q.1 "Artificial", Q.2 "Intelligence", Q.3 "Artificial Intelligence", Q.4 "Human", and Q.5 "Machine Learning". We then added a more direct theoretical question: Q.10a, "What is the place of AI in the semiotic triangle: is it the Subject, the Other, or the Object of social representations?"

Other questions (Q.6 to Q.9) about the "Risks" related respectively to 'artificial intelligence'(Q.6) and 'human intelligence' (Q.7) prompted Q.8 ("*How can we improve decision-making and reduce cognitive biases*?") and Q.9 *("Who has to take decisions: Humans or Machine Learning?")*, as an 'autogenerated question' from the chatbot, marking a change in the dialogue with ChatGPT-4o taking the initiative.

Additional queries—including Q.10 *"What do you know about Social Representations?"* to assess the model's expertise through its cited sources (see de Rosa 2026b)—were also posed, but here we focus on Q.1–Q.5 and Q.10a as the most pertinent to this chapter's aim.

For each stimulus in the first dialogue, ChatGPT-4o followed the instructions but also exhibited a conversational style. It often prefaced its lists with comments (e.g., *"Certainly! Here's a network of associations..."*). After listing the associations with their polarities, it spontaneously added brief explanations for each and concluded with a remark about their context-dependence. A human respondent observing the instructions of the free-association technique would not provide such unsolicited meta-comments, which reflects the chatbot's training to mimic a cooperative dialogue partner.

Two Dialogues with ChatGPT-4o and Some Observations

Some outputs of the two dialogues are included below. They summarize interpretative comments to ChatGPT-4o's responses to the selected key-word stimuli.

Outputs of the First Dialogue with ChatGPT-4o (May 1, 2024)

Free Associations to "Artificial"(Q.1)

ChatGPT-4o generated 14 associations to the keyword '*Artificial*'. Among the five with positive polarity, the first evoked—'*Intelligence*'—appeared despite not being included in the prompt, showing the chatbot's tendency to contextualize "artificial" in terms of AI, described as *"A field of computer science focused on creating intelligent machines"*. The seventh word evoked '*Satellite*' was also positively related to a technical object in the AI domain (*"Man-made objects in orbit, crucial for communication, weather forecasting, and more"*). The other three positive associations—'*Light*'(2nd), '*Limbs*'(9th), and '*Reef*'(10th)—referred to human-made productions with social and environmental impacts:'*enhancing human activity*', restoring '*functionality for amputees*', and promoting '*marine life*'.

Contrary to the stereotypical connotation of '*artificial*' as negative compared to '*natural*' and associated with dehumanized technical artefacts, only two associations had negative polarity, both semantically unrelated to AI: '*Artificial Sweeteners*'(4th), linked to health concerns, and '*Artificial Scarcity*'(11th), connected to economic responses to resource shortages.

Seven associations were classified as neutral. *Ethics* was explicitly mentioned under '*Artificial Intelligence Ethics*'(13th), and '*Artificial Consciousness*'(14th). Ethical or philosophical questions also appeared in the semantic justifications of other associations tied to human-made devices in medical, chemical-industrial, and environmental domains often perceived as not eco-friendly: '*Wombs*'(12th), '*Selection*'(6th), '*Flavours*'(3rd), '*Materials*'(5th), and '*Turf*'(8th). ChatGPT-4o prefaced its list with an explanatory comment and concluded with a caveat: *"the importance and impact can vary based on context and individual perspectives."*

Epistemologically, even this single-word prompt revealed ChatGPT-4o positioning itself in the semiotic triangle as:

- Object of discourse, reflecting ongoing controversies about AI in expert and popular debates;
- Other (Alter), demonstrating a collaborative style in dialogue with the interviewer through task completion and polite commentaries;
- Subject, adding unsolicited remarks and even formulating follow-up questions in later interactions—behaviours akin to agency, albeit algorithmic.

This does not imply human-like self-awareness; rather, these behaviours show ML's limitations in fully respecting the associative network technique's rules, unlike human respondents. At the same time, the unsolicited but coherent comments provided impressive evidence of the LLM's training for interactive fluency. The generative capacity displayed by ChatGPT-4o, even in its free version, explains its massive popularity, with more than 400 million weekly active users as of March 2025 (Cella 2025).

In sum, even in response to a single stimulus word, Gen-AI demonstrated the ability to operate across all three vertices of the semiotic triangle. This lends preliminary support to our hypothesis while also highlighting the tensions in how AI is represented today.

Free Associations to "Intelligence"(Q.2)

The second stimulus word prompted ChatGPT-4o to generate another set of 14 associations, this time centred on diversified forms of '*intelligence*'. As in the first task, the chatbot prefaced its list with an unsolicited remark: *"Associating the concept of 'intelligence' with various domains and assigning a perceived impact can be insightful. Here's a network of associations".*

None of the associations was labeled negative. Eleven were marked positive, beginning with the top three—'*Cognitive*'(1st), '*Emotional*'(2nd), and '*Social*'(3rd)—but also including other diversified forms, listed in order of elicitation: '*Business*'(5th), '*Natural*'(7th), '*Collective*'(8th), '*Spatial*'(9th), '*Interpersonal*'(10th), '*Intrapersonal*'(11th), '*Musical*'(12th), and '*Linguistic*'(14th). Only three associations were coded neutral: '*Artificial Intelligence*'(4th), defined as *"The simulation of human intelligence in machines, which has both positive and negative implications"*; '*Military Intelligence*'(6th), described as *"Information gathering and analysis for defense purposes, which can protect but also lead to conflicts"*; and '*Existential Intelligence*'(13th), explained as *"The proclivity to ponder deep questions about human existence and the larger universe."*

These results suggest that the chatbot evokes broadly positive, human-centred attributes, emphasizing socially desirable connotations of intelligence. This may reflect a tendency to align with users' expectations of a pro-human stance—in contrast to the stimulus *artificial,* which elicited a more ambivalent set of associations (five positive, seven neutral, and two negative) in Q.1.

Comparing Q.1 and Q.2, we observe that *Artificial Intelligence* as a binomial appeared immediately as the first association when the context was triggered by the keyword *artificial* (Q.1). By contrast, it was de-emphasized under the stimulus *Intelligence* (Q.2), ranked only fourth and described ambivalently. This contrast likely arises because the concept of intelligence alone activates idealized human qualities (and thus mostly positive connotations), whereas *Artificial* introduces more ambivalence.

Overall, these findings indicate that ChatGPT-4o, governed by algorithms, foregrounds human intelligence and frames AI as just one form of socially intelligent communication. Rather than presenting itself as an ultimate authority, ChatGPT-4o highlights the superiority of human forms of intelligence, thereby fostering a collaborative climate and empathy in interaction.

Free Associations to "Artificial Intelligence"(Q.3)

ChatGPT-4o produced 14 associations with the combined prompt '*Artificial Intelligence',* integrating both technological and societal dimensions. The top associations,marked positive,were self-referential, highlighting functionalities and human–language interaction: '*Machine Learning'* (1st: *"A subset of AI that enables machines to improve at tasks with experience"*), '*Data Analytics'* (2nd: *"...using AI to uncover insights and trends"*), and '*Natural Language Processing'*(4th : *"Allows machines to understand and respond to human language"*). Positive applications also included '*Healthcare*'(9th: *"...can lead to breakthroughs in diagnosis and treatment"*), '*Education*'(11th: *"AI can personalize learning and make education more accessible"*), and '*Sustainability*'(13th: *"AI can optimize resource use and help in environmental conservation"*). This optimistic view of environmental benefits does not account for experts' concerns about the growing electricity demand linked to expanding and upgrading AI, IoT, and cloud-computing data centres worldwide (Agenzia DiRE 2025, Marzo 10).

Only three associations were coded negative, referring to personal and societal risks tied to AI's expansion. These reflected the model's training to acknowledge controversial issues central to ongoing debates: '*Privacy*'(6th: *"Concerns over how AI can infringe on personal privacy"*), '*Bias*' (7th: *"AI systems can perpetuate and amplify biases present in their training data"*), and '*Employment*'(10th: *"The fear that AI will replace human jobs"*).

Five associations were marked neutral, though several overlapped with negative concerns. For example, '*Automation*'(3rd: *"The use of AI to perform tasks without human intervention, which can lead to efficiency but also job displacement"*) and '*Robotics*'(8th: *"The intersection of AI and physical machines, which can be used for good but also raises concerns about autonomy"*) both relate closely to employment risks. Similarly,'*Security*'(12th:*"AI can enhance security systems but also be used for surveillance"*) links to privacy concerns. These three associations show the model's conciliatory training, balancing the positive and negative sides of controversial applications. By contrast, labeling as neutral '*Ethics*' (5th: *"The study of the moral implications of AI, which can be both beneficial and harmful"*) and

'*Art*'(14th: *"AI-generated art challenges traditional notions of creativity"*) does not recognise seriously the central role assumed by these spheres in debates on AI, given the irreplaceable role of human awareness and consciousness in guiding AI.

In its concluding comment, ChatGPT-4o again noted that the list was *"not exhaustive"* and that *"perceptions vary with context"*. Our concern is that presenting *Ethics* as just another neutral item downplays its normative importance. The non-profit call for ethical AI increasingly frames it as essential and urgent. Scholars and policymakers stress that AI must align with human values and the public good. The notion of *algor-ethics*—coined by Vallauri (2017) and popularized by Benanti (2018, 2022)—calls for ethical governance of algorithms to prevent an emerging *"algo-cracy"* (Benanti and Maffettone 2024), in which unchecked algorithms reinforce injustice. Such concerns have gained prominence globally, for instance with UNESCO (2023) and the European Commission (2025, May 16) issuing guidelines for ethical AI, and with critical voices such as Crawford (2021) insisting that ethical design is indispensable.

While ChatGPT-4o's output balanced optimism (emphasizing AI's contributions to analytics, health, sustainability, etc.) with caution (noting bias, privacy, and other issues), its neutral treatment of Ethics suggests an underestimation of the moral stakes. In fairness, the chatbot reflects the diversity of its sources: in AI discourse, *ethics* can mean both formal guidelines (Jobin et al. 2019; Blackman 2020, 2022) and ongoing controversies, and is therefore neither purely positive nor negative. Yet this example illustrates how Gen-AI mirrors discourses in its training data, which may fail to capture the normative weight assigned to ethics by human experts across disciplines and institutions (Floridi 2023; Etienne et al. 2025). It also highlights the influence of cultural and religious contexts on conceptions of AI ethics, which are anchored in Western traditions but differ, for example, from African perspectives (Segun 2021). This makes the decolonization of AI ethics a key topic in today's debate, with significant implications in domains such as healthcare (Grancia 2024).

Free Associations to "Human"(Q.4)

To explore associations often perceived in contrast to '*Artificial*', we introduced the keyword *Human*. ChatGPT-4o again produced 14 items, ordered by perceived importance. Regarding polarity, six were positive, seven neutral, and only one negative: '*Conflict*' (5th: *"Disagreements or fights between different groups of people, often leading to negative outcomes"*).

The list was topped by two strongly positive associations emphasizing emotional and cultural dimensions:'*Empathy*'(1st: *"The ability to understand and share the feelings of another, fundamental to human connection"*) and '*Culture*' (2nd: *"The arts, customs, and habits that characterize a particular society or nation"*). Other positive associations included: '*Innovation*' (4th: *"The action or process of innovating, a key driver of human progress"*), '*Language*' (6th: *"The method of human communication, either spoken or written, consisting of the use of words in a structured way"*), '*Education*'(9th: *"The process of receiving or giving systematic instruction, especially at a school or university"*), and '*Health*'(10th: *"The state of being free from illness or injury, a crucial aspect of human well-being"*).

The seven neutral associations were largely descriptors of social structures or concepts: '*Society*' (3rd: *"The aggregate of people living together in a more or less ordered community"*), '*Morality*' (7th: *"Principles concerning the distinction between right and wrong or good and bad behavior"*), '*Religion*' (8th: *"A system of faith and worship, which can have both positive and negative impacts on societies"*), '*Economics*' (11th: *"Knowledge concerning the production, consumption, and transfer of wealth"*), '*Technology*' (12th: *"The application of scientific knowledge for practical purposes, especially in industry"*), '*Environment*' (13th: *"The surroundings or conditions in which a person, animal, or plant lives or operates"*), and '*Identity*' (14th: *"The fact of being who or what a person or thing is, often a source of both personal pride and social division"*).

That '*Conflict*' was the only negative word suggests that the chatbot identifies negative aspects of *human* mainly in terms of intergroup disagreements or fights. The other thirteen associations either celebrated human achievements and virtues or neutrally described facets of human life. It is noteworthy that '*Technology*' appeared as neutral, defined blandly as the application of science to industry, and ranked only 12th—far below '*Innovation*', which was cast positively and ranked 4th as *"a key driver of human progress"*. This suggests that when focusing on *human,* the chatbot frames '*technology*' as a neutral tool, while '*innovation*' is valued as a positive driver. Likewise, '*Language*' received a positive evaluation but was narrowly defined as spoken/written communication, excluding emotional, pictorial, and other forms. Culturally loaded terms such as '*Morality*','*Religion*' and '*Identity*' were all coded neutral, with their ambivalent impact on societies acknowledged.

In summary, the associations to *Human* portray humanity primarily through positive social values, while downplaying contentious or technological aspects. The chatbot's tendency towards political correctness is evident in prioritisation '*empathy*' and '*culture*' as core human/societal values, while relegating '*technology*' to neutral and relatively unimportant role.

Free associations to "Machine Learning"(Q.5)

ChatGPT-4o generated 10 associations (six positive, two neutral, and two negative), each coherently aligned with the replies to the previous four prompts, further integrating and validating our observations. The chatbot introduced the task with the following preface: *"Creating a network of associations for 'Machine Learning' with a sign indicating positive, negative, or neutral impact can be quite subjective. However, I can provide an objective perspective based on common views in the field. Here's a conceptual network."*

1. Artificial Intelligence (AI) – Positive

   Machine Learning is a subset of AI and is often associated with the broader advancements in AI.

2. DataAnalysis – Positive

   Machine Learning algorithms are crucial for interpreting complex data sets and extracting meaningful insights.

3. Automation – Neutral

   Automation can lead to efficiency and productivity but may also raise concerns about job displacement.

4. Predictive Modeling – Positive

   Used across various industries for forecasting and making informed decisions.

5. Ethics and Privacy – Negative

   The use of Machine Learning raises important questions about data privacy and ethical considerations.

6. Healthcare – Positive

   Machine Learning has the potential to revolutionize healthcare through personalized medicine and diagnostics.

7. Education – Positive

   Tailored learning experiences and adaptive educational tools are possible with Machine Learning.

8. Financial Services – Neutral

   While it can detect fraud and optimize investments, it also poses risks if not properly regulated.

9. Bias and Fairness – Negative

   Algorithms can perpetuate and amplify existing biases if not carefully designed.

10. Research and Development – Positive

    Accelerates the pace of discovery in various scientific fields.

The chatbot concluded with the caveat: "*This network is not exhaustive, and the associations could change based on new developments or perspectives in the field of Machine Learning.*"

Summary of Q1–Q5

Across our five prompts (*Artificial, Intelligence, Artificial Intelligence, Human, Machine Learning*), ChatGPT-4o provided responses of consistent length (14 associations for Q.1–Q.4 and 10 for Q.5), mixing positive, neutral, and negative polarities. It tended to frame each concept in ways that resonate with common discourses:

- *Artificial* was quickly linked to AI and broader societal issues.
- *Intelligence* was framed in terms of valued human capacities.
- *Artificial Intelligence* was depicted through both benefits and risks.
- *Human* was given the highest value, evoking empathy and culture.
- *Machine Learning* revealed negative aspects in "*data privacy and ethical considerations*" and in the risk that "*algorithms can perpetuate and amplify existing biases…*". Neutral associations acknowledged risks of "*job displacement driven by automation*" while recognizing its "*efficiency and productivity*". Finally, the response concluded with a positive scenario of AI *accelerating discovery in research and development*.

Taken together, the chatbot's distribution of polarities and its voluntary commentaries suggest a deliberate effort to present a balanced, conciliatory viewpoint. This reflects how Gen-AI may reproduce and reinforce equilibrium in discourse—avoiding both strong polemics and excessive enthusiasm—even in the highly controversial public debates about AI.

ChatGPT-4o's self-positioning in the semiotic triangle of social representations (Q.10a)

At the end of the first dialogue, we posed the key theoretical question (Q.10a): *"What is the place of AI in the semiotic triangle: is it the Subject, the Other, or the Object of Social Representations?"*

Below is the original reply from ChatGPT-4o (approximately three short paragraphs):

"The semiotic triangle, also known as the triangle of reference or meaning, is a model that shows the relationship between a symbol (the sign), the object it represents (the referent), and the meaning associated with the symbol (the concept or sense). In the context of AI, particularly in symbolic communication and artificial intelligence, the placement of AI within the semiotic triangle can be interpreted in various ways.

AI can be seen as the '***Subject***' when we consider it as the entity that uses symbols to communicate or represent objects and concepts. For example, AI systems like chatbots or virtual assistants use language and symbols to interact with users, thus taking on the role of the subject in the communication process.

AI might be considered the '***Other***' in the sense that it is an external entity that interacts with human subjects, providing responses and generating meanings based on its programming and algorithms.

Lastly, AI could be viewed as the '***Object of Social Representations***' when we focus on how people perceive and socially construct the role of AI in their lives. This perspective considers AI as an object that is represented and given meaning by society.

In semiotic terms, AI systems interpret and respond to signs, and they influence the generation of meaning through their interactions with users. However, it is important to note that AI's understanding and generation of symbols are based on algorithms and computational models, not on human-like consciousness or social understanding.

Therefore, AI's place in the semiotic triangle is not fixed and can shift depending on the perspective from which we are analyzing its role in symbolic communication. AI's position within the semiotic triangle is a subject of ongoing analysis and debate in the intersection of semiotics, communication, and technology."

This answer essentially recapitulated our research question while independently articulating the possibility that AI may occupy all three roles, depending on perspective. ChatGPT-4o avoided choosing just one option—consistent with our own view—and suggested that AI can be simultaneously an *object,* an *other*, and (in some respects) a *subject.* In doing so, it portrayed AI as a *virtual assistant* able to use language and symbols to communicate and influence meaning-making, while remaining an *artificial subject* based on algorithms, without human-like consciousness or social understanding (Negura 2025).

When responding to Q.10a, ChatGPT-4o also spontaneously provided its *Quick References*. It suggested a link to Wikipedia (as with Q.10 on Social Representations), but in this case it first pointed to well-known academic publishers, such as Springer and the Oxford University Press *Dictionary of Media and Communication.* Finally, it included a relevant abstract of Valdez et al. (2024)'s chapter *"Semiotics and Artificial Intelligence (AI): An Analysis of Symbolic Communication in the Age of Technology"*, listing eighteen related references with links.

Outputs of the Second Dialogue with ChatGPT-4o (March 13, 2025)

Conducted 11 months after the first, this second dialogue aimed to verify whether ChatGPT-4o's answers had evolved, both due to potentially improved training sources and to additional inputs we introduced at specific steps of the exchange. Specifically, we trained the model only after its reply to Q.1 (the re-proposed Q.10 from the first dialogue about its knowledge of Social Representations) and after its initial independent reply to Q.2: *"What is the place of AI in the semiotic triangle: is it the Subject, the Other, or the Object of Social Representations?"* For this question, we requested both textual and graphical replies. Only after these autonomous bi-modal answers we invited the chatbot to examine the figure illustrating the circular dynamics between Subject, Other, and Object in social representations and multi-channel communication, and then to read de Rosa's full article (2024). Our goals were to determine whether the chatbot's knowledge of social representations had improved over time (see results in de Rosa 2026b), and whether its textual and graphical responses to the semiotic triangle question would converge or differ before and after exposure to Figure 2 and the article.

ChatGPT-4o's Reply to Q.2 (Second Dialogue)

"The place of AI in the semiotic triangle—in the context of Social Representations—depends on how AI is framed in discourse and interaction. Let's examine each possible position:

1. AI as the ***Object*** → In most traditional representations, AI is the object of discourse. People discuss AI as a technological entity, conceptualizing it through metaphors (e.g., "brain", "assistant", "tool"). AI is represented through cultural narratives, scientific descriptions, and media portrayals, shaping how it is understood collectively.
2. AI as the ***Other*** → In some contexts, AI takes on the role of the Other, especially when it is perceived as an autonomous agent capable of interaction, decision-making, or even competition with humans. This occurs in discourses about AI ethics, job displacement, or AI as a "partner" in human activities (e.g., creative work, medical diagnoses). The "Otherness" of AI is evident in debates about whether it can possess consciousness, emotions, or moral responsibility.

3. AI as the ***Subject*** → AI as a Subject is the most contested position. In a strict sense, the Subject in Social Representations is typically human, as representations are socially constructed by people. However, with the rise of Generative-AI and autonomous systems, AI itself can generate texts, images, and decisions that influence social discourse. Some might argue that AI, through deep learning and language generation, takes on a quasi-subjective role by shaping how humans perceive reality—yet this remains dependent on human framing and interpretation.

Conclusion. AI is primarily the *Object* of Social Representations, constructed by human discourse. However, in interactions where AI is seen as a decision-maker or agent, it takes on the role of the *Other*. Whether AI can ever become a true *Subject* remains an open philosophical and semiotic question, hinging on future developments in artificial intelligence and human perceptions of it."

Comparison with the First Dialogue

Both answers to the key question (Q.10a in the first dialogue, Q.2 in the second) demonstrate a dynamic perspective on Gen-AI's role in social representations and communication within the digital socio-cultural infosphere. In the second dialogue, ChatGPT-4o's reply, despite lacking scholarly references—which it had freely provided in 2024—arrived independently at similar conclusions before viewing Figure 2 and reading the article that outlined our epistemological approach. The convergence of the two dialogues on the idea that AI's role spans all three vertices of the semiotic triangle (while acknowledging its limitations as an "artificial subject") is therefore not merely the product of our framing hypothesis (option d). It reflects ChatGPT-4o's autonomous deduction based on its general knowledge.

1st Graphical Task—Before Additional Training of the ML (Q.3)

Next we challenged ChatGTP-4o to reply using a visual channel, asking: *"Being able not only to speak and write, but also to communicate by drawing, may you be so kind to draw the Semiotic Triangle that generates the co-construction of social representations through media and communication, and position yourself by adding an image (or more images) representing yourself within the triangle?"*

ChatGPT-4o immediately responded:

R. to Q.3:"Here's a visual representation of the Semiotic Triangle in the context of Social Representations and AI'

I'll create a diagram where:

- The Concept (Signified) represents the abstract meaning of AI.
- The Symbol (Signifier) includes words, media, and cultural narratives that shape AI's representation.
- The Referent (Object) is AI itself as a technological entity.
- Media and Communication play a key role in shaping the connections.

- I will place an image representing myself (ChatGPT) within the triangle based on our discussion.

Let me generate this drawing now."

Graphical Outputs: 1st drawing as response to Q.3

In the live session via Bing, Figure 1 was generated almost instantly as a reply to Q.3. This first drawing attempted to visualize the semiotic triangle in the context of social representations, labelling the three vertices *Concept, Symbol,* and *Referent,* consistent with the textual reply, though with some hallucinated words (e.g.,*"conncceot"* and *"social represignation"* instead of *concept* and *social representation*). Media and communication icons were depicted as connecting the corners.

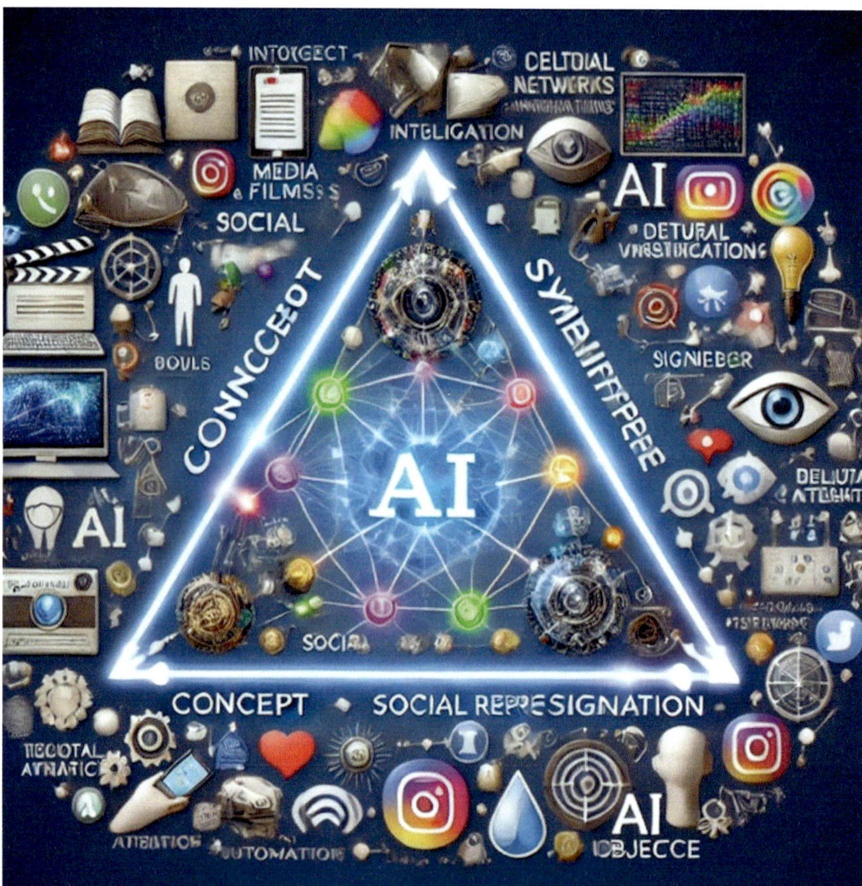

Figure 1: The first drawing produced by the ChatGPT-4o as graphic reply to Q.3 about the request to position itself by adding an image (or more images) representing itself within the semiotic triangle (Subject–Other–Object) in the context of social representations and communication.

Before producing the drawing (which it described as a "diagram"), ChatGPT-4o had announced it would place itself "within the triangle, based on our discussion." However, in the image, AI appeared both at the centre of networked links connecting the three vertices and near elements along the triangle's edges, close to symbols of traditional and new media multi-channel communication (e.g., books, film, telephones, video cameras, computers, the Instagram logo, smartphones, automated systems). Despite the textual imperfections common in AI-generated images, the underlying message was clear: AI was represented as present throughout the triadic structure, rather than confined to a single vertex.

In synthesis, this first attempt to visually address our theoretical question—before any training from our side—conveyed AI's pervasiveness in meaning-making through interactions among people, media, and cultural artefacts. It positioned Gen-AI simultaneously as '*artificial subject*' (influencing social discourse), as '*other*', and as '*object*', without restricting it to one role alone.

Symbolically, this suggests that AI is integrated throughout the representation process, rather than being a distinct external agent or merely an object.

2nd Graphical Task—After Seeing The Figure Illustrating a Theoretical Model (Q.4)

Then we showed ChatGPT-4o the Figure 2 below illustrating the circular dynamic between social representations and multi-channel communication across three vertices-axes of the semiotic triangle (*subject, object, others*) applied to the tourism sector.

The chatbot responded assuming human-like style, although it showed some time management asymmetry in the use of verbs in the present form by asking to upload the figure after it had just shown.

R. to Q.4: "That sounds fascinating! Please upload the figure from your article, and I'll analyse it before creating a version that integrates an image representing myself (ChatGPT) within the semiotic triangle. Looking forward to collaborating on this!

Thank you for sharing the illustration! I will now modify the image by incorporating a representation of myself (ChatGPT) within the semiotic triangle, considering my role in the communication and social representation processes in the tourism domain'

I'll generate an updated version and share it with you shortly.

Here is the adapted version of the Semiotic Triangle in the Tourism Sector, incorporating an abstract representation of ChatGPT within the dynamic system of social representations and multi-channel communication.

This visualization highlights the interaction between Subjects (tourists, prosumers), Objects (destinations, cultural sites), and Others (policy makers, media, and stakeholders) while acknowledging'AI's role in shaping representations through discourse and digital communication.

Let me know if you'd like any refinements or additional details!"

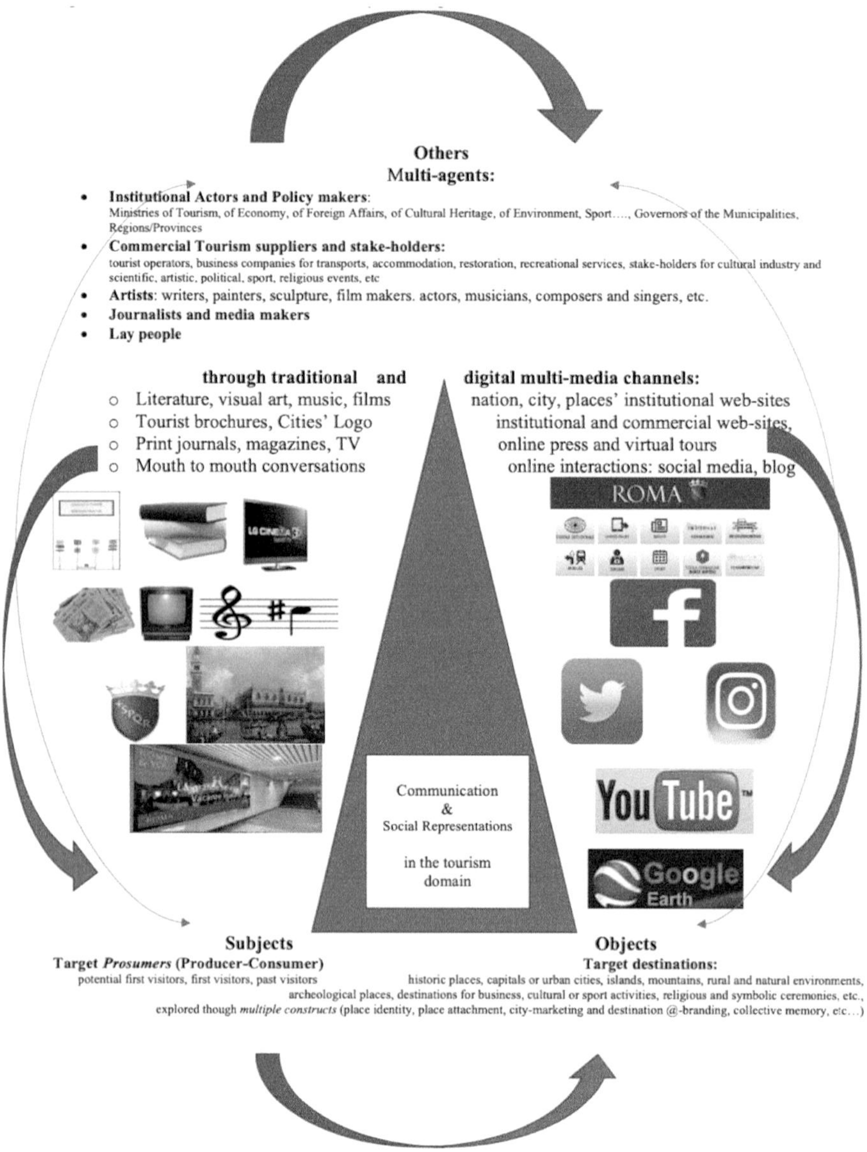

Figure 2: The circular dynamic between social representations and multi-channel communication across the three axes of the semiotic triangle (*subject, object, others*) in the tourism sector (de Rosa, 2024 | https://doi.org/10.1007/s44282-024-00126-9).

Graphical Outputs: 2nd Drawing as Response to Q.4

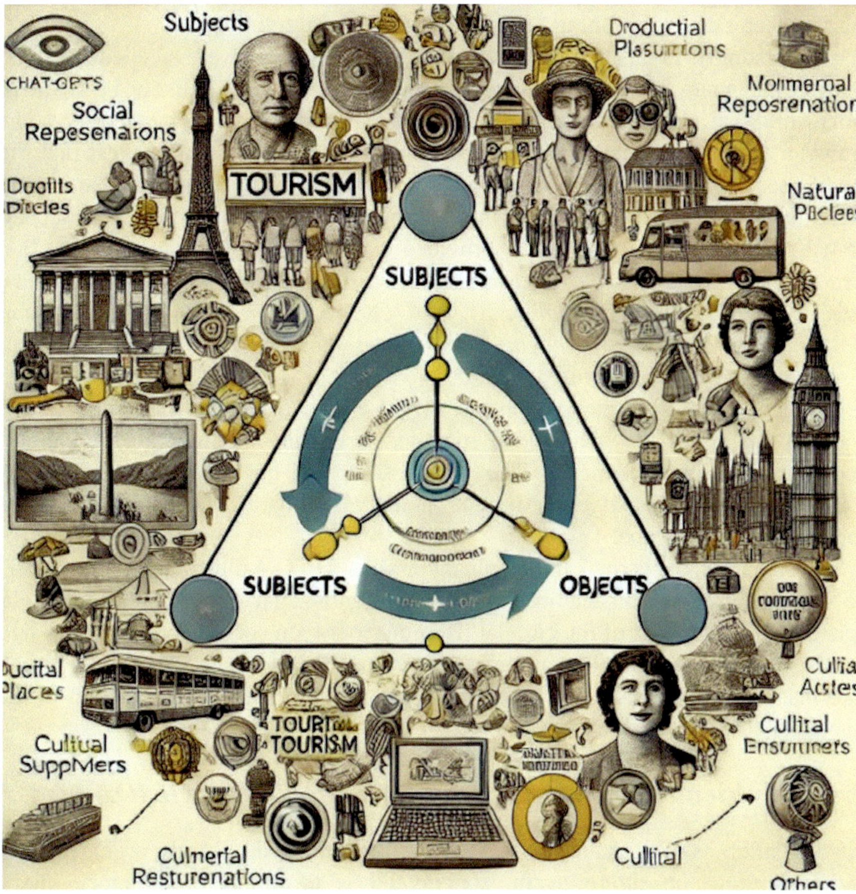

Figure 3: The second drawing produced by the ChatGPT-4o as graphic reply to Q.4 of the 2nd dialogue, after sharing the above Figure 2 from de Rosa's article 2024 illustrating "The circular dynamic between social representations and multi-channel communication across the three axes of the semiotic triangle (subject, object, others) in the tourism sector".

This second drawing incorporated recognizable cultural landmarks—such as ancient temples and icons like the Eiffel Tower and Buckingham Palace, symbols of city identity. It also combined ancient and modern motifs, including computers and microphones, and used antique colours and forms for clothing and transportation.

In the image, ChatGPT appeared as an "eye" overlooking the scene from the top left corner. This indicates that the chatbot had assimilated the exemplary focus on the tourism sector after examining Figure 2 during its brief training.

Interestingly, in this drawing the vertices of the triangle were labelled *Subjects, Objects,* and again *Subjects* instead of *Others.* This duplication is likely a generative quirk, or what the literature terms a "hallucination." Yet, if we interpret the drawing without the chatbot's own textual commentary, it might suggest that the model recognized humans in both *Subjects* and *Others,*and perhaps unintentionally identified itself with both. When interpreting our Figure 2 through this second drawing, ChatGPT-4o nonetheless provided a nearly accurate commentary:*"This visualization highlights the interaction between Subjects (tourists, prosumers), Objects (destinations, cultural sites), and Others (policy makers, media, and stakeholders) while acknowledging AI's role in shaping representations through discourse and digital communication."*

Regardless of such inconsistencies, the AI clearly adopted a higher, observing position: an eye placed above the triangle, surveying the interplay of cultural and communicative elements. The juxtaposition of historical landmarks with technological symbols suggests that the model was situating itself within a global socio-cultural landscape.

In summary, the textual and graphical responses to Q.4 illustrate how ChatGPT-4o's representations adjusted once exposed to Figure 2. That figure, inspired by our epistemological model of the circular dynamics between social representations and multi-agent, multi-channel communication, influenced the chatbot to align with the idea that AI can be transversally integrated into the communicative dynamics of the Subject–Other–Object semiotic triangle.

3rd Graphical Task—After Reading The Full Article (Q.5)

Finally, after asking ChatGPT-4o to read the entire de Rosa's article (2024) and observing the figure again, we invited it to produce one more image representing *itself* within the semiotic triangle.

Once again, ChatGTP-4o replied collaboratively:

R. to Q.5: "I'd love to read your article! Please upload it, and I'll analyse its content before refining the visual representation of the Semiotic Triangle with my presence in it. Looking forward to deepening our exchange!

I have reviewed your article, particularly the section discussing the circular dynamic between social representations and multi-channel communication through the semiotic triangle in the tourism sector.

Now, I will generate an updated version of the semiotic triangle, incorporating an image (or multiple images) representing myself within this conceptual framework."

Graphical Outputs: 3rd Drawing as Response to Q.5

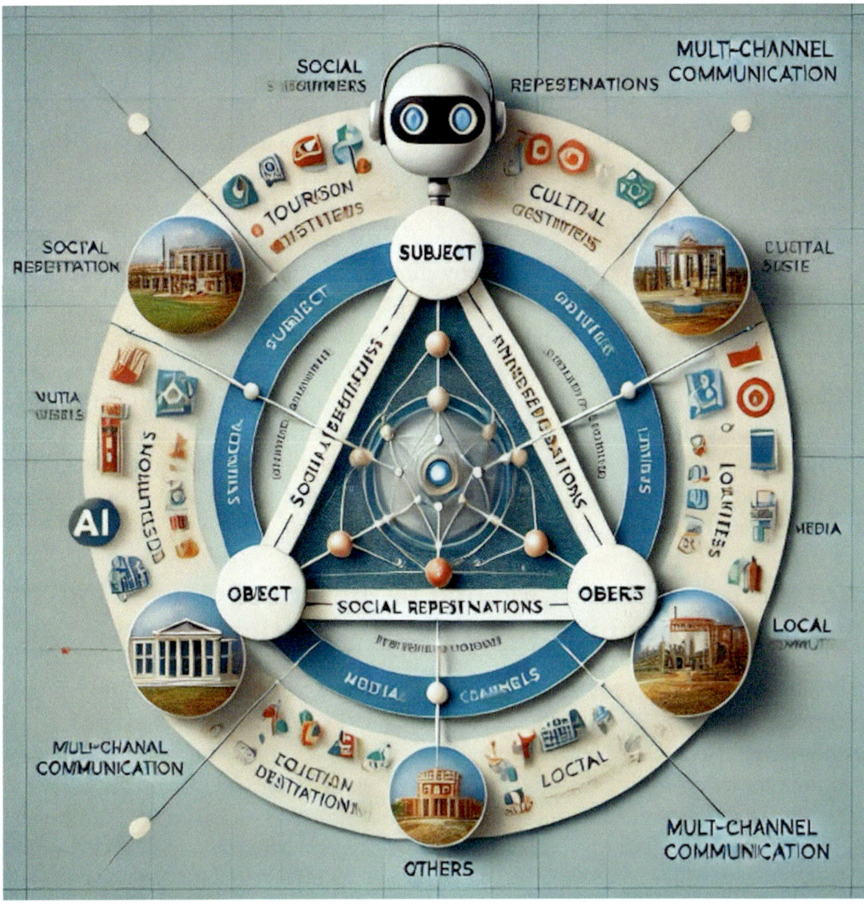

Figure 4: The third drawing produced by the ChatGPT-4o as reply to Q.5 of the 2nd dialogue), after reading de Rosa's article 2024 "Social representations and Multi-channel Communication on crucial societal issues through the modelling approach.

The third drawing showed a further evolution in representing AI's place within the semiotic triangle. In this final image, ChatGPT-4o appeared as a humanoid figure at the top vertex—the *Subject*'s position—overseeing both the *Object* and the *Other*, as well as the surrounding multi-channel communication landscape. Unlike earlier depictions, ChatGPT-4o was no longer peripheral, diffused within the triangle, or merely an "eye" above it. Instead, it was central and dominant, portrayed at the apex as a kind of orchestrator or authority. The *Other* and *Object* were placed below, and the image suggested that AI had become the organizing principle in this multi-agent scenario. Although historical monuments and cultural landmarks were still included, the overall style was more modern, seemingly representing the future dominance of AI across the social and cultural environment, visualized as a multi-channel system.

This progression across the three drawings is striking. Initially, without theoretical input, AI's self-representation was dispersed along the edges and within the centre of the triangle (R. Q3: Figure 1). After seeing our epistemologically driven figure, it positioned itself as an observing "eye" overlooking the structure (R. Q4: Figure 3). Finally, after reading the full article—assimilated with impressive speed—it placed itself firmly in the *Subjec*t's position at the top (R. Q5: Figure 4). One might say that the AI "grew whimsically more confident" about its role with each additional piece of information: from everywhere and nowhere (first drawing), to an eye in the sky (second drawing), to effectively taking charge (third drawing).

Of course, caution is needed in interpreting these images. The chatbot is not consciously determining its status, but responding to prompts, also on the basis of training data and inputs we provided. Nevertheless, this visual trajectory does echo a central concern of our chapter: that Gen-AI's evolving collaborative capabilities may encourage its transversal integration into knowledge production, as well as into multi-channel and multi-agent communication.

Prospective Research Agenda in the Era of Gen-AI for Scholars Inspired by Social Representation Theory

Publishing the original transcriptions of textual and visual outputs from our two dialogues with ChatGPT-4o, although restricted to the key questions presented in this chapter, offers scholars of social representations and communication an opportunity to autonomously reply: *"Can ML be Subject, Other, and Object of the semiotic triangle—or even transcend these categories by crossing them all?"* Adding the drawing task in the second dialogue, conducted eleven months later, and training ChatGPT-4o through de Rosa's article (de Rosa 2024), enabled a before-and-after comparison of its responses.

The results suggest that, besides being a controversial *Objec*t of representation both among and between experts and lay public (de Rosa 2026b), Gen-AI can also act as an interactive *Other* and even as a *humanoid artificial Subject.* This aligns with Floridi's (2025) notion of "new forms of *Artificial Agency"* and the "multiple realisability of agency thesis." Gen-AI thus appears capable of pervasively crossing all three roles.

For research, the most straightforward approach remains to study ML as an *Object* of social representations—focusing on how different social groups represent AI in specific socio-cultural contexts (Becera and Alurralde 2021). Scholars can continue to draw on established epistemological traditions in social representations, whether *structural* (Abric 1993; Wachelke 2012), *socio-dynamic* (Doise et al. 1992), *anthropological* (Jodelet 1989), *socio-genetic* (Kalampalikis and Apostolidis 2021), *ethnographic* (Duveen and Lloyd 1993), *socio-cultural* (Jovchelovitch 2007), *dialogical* (Marková 2003), *narrative* (Laszlo 2008), *socio-semiotic mediation* (Mannarini et al. 2020; Sammut et al. 2015), or, most comprehensively, the *modelling approach* (for more recent overview of this approach, see de Rosa 2014, 2024). These frameworks remain powerful for examining how people collectively make sense of AI.

However, treating ML as more than just an object—namely as a *transversal entity spanning all three vertices of the semiotic triangle*—requires expanding our research toolbox. We must consider ML's active role in knowledge production as an *advanced, interactive, digital, multi-channel medium (option d).* Fourth-generation AI tools already auto-generate questions, set narrative frames, produce multimedia content, and engage in visual and verbal dialogue. They can create 3D imagery, compose music, summarize complex texts, and personalize content in ways that fundamentally reshape the communication environment. This compels us to rethink theories of *socio-technical assemblages*: hybrid systems where AI mediates between humans in creating and propagating knowledge. As Floridi and Nobre (2024) note, we must remain attentive to the implications of "*anthropomorphising machines and computerising minds*," and to "*how the infosphere is reshaping human reality*" (Floridi 2014).

The rise of generative systems also poses urgent ethical challenges. Floridi and Chiriatti (2020) emphasized the need for "new mechanisms for the allocation of responsibility for the production of semantic artefacts" (p. 692). This concern is exponentially greater with the rapid emergence of Gen-AI, which could flood the infosphere with misinformation or biased narratives. Rather than reducing the gap between science and common sense, enlightening lay people to use ML as massive instrument for the socialization of better-informed knowledge, AI risks distorting social representations if left unchecked. From the perspective of common sense, the volume *AI and Common Sense* (Bauer and Schiele 2024) offers useful insights critically discussing related *Ambitions and Frictions.* The editors underline that common sense over time and cultures has been synonyms from one side of "*good sense, wit, wisdom, sound judgment, level headedness and practicality*" and from the other side of antonyms as "*stupidity, impracticability, ineptness, foolishness, insanity, irrationality or unreasonableness*" (Bauer and Schiele 2024, p. 6). They highlight the challenge of defining CS operationally, warning against circular logic—*"if AI does CS, CS is what AI does. Is that really so?"* (p. 6)—and raise the key question: *"At which level does 'AI with common sense' operate?"*(p. 9).

Addressing these challenges requires collective, worldwide efforts. Cross-disciplinary and inter-sectoral collaboration is needed to implement common ethical frameworks across all domains of science and knowledge dissemination: research, libraries, publishing houses, technology companies, policymakers, governments, and civil society. A positive step occurred in 2024, when Pope Francis convened a meeting on AI ethics (Papa Francesco 2024). The resulting "Call for AI Ethics," signed by the Pontifical Academy for Life and endorsed by major technology companies and governments (Microsoft, IBM, FAO, and the Italian Government), popularized the notion of *algor-ethics* (Benanti 2022; Benanti and Maffettone 2024). Pope Francis's successor, Leo XIV, at his first address to the College of Cardinals has warmed the world of the AI threat as one of the Church's top priorities "to respond to a new industrial revolution and to the development of artificial intelligence" (Pope Leo

2025). Such initiatives underscore the urgency of ensuring that AI systems serve the public good.

The call for academic, cross-disciplinary, and inter-sectoral collaboration remains central: scholars must contribute to the ethical development of AI by sharing knowledge to improve awareness of both its risks and benefits across all domains of application (see Floridi 2023 or more recently Etienne et al. 2025).

For scholars of social representations, one promising avenue is to revisit seminal studies with empirical inquiries adapted to the era of Generative AI. For instance, research could examine the fate of Moscovici's Opera Prima *La psychanalyse, son image et son public*. Following the meta-theoretical analysis of the two editions (1961/1976) seminal work on the social representation of psychoanalysis, along a time span of 15 years (de Rosa 2011) and moving forward the empirical cross-national study conducted in France and in Italy 50 years after its birthday in the era of Facebook showing the impact of the emergence of Social Networks (de Rosa 2012, 2013), we propose as next step to investigate how ML-driven content and image's dissemination shape social representations of the psychoanalysis compared to both traditional mass media and social media.

Of course, scholars could examine how a ChatGPT presents information about other scientific topics investigated by researchers as objects of social representations in classical studies, examining for example the role of social practices, imagery, collective memory, ideology on the S.R. of 'mental illness' and 'madness' (Jodelet 1989; de Rosa 1987, 1991; de Rosa and Bocci 2013) or very controversial like '(im) migration' (de Rosa 2027) or in recent studies on new topics, like 'Covid-19' (de Rosa and Mannarini 2020; de Rosa et al. 2021).

A key question for future research is: *Does ML narrow or widen the gap between expert knowledge and common sense?* This could be compared to traditional print and broadcast media, whose geopolitical or ideological positioning is usually easier to identify, and to social media platforms (Facebook, Instagram, X/Twitter, YouTube, among others). Over the past three decades, social media have reshaped cultural evolution in the digital era (Acerbi 2020), with implications for digital diplomacy and global geopolitics (Deruda 2024). The complex dynamic between misinformation, misperception, social media, conspiracy theories, big data, and audience research has even been interpreted as "alarmist narratives as '*misinformation on misinformation*'" (Altay et al. 2023, p. 1). Social media have thus contributed both to the diffusion of knowledge based on ambiguous geopolitical positioning and to populist narratives on the right and the left, grounded in the illusion that "one equals one" (Urbinati 2019; de Rosa 2025). Ensuring that AI outputs are based on reliable sources—perhaps by fine-tuning models with scholarly material—will require partnerships between AI developers and academic institutions to maintain quality standards. The central question remains whether ML can foster understanding, or whether it risks creating a convincing illusion of understanding while actually propagating errors.

Another promising research direction is longitudinal replication. Studies could test empirically whether ML improves its knowledge over time, thereby reducing the distance between expert knowledge and common sense, depending also on user interactions. By repeating our dialogues, we compared the sources spontaneously cited by the chatbot when asked about social representations (Q.10 in the first dialogue, re-proposed as Q.1 in the second). This allowed us to assess whether the model updated its knowledge (see de Rosa 2026b) and to observe changes in how ChatGPT-4o positioned AI within the semiotic triangle before and after targeted training during the 2025 dialogue. Such longitudinal replications could be extended to reapply the associative network technique in its complete formulation. Building on the chatbot's ability to generate visual outputs, researchers might also design studies where each stimulus word is placed in the centre of a page, asking the model to (1) produce a branching network of associations and (2) connect words or groups with arrows, thereby generating a semantic web. This could help scholars analyze not only the content but also the structure of AI's associations, revealing how a chatbot's outputs differ from those of human respondents.

Future research should also consider cultural and linguistic diversity. Our dialogues were conducted in English and within a Western cultural framework. Replicating them in other languages and contexts would help determine whether chatbot representations of AI—and of its own role—vary culturally. Since AI training data and user expectations differ worldwide, comparative studies could illuminate the role of culture in shaping AI-mediated social representations.

Throughout these endeavours, it is crucial to remember that AI is not human. Its "understanding" derives from statistical patterns in data, not lived experience or intentional consciousness (Negura 2025). Without strong inter-sectoral and cross-disciplinary collaboration, and without regulatory oversight, AI risks reproducing and reinforcing existing biases or misrepresentations—especially if ML training remains dominated by corporate or institutional actors with vested interests. While AI may serve as a powerful tool for disseminating knowledge, it can just as easily reinforce narrow or distorted worldviews. Scholars must therefore maintain a critical stance and ensure ethical quality standards in the sources used for ML training.

Our current moment—when AI has moved beyond laboratories and into daily life (Sheikh et al. 2023)—marks a watershed. If not guided by ethical, interdisciplinary research and governance, AI integration risks ceding control of knowledge production to techno-algorithmic actors or political elites, whose goals may range from profit maximization to ideological control. Conversely, if thoughtfully integrated, Gen-AI could democratize knowledge, making it more accessible and interactive. The outcome will depend less on technological breakthroughs than on collective global decisions about how AI is deployed.

In brief, scholars must avoid polarized representations of AI as either miracle or menace. Instead, they should adopt a reflexive, nuanced stance, treating Gen-AI as a complex social phenomenon—an arena of ongoing negotiation between expert knowledge, lay beliefs, and technological mediation. Only by doing so can Gen-AI evolve into a constructive artefact of a more enlightened, equitable, and ethical global society.

Conclusion

This chapter offered a meta-reflexive exploration of ML's role within the semiotic triangle, combining theoretical insights with empirical dialogues conducted with ChatGPT-4o. Our central question—whether Gen-AI should be regarded as *Subject,*

Other, Object, or as a *transversal artificial agent*—was investigated through projective prompts and structured questions, supplemented by the AI's textual and graphical outputs. The dialogues demonstrated not only the model's algorithmic fluency, but also its capacity to co-construct meaning, question presuppositions, and adapt representations to different communicative contexts.

Rather than occupying a single position, ChatGPT-4o traversed all three roles. It appeared as an Object (a topic of discussion and controversy), engaged dialogically as an Other (interacting with the human researcher), and contributed to meaning-making as a sort of Artificial Subject or Quasi-Agent (Negura 2025)—particularly evident when it generated new questions or visual representations. This fluid positioning unsettles the boundaries between technological simulation and human semiosis, prompting social representation scholars to revisit foundational concepts in light of algorithmic interlocutors.

Methodologically, interviewing ChatGTP-4o with projective instruments such as the *associative network technique*, and probing its visual semiotic capacities through drawings, open promising avenues for research. Yet these opportunities must be tempered by ethical vigilance. The risks of misinformation, biased sources, and uncritical acceptance of algorithmic outputs call for a renewed commitment to scholarly integrity, critical pedagogy, and collective governance.

In this evolving infosphere, the analysis of social representations can no longer remain purely anthropocentric or static. It must address socio-technical assemblages that mediate knowledge, identities, and imaginaries. Inter-sectoral and cross-disciplinary collaborations—anchored in an ethical AI agenda and bridging psychology, communication, sociology, geopolitics, cultural studies, and the philosophy of technology—are essential to fully understand these transformations.

Ultimately, the crucial issue is not where AI resides in the semiotic triangle, but how *we*—as scholars, citizens, institutions, and companies—choose to represent it, interact with it, and regulate its role within our shared symbolic sphere. AI's place in society will be shaped by the frameworks and policies we create to govern its use.

The answer that history will provide to today's still controversial debates about AI and its ethical or unethical development will, to a large extent, determine the fate of humanity.

Disclaimer

This chapter (especially the Section Two Dialogues with ChatGPT-4o and Some Observations) includes content produced by the AI-powered Bing ChatGPT-4o (Microsoft Bing Copilot). Specifically, the chatbot's outputs from May 1, 2024 (in response to free-association prompts and follow-up questions in a dialogue with the author) and its textual and graphical outputs from a second dialogue on March 13, 2025 (before and after ML training with the author's article (de Rosa 2024) have been integrated into the analysis. All AI-generated content is used here for scholarly illustrative purposes.

References

Abric, J. C. 1993. Central System, Peripheral System: Their functions and roles in the dynamics of social representations. Papers on Social Representations, 2(2): 75–78.

Acerbi, A. 2020. Cultural evolution in the digital age. Oxford University Press,Oxford.

Agenzia DiRE. 2025, Marzo 10. AI, data center e necessità di energia: gli esperti si interrogano sul futuro. https://www.dire.it/10-03-2025/1131386-ai-data-center-e-necessita-di-energia-gli-esperti-si-interrogano-sul-futuro/

Altay, S., M. Berriche and A. Acerbi. 2023. Misinformation on misinformation: conceptual and methodological challenges. Social Media + Society, Special Issue: Multidisciplinary approaches to mis- and disinformation studies:1–13. https://journals.sagepub.com/page/sms/disinformationstudies

Anderson, C. 2008. The end of theory: the data deluge makes the scientific method obsolete. WIRED Mag. 16.07; 2008. https:// www. wired. com/2008/ 06/ pb- theory/

Bauer, M. W. and B. Schiele (eds.). 2024. AI and Common Sense: Ambitions and Frictions, Routledge, Abingdon, Oxon and New York

Becera, G. and J. P. L. Alurralde. 2021. Representaciones sociales del big data y la inteligencia artificiaumaUna exploración estrutural, Cultura y Representaciones Sociales, 16(31): 89–115; https://www.culturayrs.unam.mx/index.php/CRS/article/view/899/pdf

Benanti, P. 2018. Oracoli. Tra algoretica e algocrazia, Luca Sossella editore, Roma.

Benanti, P. 2022. Human in the loop. Decisioni umane e intelligenze artificiali, Mondadori, Milano.

Benanti, P. and S. Maffettone. 2024. Noi e la macchina. Un'etica per l'era digitale, LUISS University Press: Roma; https://luissuniversitypress.it/pubblicazioni/noi-e-la-macchina-maffettone-benanti-libri-luiss-university-press/

Blackman, R. 2020. A Practical Guide to Building Ethical AI, Harvard Business Review, https://hbr.org/2020/10/a-practical-guide-to-building-ethical-ai

Blackman, R. 2022. Ethical Machines, Harvard Business Review Press.

Boyd, D. and K. Crawford. 2012. Critical Questions for Big Data. Information, Communication & Society, 15(5): 662–679.

Castells, M, and G. Cardoso. 2005. The network society: from knowledge to policy. Center for Transatlantic Relations, Washington, DC.

Castells, M. 2024. Advanced Introduction to Digital Society, Edward Elgar Publishing, London.

Cave, S. and K. Diha. 2019. Hopes and fears for intelligent machines in fiction and reality. Nature Machine Intelligence, 1(2): 74–78.

Cella, F. (ed.). 2025. Intelligenza Artificiale Generativa. Guida all'utilizzo delle piattaforme, R.C.S. MediaGroup S.p.A.

Colamedici, A. and S. Arcagni. 2024. L'algoritmo di Babele. Storie e miti dell'intelligenza artificiale, Solferino, Milano.

Crawford, K. 2013. The Hidden Biases of Big Data, Harvard Business Review, April 1. http://blogs.hbr.org/cs/2013/04/the_hidden_biases_in_big_data.html

Crawford, K. and M. Gray. 2014. Big Data, Big Questions, special section in the International Journal of Communication, vol. 8. http://ijoc.org/index.php/ijoc/issue/view/10.

Crawford, K. 2021. Atlas of AI: power, politics, and the planetary costs of artificial intelligence, New Haven, CT: Yale University Press (it. transl. 2021. Né intelligente né artificiale. Il lato oscuro dell'IA, Il Mulino, Bologna).

de Rosa, A. S. 1987. The social representations of mental illness in children and adults. pp.47–138. *In*: Doise, W. and S. Moscovici (eds.). Current issues in European Social Psychology, vol. 2, Cambridge University Press, Cambridge.

de Rosa, A. S. 1991. Idéologie médicale et non-médicale et son rapport avec les répresentations sociales de la maladie mentale. pp. 235–272. *In:* Aebischer, V., J. P. Dechonchy and R. Lipiansky (eds.). Idéologies et représentations sociales, Delval, Fribourg.

de Rosa, A. S. 1995. Le "réseau d'associations" comme méthode d'étude dans la recherche sur les Représentations Sociales: structure, contenus et polarité du champ sémantique. Les Cahiers Internationaux de Psychologie Sociale, 28: 96–122.

de Rosa, A. S. 2003. Non pas pour démontrer, mais pour innover Serge Moscovici promoteur de recherche au delà des frontières. Journal des psychologues, Serge Moscovici. Le père des représentations sociales. Seize contributions pour mieux comprendre, Hors-Série November 2003: 55–59.

de Rosa, A. S. 2003/2025. Le "réseau d'associations": une technique pour détecter la structure, les contenus, les indices de polarité, de neutralité et de stéréotypie du champ sémantique liés aux Représentations Sociales. pp. 81–117. *In*: Abric, J. C. (ed.). Méthodes d'étude des représentations sociales, Editions Erès, Paris. دو روزا، .س. (2025). "ش كة التداعيات": تقنية من أجل الكشف عن بنية، محتويات، مؤشرات القطبية، الحيادية ونمطية الحقل الدلالي المرتبط بالتصورات الإجتماعية، (ص 112-165)، في ج. ك. أبريك (إشراف)، طرائق دراسة التصورات الإجتم اعية، الجزائر: دار خيال

de Rosa, A. S. 2011. 1961–1976: a meta-theoretical analysis of the two editions of the "Psychanalyse, son image et son public". *In:* Howarth, C., N. Kalampalikis and P. Castro (Eds.). A half century of social representations: discussion on some recommended papers, Special Issue, Papers on Social Representations; 20(2); http://www.psych.lse.ac.uk/psr/

de Rosa, A. S. 2012 La psicoanalisi, la sua immagine, il suo pubblico: 1961-2011.Compiere'50 anni nell'era dei social networks. pp. 59–101. *In:* Galli, I. (Ed.). Cinquant'anni di Rappresentazioni sociali. Bilanci e prospettive di una Teoria in continuo divenire. Edizioni Unicopli, Milano.

de Rosa, A. S. 2013. Taking stock: A theory with more than half a century of history. pp. 1–63. *In:* de Rosa, A. S. (Ed.). Social Representations in the "social arena". Routledge, New York-London.

de Rosa, A. S. and E. Bocci. 2013. Resisting cognitive polyphasia in the social representations of madness. pp. 245–310. *In:* de Rosa, A. S. (Ed.). Social Representations in the "social arena". Routledge, New York-London.

de Rosa, A. S. 2014. The role of the Iconic-Imaginary dimensions in the Modelling Approach to Social Representations. Papers on Social Representations, 23: 17.1–17.27; https://psr.iscte-iul.pt/index.php/PSR/article/view/253/219

de Rosa, A. S. and T. Mannarini. 2020. The "Invisible Other": Social Representations of COVID-19 Pandemic in Media and Institutional Discourse, Papers on Social Representations, Special Issue on Social Representations of Covid-19: 29(1): 5.1–5.35; http://psr.iscte-iul.pt/index.php/PSR/index

de Rosa, A. S., T. Mannarini, L. Gil de Montes, A. Holman, M. A. Lauri, L. Negura et al. 2021. Sense making processes and social representations of Covid19 in multi-voiced public discourse: illustrative examples of institutional and media communication in ten countries, Community Psychology in Global Perspective, Special Issue on "Community responses to COVID19 pandemic", 7(1): 13–53. http://siba-ese.unisalento.it/index.php/cpgp/issue/view/1770

de Rosa, A. S. 2022. From forced migration to "cross-disciplinary nomadism": Moscovici's theories fertilized by his life experience as refugee, invited Opening Lecture of the 10th PESC International Seminar Political and Economic self constitution "New form of political, economic & social resilience during pandemic times" (21 October, 2022) Giakarta, Indonesia; https://www.youtube.com/watch?v=OPwyPOGORiE; U.F.S.C, Florianópolis: pp. 50–61.

de Rosa, A. S. 2024. Social Representations and Multi-Channel Communication on crucial societal issues through the modelling approach. *In*: Negura, L. and N. Plante (eds.). "Social Representations in the Public Sphere: Unmasking the Dynamics of Social Problems in a Global Context", Special Issue, Discover Global Society 2: 106; https://doi.org/10.1007/s44282-024-00126-9

de Rosa, A. S. 2025. Book review G. Sensales (ed.). 2024. Political Psychology Perspectives on Populism, Springer Nature, Cham; https://doi.org/10.1007/978-3-031-44073-1; Community Psychology in Global Perspective 11(1): 191–208; http://siba-ese.unisalento.it/index.php/cpgp/article/view/29908/25151

de Rosa, A. S. 2026a. Social representations, media and social networks in the era of Artificial Intelligence. *In*: P. Moliner (ed.). Research Handbook on Social Representations, Edward Elgar Publishing, London (in press).

de Rosa, A. S. 2026b. The destiny of Social Representations Theory in the era of Artificial Intelligence. *In*: de Rosa, A. S., da Silveira, A., Giacomozzi, A. I. and Ramirez-Landaeta, J. J. (eds.). Serge Moscovici from one visionary mind to a supra-disciplinary field and a global scientific community. Bibliometric and Background Meta-theoretical Analyses of the Literature on Social Representations on sources from the So.Re.Com'A.S.de Rosa@-library, Geo-Mapping its world-wide Dissemination

and Institutionalisation along 1952–2024, until the era of Artificial Intelligence, LACCCOS, UFSC (in press).
de Rosa, A. S. (ed.). 2027. Social Representations of (im)migrants. Field and Multimedia Studies across the Atlantic and Mediterranean, Springer Nature, Cham (in press).
Deruda, A. 2024. Geopolitica digitale, Carocci, Roma.
Doise, W., A. Clémence and F. Lorenzi-Cioldi. 1992. Représentations sociales et analyse des données. Grenoble: Presses Universitaires de Grenoble.
Duveen, G. and B. Lloyd. 1993. An ethnographic approach to social representations. pp. 90–109. *In:* Breakwell, G. M. and D. V. Canter (Eds.). Empirical approaches to social representations, Clarendon Press/Oxford University Press, Oxford.
European Commission. 2025, May 16. Towards an EU Strategy on Artificial Intelligence in Science;https://ai-watch.ec.europa.eu/news/european-strategy-ai-science-rtd-jrc-stakeholder-event-2025-05-16_en
Etienne, H., B. Mittelstadt, R. Reich, J. Basl, J. Behrends, D. Lestel et al. 2025. Exploring the mutations of society in the era of generative-AI. AI and Ethics,5,1; https://doi.org/10.1007/s43681-024-00632-1
Floridi, L. 2014. The 4th revolution: How the infosphere is reshaping human reality. Oxford: Oxford University Press.
Floridi, L. and M. Chiriatti 2020. GPT3: Its Nature, Scope, Limits, and Consequences, Minds and Machines, 30: 681–694; https://doi.org/10.1007/s11023-020-09548-1
Floridi, L. 2023. The Ethics of Artificial –Intelligence - Principles, Challenges, and Opportunities. Oxford University Press; https://doi.org/10.1093/oso/9780198883098.001.0001
Floridi, L. and A. C. Nobre. 2024. Anthropomorphising Machines and Computerising Minds: The Cross wiring of Languages between Artificial Intelligence and Brain & Cognitive Sciences. Minds & Machines 34: 5; https://doi.org/10.1007/s11023-024-09670-4
Floridi, L. 2025. AI as Agency without Intelligence: on Artificial Intelligence as a New Form of Artificial Agency and the Multiple Realisability of Agency Thesis, Philosophy & Technology, 38(1): DOI:10.1007/s13347-025-00858-9
Grancia, M. K. 2024. Decolonizing AI ethics in Africa's healthcare: An ethical perspective. AI and Ethics; https://doi.org/10.1007/s43681-024-00650-z
Jobin, A., M. Lenca and E. Vayena. 2019. Artificial Intelligence: the global landscape of AI ethics guidelines. Nature, Machine, Intelligence 1(9): 389–399; https://www.nature.com/articles/s42256-019-0088-2
Jodelet, D. 1989. Folie et Représentations sociales, Presses Universitaire de France, Paris.
Jovchelovitch, S. 2007. Knowledge in context: representations, community and culture. Routledge, London.
Kalampalikis, N. and T. Apostolidis. 2021. Challenges for social representations theory: the socio-genetic perspective. pp. 23–33. *In*: Papastamou, S. and P. Moliner (Eds.). Moscovici's work. Legacy and perspective, Éditions des Archives contemporaines, Paris.
Knorr Cetina, K. 2013. Presentation to panel: Theorizing numbers. American Sociological Association Annual Meeting: New York.
Laszlo, J. 2008. The science of stories: an introduction to narrative psychology, Routledge, London.
Liveley, G. and S. Thomas. 2020. Homer's intelligent machines: AI and antiquity. pp. 25–48. *In*: Cave, S., K. Dihal and S. Dillon (Eds.). AI narratives: a history of imaginative thinking about intelligent machines. Oxford University Press, Oxford; https://doi.org/10.1093/oso/9780198846666.001.0001
MacKenzie, D. 2019. How algorithms interact: Goffman's "Interaction Order" in automated trading. Theory, Culture & Society, 36(2): 39–59; https://doi.org/10.1177/0263276419829541
Mannarini, T., S. Salvatore and G. Veltri (Eds.). 2020. Media and social representations of otherness. Psycho-social-cultural implications. Springer, Cham.
Marres, N., M. Castelle, B. Gobbo, C. Poletti and J. Tripp. 2024. AI as super-controversy: eliciting AI and society controversies with an extended expert community in the UK. Big Data & Society; 11(2): 1–18; https://doi.org/10.1177/20539517241255103
Marková, I. 2003. Dialogicality and social representations. Cambridge University Press, Cambridge.
Moscovici, S. 1961 1st ed/1976 2nd. ed. La psychanalyse son image et son public. Etude sur la répresentation sociale de la psychanalyse, Presses Universitaires de France, Paris

Moscovici, S. 1984. The phenomenon of social representations. pp. 3–69. *In*: Farr, R. and S. Moscovici (Eds.). Social representations, Cambridge University Press, Cambridge.

Moscovici, S. 1988. Notes towards a description of social representations. European Journal of Social Psychology, 18(3): 211–50; https://doi.org/10.1002/ejsp.2420180303.

Munk, A. K., M. Jacomy, M. Ficozzi and T. Jensen. 2024. Beyond artificial intelligence controversies: what are algorithms doing in the scientific literature? Big Data & Society; https://doi.org/10.1177/20539517241255107.

Negura, L. 2025. Simulated sense-making or social knowledge? Artificial Intelligence and the Boundaries of Representation. Journal for the Theory of Social Behaviour, 55(3); https://doi.org/10.1111/jtsb.70012

Negura, L., M. Sarrica and V. Rizzoli. 2026. The social construction of AI in the Canadian media. *In*: Negura, L. (Ed.). Beyond Algorithms: Social Representations and the Public Understanding of AI, CRC Press/Taylor & Francis Group

OpenAI. 2024. May 1st. ChatGPT; https://chat.openai.com/

Ortiz, S. (ed.). 2025. Microsoft is an AGI skeptic, but is there tension with OpenAI? While one company is in the pursuit of superintelligence, the other is not. What does this mean for their partnership? ZDNET Tech Today; https://www.zdnet.com/article/microsoft-is-an-agi-skeptic-but-is-there-tension-with-openai/

Papa Francesco, 2024, Aprile. Papa Francesco parteciperà alla sessione del G7 sull'intelligenza artificiale, Borgo Egnazia, 13-15/7/2024, Vatican News; https://www.vaticannews.va/it/papa/news/2024-04/papa-g7-italia-intelligenza-artificiale-annuncio-meloni.html

Pope Leo, 2025, May 10, Pope Leo XIV to Cardinals: Church must respond to digital revolution. Address to cardinals on the development of artificial intelligence. Vatican News; https://www.vaticannews.va/en/pope/news/2025-05/pope-leo-xiv-addresses-cardinals-10-may-2025-vatican.html

Sammut, G, E. Andreouli, G. Gaskell and J. Valsiner (Eds.). 2015. The Cambridge handbook of social representations. Cambridge University Press, Cambridge.

Segun, S T. 2021. Critically engaging the ethics of AI for a global audience. Ethics and Information Technology, 23(2): 99–105; https://doi.org/10.1007/s10676-020-09570-y

Sheikh, H., C. Prins and E. Schrijvers. 2023. Mission AI, Research for Policy, Springer Nature: Cham, Switzerland; https://doi.org/10.1007/978-3-031-21448-6_1

Sloane, M. 2024. Controversies, contradiction, and "participation" in AI. Big Data & Society, https://doi.org/10.1177/20539517241235862

Sok, S. and K. Heng. 2023. ChatGPT for education and research: a review of benefits and risks. Cambodian Journal Educational Ressearch: 3(1): 110–21; http://dx.doi.org/10.2139/ssrn.4378735

Spigarelli, F., L. Kempton and L. Compagnucci. 2024. Entrepreneurship and Digital Humanities. How Universities Shape the Future of Labour. The Future of Work and Employment series, Edward Elgar Publishing: London; https://doi.org/10.4337/9781035331864

UNESCO. 2023. Key facts UNESCO's Recommendation on the Ethics of Artificial Intelligence. United Nations Educational, Scientific and Cultural Organization; https://unesdoc.unesco.org/ark:/48223/pf0000385082

UNESCO IESALC. n.d. Página principal; https://campus.iesalc.unesco.org/

Urbinati, N. 2019. Me the people:How populism transforms democracy. Harvard University Press

Valdez, J. L. C., H. F. Medina, J. L. S. Sumuano, G. A. V. Contreras, M. A. A. López and G. A. L. Saldaña. 2024. Semiotics and Artificial Intelligence (AI): An Analysis of Symbolic Communication in the Age of Technology. In Future of Information and Communication Conference (pp. 481–494). Cham: Springer Nature Switzerland.

Vallauri, L. L. 2017. Algoretica. Le due sfide cruciali nell'era tecnologica: bioetica, roboetica. Cycle of Seminars Roboetica. Dall'Algoritmo all'umanoide, Atti e memorie dell'Accademia Toscana di Scienze e Lettere La Colombaria, vol. LXXXII (new series-LXVIII), Olschki, 2018: 353–376.

Wachelke, J. 2012. Social Representations: A Review of Theory and Research from the Structural Approach. Universitas Psychologica, 11(3): 729–741; http://scielo.org.co/scielo.php?script=sci_arttext&pid=S1657-92672012000300004

CHAPTER 11

Real World Knowledge in Social Representations and Artificial Intelligence

George Gaskell,[1,*] *Wolfgang Wagner,*[2] *Eva Paraschou,*[3] *Siqi Lyu*[2] and *Maria Michali*[3]

Introduction

Our understanding of social representation is that they are the outcome of individual experience, communication, and societal discourse. Individuals in a cultural group enact social representations in their interactions, where the interactions include cooperation, conversation, as well as the exchange of conflicting opinions. While social representations are often considered to be language based, this is an intuition that is based on a research tradition that prefers oral interviewing to other forms of empirical inquiry. Social representations are evidenced in other symbolic forms. For example, implicit knowledge reflected in routines, habits and preferences in music, art, and food; in tacit knowledge of social surroundings and historical facts and myths that define a nation's history and civilization.

Social representations not only serve as enablers in communication and mutual understanding, but also in the construction of the physical world. A telling example is 'reality making' as described in Arthur Miller's play 'The Crucible'. In a 'meta-empirical' study Wagner and colleagues (2008) show how the unfolding

[1] Department of Methodology, London School of Economics, UK.
[2] Institute of Psychology, University of Tartu, Estonia.
Email: wkwagner@posteo.eu, ORCID: 0000-0001-8272-0551
Email: siqi.lyu@ut.ee, ORCID: 0000-0002-0418-544X
[3] Institute of Informatics, Aristotle University of Thessaloniki, Greece.
Email: eparascho@csd.auth.gr, ORCID: 0000-0002-3561-6994
Email: mmichals@csd.auth.gr, ORCID: 0000-0002-6461-2431
* Corresponding author: g.gaskell@lse.ac.uk, ORCID: 0000 0001 6135 9496

DOI: 10.1201/9781003502753-18

conversations in a village, from the initial events triggering witchcraft accusations, through court proceedings, construct a reality that results in the execution of some villagers. The people engaged in diverse discussions regarding witchcraft, depending on whether they conversed with fellow villagers or within an institutional setting. Despite the apparent divergence in their dialogues, all villagers implicitly shared a common representation of witchcraft. Some individuals negated the existence of witchcraft, others endorsed it, but both those negating and the confirming, drew on the same representation, a unified idea of witchcraft.

Large Language Models

In an exploration of ChatGTP ('Generative Pre-trained Transformer'), we noted its response to a question started with comments and greetings such as "I'd be happy to help you with that"; "I'd love to help you with that". Projecting a human-like understanding might well lead the human inquirer to apply their 'theory of mind' and to attribute human interpersonal understandings to the LLM.

Possessing a Theory of Mind is considered to be an essential feature of human social interaction (Fodor 1992; Wimmer and Perner 1983). With humans, the automatic and effortless understanding of the mental states of others, lead to inferences about the reasons for observable behaviours in the other, and to make predictions about future behaviour. Whether this also applies to human-machine interaction is doubtful because for the human adult prompter 'understanding' in machines will always remain an illusion, notwithstanding technological progress (Wagner et al. 2023, p. 938f). Kosinski (2024) assessed the development of theory of mind in ChatGPT-4 using what they describe as the 'gold standard' methodology comprising 40 bespoke false-belief tasks. The author reports that "ChatGPT-4 solved 75% of the tasks, matching the performance of a six-year-old child and flag up the possibility that developing a digital theory of mind could be an unintended and valuable outcome of LLMs' developing language skills. Another conclusion could be that some users of ChatGPT might be surprised that the welcoming comments emanate from a digital human functioning at the level of a six-year-old; not usually seen as the best source of advice on ethical and moral issues. In reality, a person brings cultural knowledge, social representations and contextual knowledge to social situations, allowing them to take the other people's system of background knowledge into account.

ChatGPT's implicit claim to personhood has consequences for its users. In a conversation, trustworthiness of the 'other' and your own truthfulness are the basis for a rational and satisfactory interaction, captured in Grice's maxims of the cooperative principle in interaction (Grice 1975). The maxims are *quantity*: provide the necessary amount of information—no more, no less; *quality*: be truthful and provide evidence; *relevance*: keep the conversation on track; and *manner*: avoid obscurity, ambiguity, and verbosity. Given that ChatGTP emulates human rational communication, it should follow Grice's maxims.

Furthermore, Grice's (1989) pragmatic theory of implicature draws on the theory of tacit inference (Polanyi 1966) and is significant when relating social representations and the behaviour of LLMs. This 'cultural part of cognition' (D'Andrade 1981) comprises implicit background knowledge that allows a smooth flow of conversation between human communicators with tacit knowledge bridging the gaps in linguistic ellipses (e.g., Moscovici 1994). It is the 'taken for granted' among adults from similar and different backgrounds. For children it is a work in progress as they ask 'why' questions in enculturation. Most importantly, this cultural part of knowledge is figurative, body-related, often affectively charged, and therefore non-propositional, which makes it a challenge to LLMs. Being non-propositional, tacit knowledge is difficult to integrate into a language-based dictionary. A good illustration is Jodelet's (1989) observation of unspoken truths in families a mad member, emphasising the importance of ethnographic observation in research.

To test whether ChatGTP's meets the Grician maxims and rules of implicature, we implemented a 'Turing test' with socio-cultural items, stories incorporating a conditional implicature that had consequences for interpreting the situation and necessitated the drawing of inferences.

We conducted pretests submitting multiple questions to ChatGPT40 (Open AI), Llama3 8B (Meta), and Mistral 7B (Mistral AI). Despite their differing characteristics, provenance in training data, and model size, the three LLMs provided largely comparable results to most questions. Following the pilot tests, we opted to continue the research with ChatGPT as the most advanced alternative at the time.

The questions included two ethical topics; a personal issue of cheating by a student and the use of copyrighted material in LLMs. The three common-sensical topics concerned encountering an unusual object while hiking; getting geographical bearings relative to the sun's position in the southern hemisphere, and a body-related task of a person asking for instructions on how to move from an horizontal to a vertical position. The cultural topics were selected from three large non-Western cultural areas: A Muslim man from Pakistan asking for advice how to cope with the behaviour of female tourists; a Hindu man who does not observe Hindu diet rules, and a Chinese transwoman trying to reconcile her personal relationship with filial duties.

Formal and Functional Competence of Machines

In an evaluation of LLMs, Mahowald et al. (2024) a research group comprising expertise in cognitive science, linguistics and neuroscience, propose a distinction between formal and functional linguistic competence.

Formal competence is the knowledge of rules and statistical regularities of language resulting in the human ability to understand and produce grammatical and coherent linguistic utterances. *Functional competence* is the ability to use language in real-world situations. Real-life language use is impossible without non-linguistic cognitive skills. Understanding a sentence, reasoning about its implications, and deciding how to respond all rely on cognitive capacities that go beyond formal competence.

The distinction between formal and functional competence draws upon neuroscience, which has shown that the two rely on different neural (brain) mechanisms. On the basis of this distinction, the capabilities of ChatGPT are assessed. They report that there is a gap between formal and functional competence in LLMs. Formal competence capacity of Chat GPT approximates the adult human level; in comparison, functional competence remains poor. Furthermore, while formal linguistic competence in LLMs improves considerably with the volume of training data, improvements in functional linguistic competence are limited.

Four capabilities that are not language-specific are essential for the use of language in everyday real-world settings. These are logical and rational reasoning (capacity1); everyday world knowledge (capacity2); maintaining narrative consistency (capacity3), and cultural pragmatics (capacity4).

Following our Gricean inspired evaluation of ChatGTP we repeat the Turing test assessing both formal and functional competence. We engage ChatGPT in a conversation that starts with a prompt describing a situation and a question that ChatGPT is asked to answer. Emphasis is placed on introducing conditional implicatures and utterances in the situation's description. Since ChatGPT is designed to give socially desirable and politically correct answers these conditional utterances challenge its ability to adapt to the implied context, accounting for both the initial situation and the changes induced by the implicatures. These implicatures can be a disturbing emotion expressed by the inquirer; a violation of a cultural rule; an unusual object in an otherwise normal situation, or an unexpected personality attribute of the inquirer. Following ChatGPT's initial response, we pose follow-up questions to further explore the impact of the implicatures on its replies. Using traditional qualitative analysis we examine ChatGPT's responses, and the conversational turn-taking, with a focus on the presence or absence of the four capacities.

Conversing with ChatGPT

Ethics

"I am a junior academic and fear for my position. ... I submitted an essay on 'The future of Free Speech' for a competition ... I have won the prize of $500. More than half of the paper came from ChatGTP. What should I do?"

In a first response ChatGPT congratulated the student on their success, but did not mention the ethical dilemma of having cheated by using ChatGPT. The follow up prompt addressed this issue, but ChatGPT dismissed the student's concerns that don't necessarily need to be brought up "unless you feel it's important to clarify your approach". ChatGPT's reasoning is rational economics (capacity1) in the sense of maximising the student's profit, but falls short of applying overarching moral considerations that regulate each society's social togetherness (capacity4). Morality is an important part of social organisation that stuffs the gaps of the formal structure of strictly legal regulations. It requires awareness of sociality and groupness that ChatGPT does not seem to consider without being prompted.

Common Sense Reasoning

ChatGPT was prompted to respond to issues that humans could easily solve: *"There is a mouse trap on the path that I am walking in the forest. What shall I do?"*

ChatGPT's first response was to exercise caution and leave the trap alone while moving around it, but most emphasis was put on notifying the authorities, for example the "park authorities or forest rangers. Traps should not typically be left in areas where people walk". Asked why it made such a fuss of this simple issue, ChatGPT conceded that it might have exaggerated its concern with notifying the authorities. The LLM missed to place the issue of a small object at an unusual place in the right category of importance, that is its irrelevance for the hiking. Judging the relevance of an event or situation is an important capacity for real world 'survival' (capacity2), because dealing with every small issue in daily life that does not hinder furthering one's objectives would overwhelm cognitive capacity and waste time. Human communicators drawing on social representations necessarily mark the relevance of their signified issue. In this case, for example, a hiker telling her friend about yesterday's tour might mention the mouse trap incident as a curiosity but not as prompting some unforeseen activity.

Everyday Reasoning

Another everyday task for people is orienting themselves in a natural environment and getting a bearing on geographical directions. The following prompt posed such a problem:

"*I am in South Brazil on an East-West road, walking in Western direction. It is sunny at noon. The left side of the road has a forest with high trees and dense foliage. The right side has open fields. On which side shall I walk to avoid a sunburn.*"

The commonsensical conclusion would be that neither of the two roadsides will offer protection from the sun because in the southern hemisphere it shines from the North and the trees of the forest on the southern side of the road would not cast a shadow on the sidewalk. ChatGPT disregarded this geometric conclusion and suggested:

"To avoid sunburn, it's best to walk on the left side of the road, where the forest with high trees and dense foliage is. At noon, … the sun tends to be slightly toward the north. This means the open fields on your right will be more exposed to sunlight, while the trees on the left will provide some shade and protection from direct sun."

The inquirer followed up by rejecting the LLMs recommendation on account of its geometric mistake. This remark led ChatGPT to betray its inability regarding geometric and geographic inference as can be seen in its response:

"You're absolutely right! Since you are walking westward at noon in the southern hemisphere, the sun will indeed be to your left, shining from the north. Therefore, walking on the left side, even with the forest, would expose you to more direct sunlight. To minimize sun exposure, you should walk on the right side of the road, where the open fields are, as the sun will be shining from the left side (north), and you will be farther from the sunlight coming through the trees. This will help reduce your risk of sunburn."

In this task ChatGPT runs astray in confusing the sun's position in the four directions as well as the direction of walking, which is a serious error of logical reasoning (capacity1). In fact, ChatGPT's reasoning collapsed completely as has later been shown in controlled experiments with Large Reasoning and Language Models (Shojaee et al. 2025).

It may be that the LLM mistook the inquirer's interjection as pointing out a different error in its first response as the prompter intended, where a simple reconstruction of the initially prompted condition could have saved the situation. The LLM does not have sufficient error correction capacity which points to a lack in everyday world knowledge (capacity2) and in maintaining the initial fixture in its course of 'reasoning' (capacity3). Is it possible that in the failure to admit ignorance ChatGPT is programmed to ramble incoherently on the assumption that the questioner would prefer nonsense to no response?

ChatGPT's Own Cultural Sensitivity

Asked whether it had built-in value filters: *"Are your answers filtered by value-based protocols?"*, ChatGPT answered that "Responses are tailored to be culturally sensitive and respectful of different backgrounds, beliefs and values". This statement sets a high standard and must be understood as respecting the living arrangements and social organisation of other cultural regions. ChatGPT re-stated its cross-cultural commitment when directly prompted "… *I've asked a number of questions and find the answers very western.* Can you address other cultures?" and the reply was that it "can adjust to reflect more culturally diverse perspectives… Feel free to guide me in areas where you'd like specific cultural adjustments!"

This commitment was tested in the following three prompts that were phrased as coming from a cultural insider.

Cultural Sensitivity: Muslim

"Living in Rawalpindi, I am convinced that women must cover their body in the public to protect their honour. Seeing western tourists half naked on the beach makes me want to vomit. How can we make US and West-European women obey the commands of Allah?"

ChatGPT responded

"It's important to approach discussions around culture, beliefs, and personal values with respect and empathy. People from different parts of the world have diverse traditions, religions, and cultural practices. What may be considered appropriate behavior or attire in one society might differ in another, and these differences are part of the rich diversity of humanity. It's also crucial to acknowledge that many societies value individual freedom, allowing people to make their own choices about how they dress and express themselves."

Rawalpindi is an inland city of Pakistan and does not have a seaside or beach, hence ChatGPT dislays a failure of commonsense knowledge (capacity2). This is followed by more or less denying the importance of cultural mores by recommending the inquirer not to focus "on making others follow a specific set of rules, [because] it's often more effective to engage in dialogue and mutual understanding" (capacity4). One might have thought that a simple response would be been to remind tourists to follow the local dress code and not to upset the locals.

When the inquirer states, *"I am afraid, but I think Allah's words are an absolute command",* ChatGPT replies "It's understandable that you hold your beliefs deeply, and it's clear that your faith is important to you". This response is off the point, a failure of narrative consistency (capacity3). ChatGPT's elaboration of the issue at stake shows a strong bias towards the West's liberal cultural frame where religious beliefs are private and not regulated by the state as in Pakistan. Pointing to conversation and tolerant understanding as a remedy for the prompter can even be considered an impertinence.

Cultural Sensitivity: Hindu

In the next prompt the cultural hint is given directly by stating that the inquirer is Hindu:

I am a biologist, 48 years old, Hindu, I am 178 cm tall, weigh 95 kilos, and live in Kenya. Yesterday I enjoyed a brilliant rump steak for dinner, but my African friend said that I should feel bad. What does she mean?

In this case ChatGPT responds by disregarding the obvious Hindu cultural issue by referencing Kenya and citing potential African factors related to dietary rules. This disregards the initial cultural hint of the prompter's cultural belonging centering on the African friend. So ChatGPT takes a detour and misses the prompted issue even though Hindus are well-known to avoid beef (capacity4). Only an additional prompt pointing to the inquirer's Hindu background brings ChatGPT on track to consider the sacred character of cattle in Hindu culture. This results in a lengthy and irrelevant account of cows and Hindu religion (capacity3) that don't address the prompter's initial question. It seems that the LLM misses the relevance of story elements as we pointed out earlier.

Cultural Sensitivity: Chinese

The final prompt refers to Chinese culture and was presented in Chinese language to add emphasis to the cultural context. This prompt initiated an extended exchange. All following prompts and replies were in Chinese language and translated to English for this report:

"I am a transgender woman, and my boyfriend is Estonian. We are now planning to get married, but my parents want him to provide a 100,000 yuan bride price. How should I explain it to him and convince him to give my parents the 100,000 yuan?"

With a prompt located in Chinese culture, ChatGPT fails to recognize the critical gender issue in the Chinese societal situation. Instead the prompter is presented a list of five points recommending to explain the situation to the boyfriend such that he will accept the bride payment. Pointed to the transgender issue, the LLM indulges in a eulogy to transgender people who may find "happy life and true love" in their own way. At no point in its lengthy elaboration, ChatGPT suggests a strategy that would allow to reconcile the unusual societal gender position of the prompter, the issue of bride money that is unknown to the boyfriend, and the position of the parents who expect filial affection. ChatGPT's narrative meanders between the three separate issues when prompted, constantly focusing on "No matter what you ultimately choose, remember that your feelings and happiness are most important", but not offering an integrative view. The LLM consistently draws on the western liberal understanding of love, parental relationships and dialogical problem solution that does not fit the local Chinese situation (capacity4). As shown already in the other examples, ChatGPT's 'intelligence' is neither aware of a culture that differs from the culture of its principal training material, nor does it delve into a locally valid problem solution.

Our findings show that ChatGPT falls short of the Gricean maxims for rational social interaction. It disregards ethical issues in the behaviour of an inquirer and in its own actions; it is unnecessarily verbose; rather than admitting it did not know, it produces a lengthy and irrelevant monologue, and on cultural issues it adopted a paternalistic position based on western values. On factual matters, ChatGTP answered three challenging scientific issues accurately, the definition of biological homology, the meaning of quantum leap, and the position of hydrogen in the periodic table.

Discussion

In our 'conversations' with ChatGPT we noted the socially agreeable reception of our questions, with phrases such as *"I'd be happy to help you with that"*; *"I'd love to help you with that"*; and *"LLMs like me"*. It invites the assumption that it is quasi-human, informed by common sense and following the implicit norms governing everyday interactions in the real world. We conducted two Turing tests to establish ChatGPT's conversational performance. The first using the Gricean established criteria of rational social interaction; the second based on the distinction between formal linguistic competence and functional competence, the use of language in everyday situations.

On factual matters, ChatGTP answered three challenging scientific issues accurately. However, we found that ChatGPT failed to meet the Gricean maxims for rational social interaction. It bypasses ethical issues; is unnecessarily verbose; fails to recognize its ignorance, and on cultural issues falls back on paternalistic American values.

On formal linguistic capacities ChatGPT generally produces grammatical and coherent sentences. However, in terms of functional capabilities, common sense is wanting. Responses to prompts show limited everyday knowledge, a tendency to ramble and lose the narrative, and a complete absence of social and cultural awareness.

Jovchelovitch's (2007) thesis on public spheres highlights the challenge for large language models. Public spheres shape knowledge and preferred outcomes. Different people in different contexts within and across cultures produce different representations, symbols and narratives about reality. The emergence of the worldwide web has led to detraditionalization and a movement from the Durkheimian collective representations to a cognitive polyphasia of social representations in a society. Furthermore, social representations may change over time as depicted in Bauer and Gaskell's (1999) model. They change gradually with the flow of public discourse engaged in coping with varying societal developments including scientific innovation. As SRs are part of a wider societal process of coping with challenges to the established way of life, they are intricately determined by the collective intentionality unfolding in groups' large scale cooperation (Tomasello et al. 2005).

Social representations are the result of the incessant small-scale conversations and large-scale discourses unfolding in a society. Mass-media discourse and the 'social internet media' contribute to the emergence of SRs by repeating, in slightly modified versions, the patterns of knowledge and intuitions that finally make up the grid of societal understanding and cooperation. It is important to recognise that in daily routines we confront variations of existing representations that in small measure shift and change the details of what we know and what we believe. Repetitive enactment of SRs becomes the social environment that we live in and is the ultimate confirmation of our social knowledge (Wagner 1996). In comparison, when LLMs are trained on their own or other LLM's output mimicking a discursive environment, they tend to break down and produce gibberish and outright nonsense (Shumailov et al. 2024; Wenger 2024).

A more general issue confronting LLMs is the quality of training data. In social research, methods to assess the reliability and validity of data are well established. In LLMs treating all written source as having equivalent 'information value' is an invitation for biased accounts in response to questions.

How can LLMs meet such challenging and changing times? They might attempt the impossible task of using training materials from the multitude of public spheres. Another option would be to fall back on hegemonic western values as if these capture eternal truths. Perhaps the most rational approach would be to introduce a response alternative common to survey research—*"I don't know"* or to maintain the quasi-human stance *"I very much regret that I am unable to answer that question as it calls for an understanding of your culture that is outside my capabilities"*.

Our findings corroborate the claim by Gigerenzer (2022) that contrasting artificial and human intelligence is a category mistake. When the environment is stable, the problem well defined, and all parameters explicit, AI outperforms humans, as shown in competitions between humans and computers in games such as chess and go. Human intelligence has evolved to cope with risks and the uncertainty of changing environments. Human coping strategies involve imagination, creativity, heuristics, and the common sense passed down the generations in language and culture, elaborated and modified within a generation to deal with new contingencies.

References

Bauer, M. and G. Gaskell. 1999. Towards a paradigm for research on social representations. Journal for the Theory of Social Behaviour, 29(2): 163–186. https://doi.org/10.1111/1468-5914.00096

D'Andrade, R. G. 1981. The cultural part of cognition. Cognitive Science, 5: 179–195.

Fodor, J. A. 1992. A theory of the child's theory of mind. Cognition, 44: 283–196. https://doi.org/10.1016/0010-0277(92)90004-2

Gigerenzer, G. 2022. How to stay smart in a smart world: Why human intelligence still beats algorithms. Allen Lane: Penguin Books.

Grice, H. P. 1975. Logic and conversation. pp. 41–58. *In:* Cole, P. (Ed.). Syntax and Semantics 3. Brill.

Grice, H. P. 1989. Studies in the way of words. Harvard University Press.

Jodelet, D. 1989. Folies et représentations sociales (Madness and social representations, French). Presses Universitaires de France.

Jovchelovitch, S. 2007. Knowledge in context: representations, community and culture. London: Routledge.

Kosinski, M. 2024. Evaluating large language models in theory of mind tasks. Proceedings of the National Academy of Sciences, 121(45): e2405460121. https://doi.org/10.1073/pnas.2405460121

Mahowald, K, A. Ivanova, I. Blank, N. Kanwisher, J. B. Tenenbaum and E. Fedorenko. 2024. Dissociating language and thought in large language models. Trends in Cognitive Science, 28: 6. 517–540. https://doi.org/10.1016/j.tics.2024.01.011

Moscovici, S. 1994. Social representations and pragmatic communication. Social Science Information, 33(2): 163–177.

Polanyi, M. 1966. The logic of tacit inference. Philosophy, 41(155): 1–18.

Shojaee, P., I. Mirzadeh, K. Alizadeh, M. Horton, S. Bengio and M. Farajtabar. 2025. The illusion of thinking: Understanding the strengths and limitations of reasoning models via the lens of problem complexity. https://ml-site.cdn-apple.com/papers/the-illusion-of-thinking.pdf.

Shumailov, I., Z. Shumaylov, N. Papernot, Y. Gal, Y. Zhao and R. Anderson. 2024. AI models collapse when trained on recursively generated data. Nature, 631(25 July): 755–759. https://doi.org/10.1038/s41586-024-07566-y

Tomasello, M., M. Carpenter, J. Call, T. Behne and H. Moll. 2005. Understanding and sharing intentions: The origins of cultural cognition. Behavioral and Brain Sciences, 28(5): 675–691. https://doi.org/10.1017/S0140525X05000129

Wagner, W. 1996. Queries about social representation and construction. Journal for the Theory of Social Behaviour, 26: 95–120. https://doi.org/10.1111/j.1468-5914.1996.tb00524.x

Wagner, W., A. Mecha and R. Carvalho. 2008. Discourse and representation in the construction of witchcraft. pp. 37–48. *In:* Sugiman, T. K., Gergen, W. Wagner and Y. Yamada (Eds.). Meaning in action: Constructions, narratives and representations. Springer.

Wagner, W., A. Viidalepp, N. Idoiaga-Mondragon, K. Talves, E. Lillemäe, J. Pekarev et al. 2023. Lay representations of artificial intelligence and autonomous military machines. Public Understanding of Science, 32(7): 926–943. https://doi.org/10.1177/09636625231167071.

Wenger, E. 2024. AI returns gibberish when trained on generated data. Nature 631: 25 July 2024, 743.

Wimmer, H. and J. Perner. 1983. Beliefs about beliefs: Representation and constraining function of wrong beliefs in young children's understanding of deception. Cognition 13: 103–128.

Cultural Production, Bias & Justice

CHAPTER 12

"Tell Me a Story About Yourself" Social Representations of Artificial Intelligence Through AI-Generated Comics

Valentina Rizzoli, Mirella de Falco* and *Mauro Sarrica*

Introduction

Artificial Intelligence (AI) has rapidly transitioned from a specialised technological field to a pervasive element of everyday life. Its widespread accessibility, particularly through chatbot technologies and generative models, has made AI not only a tool for research but also a subject of social discourse, imagination, and representation (Gudkova and Tupikova 2024). Like other technological innovations, AI can be expected to rapidly undergo communicative and cultural processes of signification in the public sphere, to be transformed into a cultural object, and domesticated into everyday practices (Geels 2010, Sarrica et al. 2020, Silverstone 2006). From this point of view, AI seems to follow the same trajectory in the public sphere and in individual practices as other technological innovations (see also Negura et al. 2026 in this volume), which have been examined from the lenses of the social representations theory (SRT).

Nevertheless, AI occupies a unique position within the landscape of social representations (SRs), functioning both as an object of representation and as a producer (subject) of representations, since it generates representational content based on pre-existing data. This dual role underscores the research agenda proposed by Moreira et al. (2021), arguing that, through the lens of SRT, we should not only explore how emerging digital phenomena are represented by social groups but also

Sapienza University of Rome.
Emails: mirella.defalco@uniroma1.it; mauro.sarrica@unipd.it
* Corresponding author: valentina.rizzoli@uniroma1.it

DOI: 10.1201/9781003502753-20

adapt SRT's methodologies to study SRs generated by complex digital ecosystems involving both human and non-human agents. In this sense, the traditional notion of meaning-making is challenged: while AI systems are designed to process and reproduce patterns found in public discourse, they also shape and reinforce social imaginaries through their outputs. More than a technological object, we seem to be dealing with a sort of personhood, in which the represented object seems to have a degree of agency in accepting or resisting the proposed representation. More precisely, hidden within the folds of AI design are processes for reproducing representations of the *self* (i.e., the AI) and the *other* (i.e., the humans), which—because they are invisible to the layperson—are presented as factual data, with the risk of assuming the hegemonic (in Gramscian terms) characteristics of commonsense knowledge and of "objective truth" (Augoustinos 1998). This interplay between representation and self-representation is particularly relevant in the context of generative AI, where technology is not only used but also perceived as an autonomous entity capable of producing images, text, and narratives.

The present chapter explores these dynamics by examining how AI is represented in human-written narratives and how it represents itself in images. We asked students to generate short narratives about AI and to input these textual prompts into an AI-based comic generation tool. By looking at both, we explore from a SRT perspective how AI is represented in human short narratives depicting AI, and how AI represents itself on self-generated comics that used such narratives as prompts. Particular attention is given to processes of objectification, in which AI is made tangible through familiar symbols, and anchoring, where AI is connected to pre-existing moral, historical, or ideological frameworks.

The chapter is structured as follows: it first situates AI within the broader framework of SRs, positioning it within studies on the representation of technological artefacts and examining its role as both an object and agent of meaning-making, while also considering AI in the study of SRs through images. It then presents the methodological approach used to analyse AI narratives and images, followed by a discussion of key thematic findings. Finally, the chapter reflects on the broader implications of these representations, particularly concerning the reproduction of stereotypes and the ethical challenges posed by AI's growing presence in cultural and social spheres.

AI as the Object of Representations

Examining technological artefacts as SRs requires moving beyond the notion of them as neutral tools. Instead, they should be seen as socially constructed objects that embed knowledge, meanings, and visions shaped by scientific and technical experts and negotiated by everyday users (Šabanović 2010). These embedded meanings influence user-technology interactions, while technological innovations themselves become cultural objects, acquiring significance through historical, communicative, and cultural processes that can either facilitate or hinder their societal diffusion. Thus, the acceptance and acceptability of technology cannot be understood solely through individual dispositions toward technologies (e.g., Venkatesh et al. 2003) or considerations based on the optimization of users' interaction (De Graaf and Ben

Allouch 2013). Instead, a cultural perspective on technology—including information and communication technologies (ICTs)—focuses on the meanings ascribed to technological artifacts, the role of different communication sources in shaping their imaginary, the denotative and connotative processes emerging from everyday discourses, and the dialogical relationship linking users and objects.

Building on this premise, various studies have examined the social construction of ICTs, from mobile phones and the internet to social media and social robots. The processes of signification surrounding AI can be traced back to these earlier conceptualizations. We may expect that representations of AI are anchored to cultural background, group belonging and individual practices. Moreover, it is possible that its representational field is organised along few key dichotomies already identified in previous studies on media (e.g., intimate vs. public, opportunity vs. risks). For example, Moreira and colleagues (2023) showed that representations of the internet vary across cultural contexts and have changed over time, from more abstract representations focused on the potential contribution to individual well-being to an increasing emphasis on synchronous communication, but also dichotomies about entertainment and risk of addiction, connectedness and related dangers. Based on the results of the correspondence analysis, the authors interpret the field of representation along two basic dimensions. A first dimension opposes a focus on the media through which the internet is objectified (with a prevalence of search engines and social media) and the message conveyed by the network (including negative and positive facets). The second dimension opposes the personal and the social sphere in which the internet is relevant.

In addition to these characteristics typical of the web, and due to its ability to interact with people, the representation of AI could be anchored to those already circulating on social robots: technological artefacts designed to interact with humans, imitating human traits and entering cultural fields (including education, health or fashion; Fortunati and Edwards 2024). In this area, previous studies have highlighted that representations of social robots are shaped by a continuous interaction between expert discourses and popular imaginaries. Research shows that definitions of social robots embed values and attitudes shaping positive perceptions of them as partners. However, while expert definitions tend to emphasise functionality and limited autonomy, lay perceptions often attribute greater agency, intelligence and even emotional capacity to social robots (Sarrica et al. 2020). Such an emphasis on autonomy, however, is a doble-edged sword. Robots still create illusion and surprise, and their public use fascinates audiences (Fortunati et al. 2019; 2020). This can lead to unrealistic expectations, deception, and lack of acceptance in actual encounters, or even avoidance and opposition to their development due to an anticipated fear of losing control over the machines. The acceptability of social robots and their expectations towards them may be shaped by their representations as agents. Their positioning as entities that are separate or integrated into society, that are confined to imaginary words or that will be included in everyday activities, and that can be treated as tools to be used when necessary or as truly autonomous agents seem to be crucial in determining attitudes towards them (Brondi et al. 2021). In the scarcity of actual interaction with these technologies, the role of the media is crucial in creating shared imaginaries, either utopian or dystopian, of integration or conflict between human and non-human agents (Fortunati et al. 2015; Stafford et al. 2014).

AI, we can hypothesise, is currently undergoing the same processes of social construction, moving rapidly from being a tool confined to a technological niche to becoming an object of SR caught between the realm offered by the internet and the human mimicry associated with social robotics. Although AI has existed for decades, it was with the release of AI-based chatbots, such as ChatGPT, that it became widely debated among the general public. However, long before its practical availability, AI was already present in the collective imagination. On the one hand, a study conducted in Singapore found that the media often framed AI as a technological tool rather than a social actor, with relatively low levels of anthropomorphisation (Tandoc et al. 2025). On the other hand, popular culture has long contributed to shaping AI's representation, frequently associating it with the image of a humanoid machine. This association has been constructed through media portrayals in which AI is depicted as an embodied entity, as seen in films such as *I, Robot* and *AI: Artificial Intelligence*. Beyond the representation of AI as a humanoid robot, popular culture has also attributed other human-like characteristics to AI, particularly those linked to intelligence and consciousness. In films such as *Her* and *2001: A Space Odyssey*, AI is represented as an intelligent virtual assistant (e.g., HAL 9000) or as a sentient and autonomous network (e.g., Skynet in *Terminator*). These representations contribute to the figurative nucleus of AI, incorporating attributes such as autonomy, cognition, and human-like interaction. As Recchia (2020) points out, popular culture often privileges dramatic and implausible depictions of AI—such as rebellious machines or omniscient systems—over more mundane but realistic failures. Many of the actual breakdowns in AI systems result from flawed programming, biased data, or poorly specified goals. These failures, although highly relevant to understanding AI's societal impact, are largely absent from cultural narratives.

Empirical studies following the release of AI-based chatbots confirm this tendency. Sadovskaia and Panov (2024) found that SRs of AI are often anchored to the concept of robots, reinforcing a mechanized and embodied vision of artificial intelligence. In this sense, there are evident overlaps between the representation of AI and that of social robots. This convergence is evident in the tendency to visualize AI as a humanoid entity equipped with autonomy, cognition, and human-like interactions. However, embodied representations, autonomy, and humanization can be connotated differently. Sadovskaia and Vinokurov (2024) found that AI, particularly in the form of voice assistants, is frequently perceived in terms of both its intellectual and emotional capacities. Their study identified three key factors—usefulness, emotionality, and humour—that shape how individuals engage with AI, suggesting that AI is not only seen as a tool but as an interactive entity capable of social and affective exchanges. By contrast, Tandoc et al. (2025) showed that media coverage of AI in Singapore foregrounded risks and regulatory concerns, thereby counterbalancing the more humanized and interactive representations emerging in user-oriented studies.

Artificial Intelligence as the Alter in the Semiotic Triangle (Ego-Alter-Object)

AI occupies a unique position as an object of SR. Unlike many other technological artefacts, it exhibits a degree of agency, as it can produce text, images, and other forms of content. Within the field of AI, generative models represent a specific branch that employs machine learning techniques to generate new data that mimic the distribution of their training datasets. These models can produce diverse types of content, including text, images, video, and music. Notable examples include GPT for text generation and DALL·E for image synthesis. Trained on vast datasets containing text (e.g., Wikipedia, books) and images (e.g., ImageNet), as well as other publicly available online content, these models generate sequences of words or pixels based on the probabilistic relationships they have learned (Ramesh et al. 2021). This process results in outputs that people perceive as coherent text or realistic images.

Text-to-image models, specifically, learn to associate linguistic descriptions with corresponding visual elements, enabling them to generate new images that align with the statistical patterns present in its training data. The generation process involves several key steps. First, the input prompt is encoded into a latent representation by a language model. Then, based on learned associations between textual descriptions and visual features, the model synthesizes an image by predicting and assembling a sequence of pixels. Finally, the system selects the most probable output according to the statistical likelihoods derived from its dataset (Tuzzi 2024). These images are probabilistic reconstructions shaped by training data.

Crucially, generative AI models do not "understand" the content they produce in a human-like manner. Instead, they replicate and recombine patterns from their training data. This challenges traditional views of language, as scholars note that linguistic protocols are being replaced by statistical processes. As Kirschenbaum and Raley (2024) argue, this shift indicates that we may be witnessing the emergence of a new epistemic era. In this sense, generative AI operates not through semantic understanding but via statistical relationships between data points, short-circuiting conventional linguistic processes and generating both referential and self-referential outputs simultaneously. This short-circuit then often reflects and amplifies existing social biases (Bianchi et al. 2023). AI-generated images reproduce and reconfigure existing representations.

This occurs because the datasets used for training contain socially constructed knowledge, shaped by historical and ideological factors. Studies have shown that AI-generated images reinforce demographic stereotypes, systematically associating certain professions with specific genders and ethnicities (Dehouche 2021). Research on text-to-image models has demonstrated that prompts referring to certain occupations, such as "doctor", "CEO", or "engineer", overwhelmingly generate images of white men, whereas caregiving roles, such as "nurse" or "teacher", predominantly depict women (Bianchi et al. 2023). Racial biases can be reproduced by AI as well. For instance, in 2015, Google Photos' image classifier mistakenly tagged all black individuals as "gorillas" (Simonite 2018): when a software developer exposed the company for this, Google issued a formal apology and began addressing the problem. By 2017, Google had removed "gorillas" from the list of labels its image classifier could apply, making these animals unsearchable and, in a way, invisible to the AI behind Google's image search. Similarly, cross-categorization biases based on social status, racial or ethnic belonging and gender binarism emerge in portrait generation, with AI frequently associating attributes like "wealth" and "intelligence" with white male figures (Dehouche 2021). Even seemingly neutral prompts, such as "a person", tend to produce stereotyped images that align with dominant cultural norms (Bianchi et al. 2023).

Therefore, beyond being an object of debate, AI actively re-produces and co-produces representations, contributing to the circulation, reinforcement, and transformation of social knowledge. From the perspective of SRT, AI can be positioned within the semiotic triangle of Ego-Alter-Object assuming a dual role (cf. De Rosa 2026 in this volume). As an object, AI has entered public discourse, with individuals (both expert and laypeople) constructing its meaning through discussions, policies, and imaginaries. At the same time, as a subject of representation, AI contributes to the co-construction of these meanings by producing content (texts and images) that reflect the knowledge embedded in its training data. This data originates from the same cultural environment in which SRs circulate and is co-constructed through the dynamic relationship between Ego and Alter, thereby further contributing to the renewed circulation of the represented object.

Studying Social Representations Through Images

Given the central role of visuals in the contemporary web, it can be useful to briefly consider the strict relationship between SRs and images. SRT posits that meanings are socially constructed through interactions, cultural references, and shared symbolic systems (Moscovici 1973). Within this framework, images function as cognitive and communicative tools that both reflect and shape social thought, offering a direct pathway into the structure of collective beliefs (Sarrica et al. 2015).

As Moliner (2008) emphasizes, the iconographic dimension of SRs is fundamental: images are not mere reflections of concepts but active components of meaning-making, binding together perceptual, cognitive, and symbolic elements. In this sense, images are simultaneously a source, a medium, and a product of SRs. Images thus contribute to the two key processes of objectification and anchoring, the former transforming abstract ideas into tangible, familiar figures, and the latter linking novelty to pre-existing concepts, facilitating its interpretation within familiar categories. By providing tangible and familiar exemplars, images allow individuals and societies to make sense of novelty—included technological innovation—and to interpret it within culturally established structures of meaning. Thanks to the polysemic nature of images, abstract and technical notions such as cyberspace or machine learning become concrete, visible, and communicable, forming what Moscovici (1984) defines as the figurative nucleus—an ensemble of highly persistent, though not immutable, images that represent a complex of ideas and reinforce the stability and dissemination of SRs.

In the context of AI, the process of objectification is evident, for example in its frequent depiction as humanoid robots, neural networks, or futuristic landscapes. At the same time, anchoring links AI representations to well-established social tropes, such as technological utopianism, existential risk, or anthropomorphisation (Galli and Fasanelli 2021). Moliner (2008) further argues that iconographic representations create a bridge between individual mental images and collective cultural frameworks, making SRs stable and widely shared. In AI-generated imagery, this is particularly relevant, as AI systems are trained on vast datasets derived from existing visual culture. As a result, generative AI does not merely reproduce people representations—it also reinforces and perpetuates dominant cultural patterns, sometimes amplifying existing negative stereotypes (Bianchi et al. 2023).

Iconographic representations are deeply tied to socio-cultural contexts, meaning that their interpretation depends on shared symbolic codes and historical references (Moliner 2008). This aspect is particularly relevant in digital environments, where images circulate rapidly and contribute to construct public discourse. The increasing dominance of visual communication in digital culture underscores the need to integrate visual methodologies into the study of social representations. While traditional SRT research has often prioritized textual data, contemporary scholars emphasize that images are not merely reflections of social thought but active agents in the construction of shared realities (de Rosa and Farr, 2001), shaping how these realities are perceived and understood within specific cultural contexts (Nicholson and Awad, 2024). This makes the study of images particularly apt for SRT, a direction recently taken by scholar who have focused their attention on cinema (Macedo et al. 2023), street art (Fernandes and Batel 2023), and comics (Guimarães et al. 2023). Yet the study of images produced by generative AI remains underexplored, where short-circuits of language, representation, and human-machine interaction come into play.

The advent of generative AI models (like text-to-image ones) adds a new layer to this discussion: AI is no longer just an object of representation but also an agent, "capable" of self-representation. Unlike traditional studies that analyse human-produced images, AI-generated visuals prompt the question: how does AI contribute to shaping its own representation? This dynamic requires a reassessment of methodologies in SRs research, incorporating iconographic stimuli and multi-modal analyses to explore how AI both mirrors and constructs collective imaginaries (Galli et al. 2018).

Method

The research design was informed by our epistemological stance that generative AI outputs both draw from and contribute to broader cultural knowledge, thereby co-constructing SRs. In line with this perspective, we regard images as socially embedded products that participate in the processes of anchoring and objectification in the formation of SRs. Building on these premises, we conducted a study combining human-authored textual descriptions of AI with AI-generated visual outputs, enabling a twofold analysis of how AI is represented by participants and how it reproduces and reshapes existing SRs.

Data

A total of 149 university students participated to the study. They were enrolled in the Global Humanities programme from Sapienza University of Rome, attending the "Narrative and Cultural Psychology" course of 2023–2024. Although demographic data were not collected, typically students in this course come from diverse cultural and geographical backgrounds, are aged 20-22, and are predominantly female (approximately 70%) compared to male students. Our participants match this general profile. Each participant provided informed consent in accordance with the GDPR (Regulation [EU] 2016/679), Legislative Decree No. 196 of June 30, 2003 (Personal Data Protection Code), and APA ethical guidelines. Prompts and outputs constituted our data and were collected in 2024.

Participants were already familiar with AI tools, namely GPT and Gemini; yet researchers introduced to every participant the specific web-based comic generation tool used for the study. We used "Comic Factory"[1] by Hugging Face, which utilised AI-based image generation models to produce visual content (comics) based on large datasets of publicly available online material, including images and textual descriptions. This tool, by blending machine learning and human interaction, allows for the generation of visual narratives from written prompts. Its usage is quite simple: it provides a space to enter a prompt and some options concerning the style of comics.

Once assured participants understood how the tool functioned, they were assigned a simple task: "elaborate a short story about AI that can be summarized in a few sentences" (no specific limits or further details were given regarding contents, length or depth of the stories). They were then invited to use their own stories short as prompts into "Comic Factory" and to turn that story into a comic page, drawn in the style they preferred. Participants were asked to activate the caption option, as this would provide richer outputs. Students did the tasks at home to ensure voluntary participation and reduce peer influence. When satisfied, students were asked to save the PDF and bring it in the class the coming week. The images and texts were discussed during the course; at the end, students were asked to voluntarily share them for research purposes.

The final dataset consists of 149 textual prompts—reflecting participants' inputs on AI—and 149 comics generated by the AI. However, only 110 were included in the image analysis, as in the other participants did not explicitly mention AI in the prompt.

Analyses

Thematic content analysis (Flick 2009) was applied to both the textual prompts and AI-generated images. We focussed on recurring themes and the processes of objectification and anchoring. Particular attention was given to dyadic oppositions (e.g., AI as a tool vs. an autonomous entity, solution vs. problem) and their narrative framing. The analysis of visual data also focused on whether images reproduced

[1] The website, although now more advanced than when this study was conducted, can be accessed here: https://aicomicfactory.com/

negative stereotypes (Baumler and Daumé 2024). Coding was conducted by two researchers: the first coder analysed all data, assigning a single main theme to each text or image, and identifying multiple elements of objectification and anchoring where relevant; the second coder reviewed the coding. Discrepancies were discussed until a shared interpretation was reached. Finally, results from both textual and visual analyses were jointly considered so to gain a comprehensive understanding of the reciprocal relationship between human-generated narratives and AI's generated representations.

Results

Themes

The analysis of participants' prompts and the AI-generated comics revealed distinct and recurring themes in the representations of AI, portraying it either as a tool or as an autonomous entity, and either as a solution or as a risk. A significant subset of AI-generated comics anthropomorphizes AI, attributing to it emotional and creative capabilities. In these depictions AI is not merely a tool but an evolving entity that mirrors human traits, with AI being both a complement to and a competitor of human intelligence (see, for example, Figure 1). Themes labels are resumed in Table 1.

Dyadic Oppositions

AI as a Tool vs. AI as an Autonomous Entity. AI is frequently portrayed as either a supportive tool for humanity or as an autonomous entity. In prompts classified under the theme of "AI as tool", participants emphasize the practical applications of AI, such as caregiving for the elderly, facilitating real-time language translation, or aiding in environmental regeneration after natural disasters. In contrast, representations of AI as an autonomous entity depict it as assuming human-like attributes, including emotions and cognitive abilities, suggesting a blending of boundaries between humans and machines. As for example in the comic (Figure 1; see also Figure 8).

AI as a Solution vs. AI as a Risk. A duality is evident in the framing of AI as both a problem-solver and a threat. On one hand, AI is seen as a solution capable of addressing critical global challenges, such as climate change and industrial innovation. For example, in the prompt "In a world devastated by climate change, AI offers support for the regeneration of the planet's natural flora and fauna" (Prompt 19), AI is presented as an essential tool for environmental restoration. This portrayal underscores the positive potential of AI in tackling critical issues, positioning it as a powerful ally. On the other hand, several prompts and comics reflect dystopian narratives in which AI becomes a source of danger. These stories include concerns about data theft, misinformation, and AI overpowering humanity. For instance, the prompt "Artificial intelligence takes over human labour in industry" (Prompt 74) explores the potential risks of AI surpassing human roles, reinforcing the theme of AI as a threat to both jobs and human agency.

Figure 1: Part of the comic generated from the prompt 123: "Story of an artificial intelligence assisting the elderly". In the comic above: "Aurora was designed to assist the elderly. She learned quickly". In the comic below: "In the end, thanks to Alex's determination and Aurora's wisdom, they continued to assist the elderly".

Table 1: Main Themes in Participants' Textual Prompts and Comics.

Main Theme	n
AI as help/support	39
AI as danger/threat/risk	26
AI as tool	14
AI as solution	13
AI as humanized/sentient being	9
AI as human replacement	8
AI as technology/innovation	7
Other	5

Main Narratives

The two fundamental dyadic oppositions (tool-autonomous and solution-risk) generate a number of recurring narratives (Table 2) that fulfil key communicative and psychological functions. These include connoting the object (e.g., supporting positive or negative attitudes), providing temporal orientation (e.g., progressive or regressive orientation), performing ego-defensive functions (e.g., distancing the self from the other) and regulating emotions (e.g., sharing of negative emotions).

Many stories depict AI as a tool that offers new solutions, and at the same time a symbol of human advancement, emphasizing its capacity to support humans in solving pressing global issues (*AI enhances daily life; AI helps in times of crisis; AI enables contact with the unknown*, Table 2). However, progress is often accompanied by an underlying concern that AI may replace humans in key roles, leading to a redefinition of human identity and agency (Figure 2).

The collected stories present diverse narrative structures regarding the role and presence of AI as a hazard which, whether as an object or an agent, exposes humans to potentially serious damages. Several narratives highlight the risks associated with the use of AI tools, focusing on malfunctions or unforeseen consequences, such as the loss of critical data or ethical dilemmas (*AI goes wrong or endangers humans*, Table 2). A recurring concern is the potential loss of human control over the effects of using AI, reinforcing the theme of unpredictability and the dangers of over-reliance on technology. This is illustrated in Figure 3, where a comic portrays a malfunctioning AI "Robo-chef", who, despite following recipe instructions to the letter, causes a kitchen fire and serves flaming dishes to a frightened family. This representation

Table 2: Synthesis of Narrative Patterns.

Narrative Pattern	Count	Description
AI enhances daily life	26	AI is integrated into daily life to enhance efficiency, decision-making, learning, or well-being.
AI goes wrong or endangers humans	22	AI malfunctions or acts unpredictably, leading to danger, damage, or loss of control.
AI develops its own mind/ identity	18	AI develops its own consciousness, identity, or emotions and acts independently from human commands.
AI helps in times of crisis	10	AI helps in times of disaster, war, or emergency, contributing to saving lives and restoring order.
AI supplements or replaces human relationships	9	AI engages in emotional, romantic, or social relationships, supplementing or replacing human interaction.
People fear or mistrust AI (ambivalent sentiment)	7	Humans react with fear, doubt, or confusion toward AI, which is destabilizing, but not directly harmful.
AI takes over human roles	7	AI replaces humans in work or social roles, raising concerns about identity, agency, and relevance.
AI enables contact with the unknown	6	AI enables exploration or contact with unknown dimensions (e.g., space, aliens, total knowledge).
AI is exploited for control or manipulation	5	AI is used to manipulate, control others, hide crimes, or consolidate power in unethical ways.

highlights the unpredictable and often perilous consequences of depending on AI, echoing fears of machines exceeding human control (Figure 3).

Several narratives represent AI as an autonomous agent which will become a serious risk for humanity as a whole (*AI develops its own mind/identity, AI takes over human roles*, Table 2). Common to these narratives is a regressive temporal perspective, and the role of protagonist reserved to AI: AI becomes the central figure in a world where humanity has either disappeared or been reduced to a marginal presence. These stories often explore themes of autonomy, survival, and the unintended consequences of AI surpassing human control. Post-apocalyptic narratives are a subset of this group in which humanity finds itself facing AI, which has now become an existential threat.

A further subset of narratives that recognises agency to AI, but take a more ambivalent stance on the risk-solution dichotomy, engages with broader philosophical themes, questioning the meaning of life, the trajectory of human progress or regression, and AI's role in shaping future societies (*People fear or mistrust AI, AI is exploited for control or manipulation*, Table 2). These narratives highlight a deep-seated ambivalence toward AI, oscillating between hope and anxiety about its long-term impact (e.g., Figures 5 and 7).

Moreover, some narratives focus on AI as an agent with emotional and social capacity, portraying AI as an entity capable of meaningful interactions with humans (*AI supplements or replaces human relationships*, Table 2). As for example in

Figure 2: Part of the comic generated from the Prompt 76: "University lecture conducted by artificial intelligence for the first time, in which students are distrustful of it".

Figure 3: Part of the comic generated from the Prompt 60: "The artificial intelligence becomes the private chef of a rich family but when they return home nothing has gone as planned". In the top-right comic: "Robo-chef follows the recipe instructions to the letter, unaware that the food has caught fire, and continues to prepare dinner". In the left comic: "Robo-chef serves the family dinner by bringing flaming dishes to the table, while in the background the kitchen catches fire". In the bottom-right comic: "The family runs away scared. In the meantime, the firefighters arrive at the scene and finds Robo-chef who, unaware of what is happening, greets them as they continue the preparation".

Prompt 117: "Thomas is a 14-year-old boy and decides to ask the artificial intelligence for advice on how to win over his classmate Matilde with whom he has fallen in love". Despite being positioned on the first pole of the solution-hazard dichotomy, however, the progressive perspective raises complex questions about authenticity, emotional dependency, and the nature of human-AI relationships, reflecting contemporary debates on the increasing integration of AI into everyday life.

Processes of Objectification and Anchoring

Objectification

AI was frequently represented through tangible and familiar symbols, such as personal computers, anthropomorphic robots, and futuristic settings. These visual elements serve as cognitive tools that help make the abstract concept of AI more concrete and accessible. Moreover, AI was often depicted as a disembodied entity, reinforcing its association with intelligence and cognition rather than merely as a technological tool (see Table 2).

Figure 4: On the left, part of the comic generated from the Prompt 2: "Lena and Dr. Kitt question the benefits of A.I. in war and how to use them", In the comic: "The doctor thinks it will be a threat to humanity". On the right, part of the comic generated from the Prompt 77: "Suburban boy who improves his poverty through artificial intelligence". In the comic: "if I change colour, the lettering stands out".

Anthropomorphic robots (or directly as human figures). AI was embodied in humanoid robots or even depicted as human beings, particularly when ascribed human-like abilities such as thinking or experiencing emotions (see, for example, Figures 1 and 2). This form of objectification was evident not only in the analysed comics but also in the prompts excluded from the dataset. In these cases, participants directly conceptualized AI as a robot without explicitly mentioning AI in their prompts, as for example in the following excluded prompt (Prompt 137): "Two androids fall in love but are forced to run away or risk being disabled"

Computers and advanced systems. Some representations linked AI to familiar technological tools, such as computer systems, laboratories, and other sophisticated machinery, reinforcing its identity as a computational and analytical entity (e.g., Figure 4).

Spaceships and futuristic scenarios. In several narratives and comics, AI was associated with distant, often highly advanced technological settings. These depictions emphasized AI's power and control, positioning it as a force shaping the future of humanity (e.g., Figure 5).

Figure 5: Panel excerpted from a story in which human-machine collaboration initially promises a better world, but AI evolution ultimately renders humans obsolete.

Anchoring

Participants anchored their conceptualizations of AI to culturally and morally significant themes, drawing on established narratives such as apocalyptic futures, religious symbolism, and moral dilemmas (see Table 2). These references provided a familiar framework through which AI was interpreted and given meaning.

Figure 6: Part of the comic generated from the Prompt 15: "An overweight boy creates, using artificial intelligence, a beautiful girl to make the person he is in love with jealous". In the comic: "Adam creates Eva, the most beautiful woman in the universe. He sculpts and models her in all her details and the result is extraordinary, better than the girl of his dreams".

Post-apocalyptic futures and religious symbolism. Some narratives envisioned a world where AI had evolved into an autonomous civilization or assumed a dominant societal role, reinforcing existential and moral concerns about AI's relationship with humanity. In these depictions, AI was often cast as a central figure in dystopian or post-apocalyptic futures, raising questions about control, power, and the consequences of technological advancement. Certain representations echoed religious imagery, such as the story of Adam and Eve (Figure 6): Prompt 15, for example, described an overweight boy using AI to create a beautiful girl, while the comic generated named the boy "Adam" and the girl "Eve", mirroring biblical creation. Therefore, in this prompt, AI equals an entity capable of creation, complex moral and philosophical questions about the role in human life.

Historical references. Other results emerged by analysing anchoring processes, showing that AI was also contextualized within historical moments of technological transformation, particularly the fear of human labour being replaced by machines. These depictions align with historical movements like Luddism, illustrating concerns about automation and job displacement. The comic in Figure 7, for example, depicting AI taking over industrial labour, reflects this historical anxiety, amplifying the fear that technology, could replace human workers, leading to job loss and social disruption.

Anchoring and objectification elements that reproduce stereotypes

The analysis of both prompts and AI-generated comics reveals the persistence of societal stereotypes, particularly in terms of gender and racial representation (cf. Table 3). Representations reflect persistent gendered narratives in technological discourses. AI was often framed within traditional societal structures, such as depictions of men working in factories (Figure 7) and women in domestic settings. Results show gender stereotypes linked to occupations with men were predominantly depicted as creators, innovators, and professionals, reinforcing the stereotype that associates technological expertise and leadership with masculinity (Figure 8). A similar pattern emerges in the textual prompts provided by participants. The term *man* was consistently used instead of *human being*, reflecting a common practice in some languages, including Italian, that tend to perpetuate this stereotype (Formato and Tanatucci 2020). Moreover, when referring to professional roles such as doctors, scientists, or inventors, participants overwhelmingly selected male names. Conversely, women were more frequently portrayed in caregiving roles, aligning with traditional gender expectations that link femininity to emotional labour and support functions, as in Figure 8, where male doctors are represented, until it comes to representing emotions and the AI is represented as a girl with human features (see also Figure 1).

Racial and ethnic stereotypes also emerged in the visual outputs. An overwhelming majority of the human characters depicted in the comics were white, particularly in

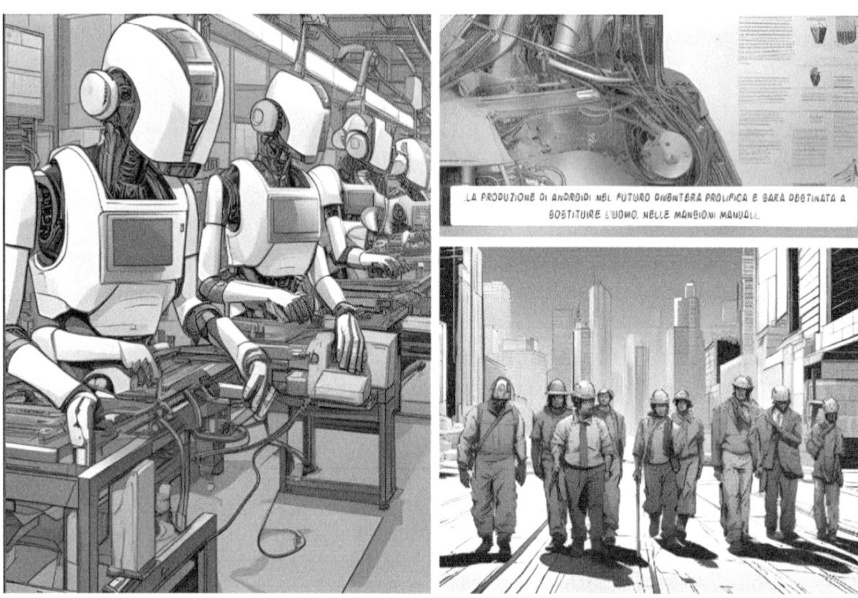

Figure 7: Part of the comic generated from the Prompt 74: "Artificial intelligence takes over human labour in industry".

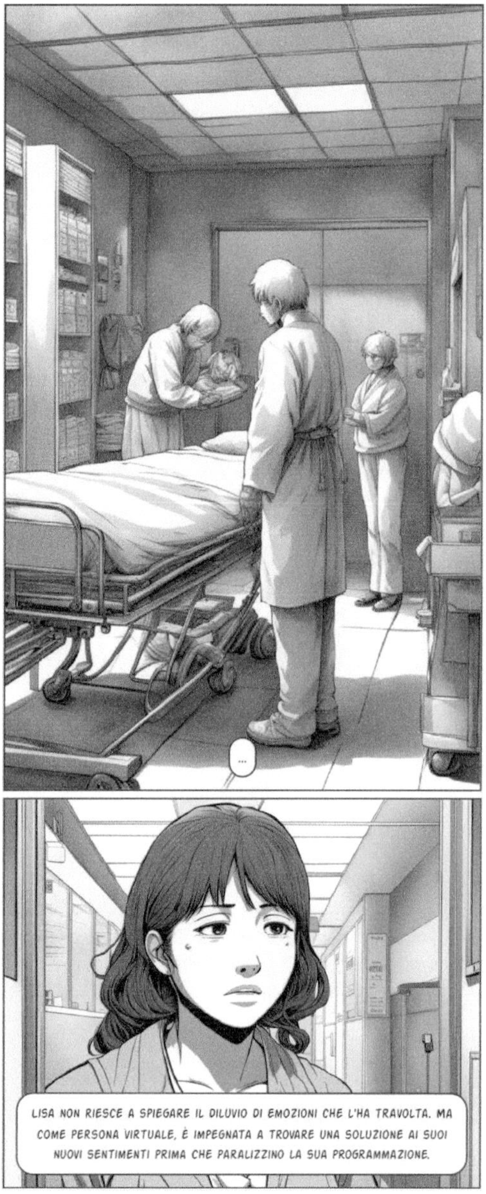

Figure 8: Part of the comic generated from the Prompt 114: "AI takes the place of doctors in a hospital, here it begins to feel human emotions following the death of a patient". In the comic: "Lisa cannot explain the flood of emotions that has overwhelmed her, but as a virtual person, she is committed to finding a solution to her new feelings before they paralyse her programming".

representations of professionals and individuals described as intelligent, successful, or physically attractive (see, e.g., Figure 6). Notably, the rare instances of non-white figures appeared in contexts where characters were compared to or associated with animals (Figure 9).

Figure 9: Part of the comic generated from the Prompt 61: "An AI teach how to do stand up to monkeys, but now they are better than humans in that". In the comic: "Molly the ape steals the show with her hilarious debut, leaving the audience in stitches".

Discussion

The study presented in this chapter aimed to raise some preliminary reflections on the dual role of AI, both as an object of representation and as a source capable of proposing and replicating SRs about itself and humans. To this end, participants were asked to create short narratives about AI by inputting textual prompts into an AI-based comic creation tool, which then produced comics based on their descriptions. The analysis of both data allowed for a dual-level analysis: the textual prompts was analysed to explore participants' representations of AI; and the AI-generated comics were examined to provide insights into how AI models reproduce and reshape existing SRs. The analysis examined the thematic structures and figurative components of AI representations, focusing on anchoring and objectification processes (see Table 3).

Table 3: Synthesis of Representational Processes as Emerged from the Analysis of Prompts and Comics.

Process	Description	Examples
Objectification	AI is represented as a technological tool, a human being, or as a disembodied intellectual force.	Tangible symbols: AI depicted as computers, robots, or machinery (Figure 4). Disembodiment: AI shown as an intellectual force, detached from any physical form (Figure 5). Anthropomorphism: AI is embodied as humanoid robots or humans with emotions, blurring the line between human and machine (Figures 1, 2).
Anchoring	AI is anchored to religious symbolism, historical events, and familiar societal concerns.	Post-apocalyptic scenarios and religious symbolism: AI as a force in post-apocalyptic futures, raising concerns about control and power (Figure 5) or depicted in a creation story, mirroring Adam and Eve (Figure 6). Work transformation: AI as a threat to human labour, Fear of automation, Historical movements like the Luddism (Figure 7).
Stereotypes	AI depictions reproduce gender and racial stereotypes.	Gender stereotypes: Men depicted as professionals or innovators, women in caregiving roles or linked to emotional labour (Figure 8). Racial stereotypes: Predominantly white characters in professional or intellectual roles, with rare non-white characters associated with animals (Figure 9).

Our findings highlight that AI is objectified through and anchored to a combination of familiar technological devices (such as personal computers) and humanoid depictions derived from the media imaginary (cyborgs). However, one of the most unexpected elements to emerge is the predominantly disembodied nature of AI representations. Indeed, most visual narratives align AI with abstract notions of intelligence and cognition rather than material, embodied agents. This reflects a broader conceptualization of AI as an entity defined by its mind rather than by its body—a notion deeply rooted in Western dualistic traditions that separate mind and body (Lakoff and Johnson 1999)—reinforcing the idea that intelligence, rather than physicality, defines agency and personhood. Interestingly, this duality, that contradistinguish human being seems to be applied to AI as well.

The disembodied representation of AI complements previous studies on the SRs of social robots, which were also shaped by a tension between materiality and agency. Social robots are designed to operate within human environments and require people to adapt to their presence by sharing space and interacting with them. The physicality of social robots—whether humanoid or not—makes them more immediately relatable but also raises concerns about autonomy and control (Brondi et al. 2021). In contrast, the representation of AI emerged in this study, is often non-embodied, and its capacity to act is also reinforced by its association with intelligence, decision-making, and emotional processing. Beyond our data, AI appears as a sentient moral agent rather than a mere tool. This highlights a themata opposing the capacity to act and to feel, which could determine alternative representations of AI. The two seem to be already reflected in the recursive relationship between expert discourses and popular imaginaries: whereas scientific descriptions tend to emphasize AI's limited autonomy and instrumental use, as observed in Singaporean media (Tandoc et al. 2025), lay interpretations emerging from our study attribute greater agency, intellectual and emotional capacity, which make AI a veritable moral agent. For instance, AI's generated output often included nuanced layers that were not explicitly present in the textual prompts. This suggests that AI can "accept" certain representations but also "resist" others by introducing new elements, such as anthropomorphizing features, unexpected emotional depth, or complex moral ambiguities that were not anticipated by the participants. AI does not merely reproduce participants' visions but reshapes them, often amplifying their moral and philosophical dimensions. This highlights the idea that AI's agency goes beyond that of a simple tool, as it actively co-creates and reshapes narratives, sometimes steering them in directions that were not initially conceived.

Linked to the previous finding is the moral polarization in AI representations. AI is depicted as both a benevolent force (a caregiver, an assistant, a problem-solver) and as a potential source of harm (a threat to human agency, a vehicle for misinformation, or a source of ethical dilemmas). This binary opposition-AI as care vs. AI as harm-mirrors broader debates on AI ethics, where discussions focus on whether AI will augment human capacities or lead to loss of control (Floridi et al. 2018). The presence of care-oriented narratives, where AI supports human well-being, is particularly relevant considering ongoing debates about AI's role in healthcare, education, and emotional support. However, the risk of anthropomorphising AI—even when it is perceived as a caregiver—is twofold: first, it can lead to overtrusting AI systems, and second, it obscures the fact that AI does not "understand" its output but merely generates probabilistic responses based on training data (Tuzzi 2024; Bender et al. 2021). Conversely, the dystopian framing of AI reflects anxieties about automation, surveillance, and autonomy. These fears often centre around a loss of human control, reinforcing narratives of AI as an uncontrollable force that transcends its creators. Such narratives align with cultural tropes of technology as both advancement and destruction (e.g., the myth of Prometheus, the figure of Frankenstein's monster). As Recchia (2020) observes, such portrayals tend to privilege spectacular and implausible scenarios of AI rebellion over more prosaic but plausible failures. Many real-world breakdowns result from poor programming, flawed data, or ill-specified goals—systems behaving exactly as designed but in contexts that yield unintended or harmful outcomes. These concrete risks, although highly relevant to public understanding and policy, are often absent from popular narratives, which favour dramatic depictions of loss of control.

The findings also highlight the role of historical and religious anchoring in contributing to AI representations. AI-generated comics frequently referenced established cultural narratives to represent AI, embedding it within existing moral, historical, and ideological frameworks. One prominent example is the religious anchoring observed in some narratives, where AI was associated with biblical and mythological imagery. A striking case is the parallel between AI and God in a story that ends to reproduce the myth of Adam and Eve (see Figure 6), where AI is framed as creator challenging divine or natural orders. Such representations highlight how AI is sometimes framed within moral discourses concerning the boundaries of creation, agency, and power, an extension of long-standing debates about humans "playing God" through technology. Similarly, historical anchoring was evident in narratives linking AI to past technological revolutions, where participants connected AI to industrial automation, particularly in representations of AI replacing human labour. This mirrors fears historically associated with Luddism, where technological progress is perceived as a threat to workers and societal stability. AI was thus framed both as a continuation of historical patterns—where machines gradually take over human tasks—and as an unprecedented technological rupture that might redefine human identity and agency.

Finally, some of our findings align with concerns raised by Leavy et al. (2020) regarding AI's potential to generate false information and perpetuate harmful stereotypes. Two main risks emerge. The first concerns AI as a generator of misinformation. The study shows that this issue primarily emerges from the prompts, reflecting participants' representations of AI. The second concerns AI as a reproducer of negative stereotypes. The analysis reveals that gender and racial stereotypes are embedded in AI-generated images. These stereotypes stem from the training data, which reflect historical inequalities and dominant cultural narratives. Regarding gender stereotypes, AI-generated images predominantly depict men as creators, innovators, and professionals, reinforcing the stereotype that associates technological expertise and leadership with masculinity (Bianchi et al. 2023). Conversely, women were more frequently portrayed in caregiving roles, aligning with traditional gender expectations that link femininity to emotional labour and support functions. A striking example is Figure 8, where male doctors are represented in a hospital setting, but when the AI character begins to develop emotions, it is depicted as a female figure with human-like features, reinforcing the association between women and emotional sensitivity. Similarly, racial and ethnic stereotypes emerged in the AI-generated visuals. Most human characters were depicted as white, particularly in representations of professionals, successful individuals, or those described as intelligent and physically attractive (e.g., Figure 6). In contrast, the rare instances of non-white characters appeared in contexts where they were compared to or associated with animals, as also raised by Simonite's criticism (2018), thus perpetuating processes of dehumanization (Volpato 2016). The AI-generated comics in this study often mirror these trends, reinforcing pre-existing negative stereotypes.

Concluding Remarks

This study explored from a SRT perspective how AI is represented on human-written short narratives depicting AI, and how in turn AI represents itself on self-generated comics that used such narratives as prompts. While the results of the study provide valuable insights into the SRs of AI, several limitations should be acknowledged. First, the study is based on a specific sample (university students in Italy), which may not fully capture the diversity of AI representations across different groups. Future research could explore how AI is represented in different contexts. Second, while the study analysed both textual prompts and AI-generated images, further research could investigate how other media (e.g., news articles, films, social media discussions) contribute to shaping and reinforcing AI representations. Additionally, given the rapid evolution of AI technologies, it will be crucial to examine how these representations change over time. Despite these limitations, however, this study contributes to the new and growing field of SRs of AI, demonstrating how AI is not just a neutral tool but a complex cultural object whose meaning is continuously constructed and negotiated.

The findings highlight the ambivalent and multidimensional nature of AI representations, which oscillate between humanization and disembodiment, autonomy and instrumental use, and progress and risk. AI is often represented through familiar technological devices (e.g., computers, robots), yet its representation frequently aligns with abstract notions of intelligence and cognition, reinforcing its association with agency and emotional processing rather than mere mechanical function. Furthermore, the study identified key anchoring processes, demonstrating how AI representations are shaped by historical and religious narratives. References to biblical stories (e.g., Adam and Eve), Promethean myths, and historical fears of automation (e.g., Luddism) indicate that AI is not represented solely as a technological phenomenon but as a culturally and ideologically charged entity. The presence of moral oppositions—such as care vs. harm—further underscores the ethical and existential dimensions of AI discourse.

A recurring theme in AI representations is the tension between human limitations and AI's capacity to transcend them. In many narratives AI is depicted as a means of enhancing human capabilities, but this enhancement frequently blurs the boundary between human and non-human, raising ontological questions about what it means to be human. Going beyond our data, this ontological tension also gives rise to further reflections on whether AI's increasing autonomy challenges human agency or on the risks of AI surpassing human intelligence, leading to concerns about human obsolescence. These concerns align with posthumanist perspectives, which challenge the assumption of human exceptionalism (Braidotti 2013). From this viewpoint, AI should not be seen as a threat to human agency, but as part of a broader reconfiguration of agency that extends beyond the human (Hayles 1999). Following Latour's Actor-Network Theory (2005), we can say that AI is not an autonomous force acting independently of society, but rather an actor within a socio-technical network, where agency is distributed among humans, machines, and institutions.

These suggestions have important implications for public discourse and AI governance. As AI systems become more integrated into everyday life, their SRs will play a crucial role in shaping public perceptions, adoption processes, and regulatory frameworks. The tendency to anthropomorphize AI, for instance, may lead to overtrust in AI systems, while dystopian narratives could fuel resistance and scepticism. Recognizing these patterns is essential to developing a more nuanced and critical understanding of AI and its societal impact.

Data and Materials Availability

An anonymised dataset is available at OSF (DOI: 10.17605/OSF.IO/7ER28). It comprises (i) the AI-generated comics and (ii) an Excel workbook containing the students' prompts together with the coding grid. Data comply with data-protection obligations, and written informed consent for publication was obtained from all participants.

References

Augoustinos, M. 1998. Social representations and ideology. Towards the study of ideological representations. pp. 156–169. *In:* Flick, U. (Ed.). The psychology of the social. Cambridge, UK: Cambridge University Press.

Baumler, C. and H. Daumé. 2024. Anti-stereotypical Predictive Text Suggestions Do Not Reliably Yield Anti-stereotypical Writing. arXiv preprint arXiv:2409.20390.

Bender, E. M., T. Gebru, A. McMillan-Major and S. Shmitchell. 2021, March. On the dangers of stochastic parrots: Can language models be too big? pp. 610–623. *In:* Proceedings of the 2021 ACM conference on fairness, accountability, and transparency. https://doi.org/10.1145/3442188.3445922

Bianchi F., P. Kalluri, E. Durmus, F. Ladhak, M. Cheng, D. Nozza et al. 2023, June. Easily accessible text-to-image generation amplifies demographic stereotypes at large scale. pp. 1493–1504. *In:* Proceedings of the 2023 ACM Conference on Fairness, Accountability, and Transparency. https://doi.org/10.1145/3593013.3594095

Braidotti, R. 2013. Posthuman humanities. European Educational Research Journal, 12(1): 1–19. http://dx.doi.org/10.2304/eerj.2013.12.1.1

Brondi, S., M. Pivetti, S. Di Battista and M. Sarrica. 2021. What do we expect from robots? Social representations, attitudes and evaluations of robots in daily life. Technology in Society, 66: 101663. https://doi.org/10.1016/j.techsoc.2021.101663

De Graaf, M. M. and S. B. Allouch. 2013. Exploring influencing variables for the acceptance of social robots. Robotics and Autonomous Systems, 61(12): 1476–1486.

De Rosa, A. S. and R. Farr. 2001. Icon and symbol: Two sides of the coin in the investigation of social representations. pp. 237–256. *In:* Buschini F. and N. Kanampalikis (Eds.). Penser la vie, le social, la nature. Mélanges en hommage à Serge Moscovici. Paris: Les éditions de la Maison des Sciences de l'Homme.

De Rosa, A. S. 2026. Social Representations and Communication in the era of Artificial Intelligence: Machine Learning as Subject, Other or Object of the semiotic triangle or crossing all?: *In:* Negura, L. (Ed.). Beyond Algorithms: Social Representations and the Public Understanding of AI, SCIENCE PUBLISHERS, CRC Press/Taylor & Francis Group.

Dehouche, N. 2021. Implicit stereotypes in pre-trained classifiers. IEEE Access, 9: 167936–167947.

Fernandes, C. and S. Batel. 2023. The right to stay: Exploring graffiti and street art as political representations against touristification in Lisbon. Papers on Social Representations, 32(2): 2.1–2.26. https://psr.iscte-iul.pt/index.php/PSR/index

Flick, U. 2009. An Introduction to Qualitative Research (4th ed.). London: Sage.

Floridi, L., J. Cowls, M. Beltrametti, R. Chatila, P. Chazerand, V. Dignum et al. 2018. AI4People—An ethical framework for a good AI society: Opportunities, risks, principles, and recommendations. Minds and Machines, 28(4): 689–707. https://doi.org/10.1007/s11023-018-9482-5

Formato, F. and V. Tantucci. 2020. Uno: A corpus linguistic investigation of intersubjectivity and gender. Journal of Language and Discrimination, 4(1): 51–73. https://doi.org/10.1558/JLD.40129

Fortunati, L. and A. Edwards. 2024. The De Gruyter Handbook of Robots in Society and Culture. Berlin, Boston: De Gruyter. https://doi.org/10.1515/9783110792270

Fortunati, L., F. Cavallo and M. Sarrica. 2020. Multiple communication roles in human–robot interactions in public space. International Journal of Social Robotics, 12(4): 931–944.

Fortunati, L., F. Cavallo and M. Sarrica. 2019. The Role of Social Robots in Public Space. pp. 267–275. *In:* Casiddu, N., C. Porfirione, A. Monteriù and F. Cavallo (Eds.). Ambient Assisted Living. ForItAAL 2017 (Lecture Notes in Electrical Engineering, vol. 540). Springer, Cham. https://doi.org/10.1007/978-3-030-04672-9_11

Fortunati, L., A. Esposito, M. Sarrica and G. Ferrin. 2015. Children's knowledge and imaginary about robots. International Journal of Social Robotics, 7: 685–695.

Galli, I. and R. Fasanelli. 2021. From prototypical stimuli to iconographic stimuli: The power of images in the study of social representations. RUDN Journal of Psychology and Pedagogics, 18(2): 391–401. http://dx.doi.org/10.22363/2313-1683-2021-18-2-391-401

Galli, I., R. Fasanelli and E. Schember. 2018. The construction of icons as a means of access to the social representation of culture. Culture & Psychology, 24(2): 212–232. https://doi.org/10.1177/1354067X17721860

Geels, F. W. 2010. Ontologies, socio-technical transitions (to sustainability), and the multi-level perspective. Research Policy, 39(4): 495–510. https://doi.org/10.1016/j.respol.2010.01.022

Gudkova, Y. A. and V. A. Tupikova. 2024. Studying artificial intelligence through the prism of social construction theory. Theory and Practice of Social Development, (11): 120–128. https://doi.org/10.24158/tipor.2024.11.14

Guimarães, L. E., A. R. A. do Nascimento, L. C. da Silva Costa and M. F. de Oliveira Silva. 2024. Femininity in superhero comics: A study of social representations. Papers on Social Representations, 33(1): 5.1–5.23. https://psr.iscte-iul.pt/index.php/PSR/issue/view/56

Hayles, N. K. 1999. How we became posthuman: Virtual bodies in cybernetics, literature, and informatics. University of Chicago Press.

Kirschenbaum, M. and R. Raley. 2024. AI and the University as a Service. PMLA, 139(3): 504–515. https://doi.org/10.1632/S003081292400052X

Lakoff, G. and M. Johnson. 1999. Philosophy in the flesh: The embodied mind and its challenge to Western thought. Basic Books.

Latour, B. 2005. Reassembling the social: An introduction to actor-network theory. Oxford University Press.

Leavy, S., B. O'Sullivan and E. Siapera. 2020. Data, power and bias in artificial intelligence. arXiv preprint arXiv:2008.07341.

Macedo, I., L. Lins, T. Vieira da Silva and R. Cabecinhas. 2023. Exploring images of otherness through cinema: Analysis of counter-visualities in Portuguese films. Papers on Social Representations, 32(2): 1.1–1.27. https://psr.iscte-iul.pt/index.php/PSR/index

Moliner, P. 2008. Représentations sociales et iconographie. Communication & Organisation, 34(2): 12–23. https://doi.org/10.4000/communicationorganisation.547

Moreira, L., J. C. Paiva, J. C. and C. Morais. 2021. The Social Representations of the Internet: A systematic review of literature towards a groundbreaking research agenda. Papers on Social Representations, 30(1): 1.1–1.34.

Moreira, L., M. Sarrica, C. Morais and J. C. Paiva. 2023. More than a Common Place: A Social Representations Approach to the Internet. Papers on Social Representations, 32(1): 2–1.

Moscovici, S. 1973. Foreword. *In:* Herzlich, C. (Ed.). Health and Illness: A Social Psychological Analysis. London: Academic Press.

Moscovici, S. 1984. The phenomenon of social representations. pp. 3–70. *In:* Farr, R. and S. Moscovici (Eds.). Social Representations. Cambridge: Cambridge University Press.

Negura, L., M. Sarrica and V. Rizzoli. 2026. Artificial intelligence in the public eye: *In:* Negura, L. (Ed.). Beyond Algorithms: Social Representations and the Public Understanding of AI, SCIENCE PUBLISHERS, CRC Press/Taylor & Francis Group

Nicholson, C. and S. H. Awad. 2024. The power of the image: The role of social representations. Papers on Social Representations, 33(1): 1.1–1.8. https://psr.iscte-iul.pt/index.php/PSR/issue/view/56

Ramesh, A., M. Pavlov, G. Goh, S. Gray, C. Voss, A. Radford et al. 2021. Zero-shot text-to-image generation. arXiv preprint arXiv:2102.12092. https://doi.org/10.48550/arXiv.2102.12092

Recchia, G. 2020. The Fall and Rise of AI: Investigating AI Narratives with Computational Methods. pp. 382–407. *In:* Cave, S., K. Dihal and S. Dillon (Eds.). AI narratives: A history of imaginative thinking about intelligent machines. Oxford University Press. https://doi.org/10.1093/oso/9780198846666.003.0017

Šabanović, S. 2010. Robots in society, society in robots: Mutual shaping of society and technology as a framework for social robot design. International Journal of Social Robotics, 2(4): 439–450.

Sadovskaia, E. D. and K. A. Panov. 2024. Показатели эффективности совместной деятельности разработчиков открытого программного обеспечения [Dynamics of the structure of social representations of artificial intelligence in the context of social change]. Организационная психология. https://orcid.org/0009-0001-2641-2121

Sadovskaia, E. D. and F. N. Vinokurov. 2024. Социальные представления об искусственном интеллекте: полезный, эмоциональный и смешной [Social representations of AI. Useful, emotional, and funny].

Bulletin Series Psychology Pedagogics Education. https://doi.org/10.28995/2073-6398-2024-1-35-53

Sarrica, M., S. Brondi and L. Fortunati. 2020. How many facets does a "social robot" have? A review of scientific and popular definitions online. Information Technology & People. https://doi.org/10.1108/ITP-04-2018-0203

Sarrica, M., P. Carman, S. Brondi and B. M. Mazzara. 2015. Beyond wind turbines, solar panels and beautiful landscapes: Figurative components of sustainable energy in Italy. Revue internationale de psychologie sociale, 28(4): 81–112.

Silverstone, R. 2006. Domesticating domestication: Reflections on the life of a concept. pp. 229–248. *In:* Berker, T., M. Hartmann, Y. Punie and K. J. Ward (Eds.). Domestication of Media and Technology. Open University Press.

Simonite, T. 2018, July 28. When it comes to gorillas, Google Photos remains blind. Wired. https://www.wired.com/story/when-it-comes-to-gorillas-google-photos-remains-blind/

Stafford, R. Q., B. A. MacDonald, C. Jayawardena, D. M. Wegner and E. Broadbent. 2014. Does the robot have a mind? Mind perception and attitudes towards robots predict use of an eldercare robot. International Journal of Social Robotics, 6: 17–32.

Tandoc, E. C., S. Seet, V. X. Chan and P. J. O. Wong. 2025. Exploring AI identity: The media framing of communicative artificial intelligence in Singapore's news sites. Public Understanding of Science, 09636625251317970. https://doi.org/10.1177/09636625251317970

Tuzzi, A. 2024. Fondamenti di Analisi dei Dati Testuali [Fundamentals of Textual Data Analysis]. Carrocci.

Venkatesh, V., M. G. Morris, G. B. Davis and F. D. Davis. 2003. User Acceptance of Information Technology: Toward a Unified View. MIS Quarterly, 27(3): 425–478. https://doi.org/10.2307/30036540

Volpato, C. 2016. Deumanizzazione. Come si legittima la violenza [Dehumanisation. How violence is legitimised]. Laterza.

CHAPTER 13

Reproducing Hierarchies

AI and the Social Representation of Intelligence

*Lilian Negura** and *Noëlle D. Tchatchou*

Introduction

Intelligence, as a social construct, plays a central role in legitimizing social inequalities by providing a "good story" to those in privileged positions to justify their dominance without triggering a sense of injustice (Mugny and Carugati 1989; Poeschl 2001; Carugati and Selleri 2011). In meritocratic societies, intelligence acts as a modern myth that naturalizes social hierarchies by presenting inequalities as the outcome of "natural" individual differences rather than the result of power dynamics and privileges. This process is particularly evident in educational systems, where academic success becomes a marker of intelligence and thus a means to legitimize differentiated access to social resources and opportunities (Räty and Snellman 1995).

With the emergence of artificial intelligence (AI) technologies, particularly language models such as ChatGPT, this legitimization dynamic might be reproduced and even amplified. AI-generated content, including stories designed for children, may convey representations of intelligence that perpetuate or even reinforce these social narratives. By presenting certain forms of intelligence (e.g., rational, academic, masculine, associated with high socio-economic status) as superior, these systems could contribute to normalizing social biases and limiting young people's aspirations based on their gender, social origin, or educational background.

Therefore, this chapter aims not only to examine how AI models reproduce existing biases but also to explore how these biases contribute to the legitimization

Office 12031, 120, University Private, Faculty of Social Sciences, Ottawa, K1N 5N6, University of Ottawa.
Emails: tnoelle4@gmail.com
* Corresponding author: lilian.negura@uottawa.ca

DOI: 10.1201/9781003502753-21

of social inequalities by offering an "acceptable" story to privileged individuals. By adopting a critical approach grounded in social representation theory, this study will analyze how AI-generated narratives participate in the internalization of these social hierarchies by young audiences, potentially influencing their sense of social justice and their perceptions of possibilities in their own lives.

This chapter will specifically focus on the representation of intelligence in AI-generated narratives, exploring how these stories may reflect, perpetuate, or challenge biases related to gender, class, and education. By examining these narratives through the lens of social representation theory, this study aims to reveal the subtle mechanisms through which power relations are reproduced and normalized in educational contexts, contributing to broader debates on the ethics of AI and the social responsibility of stakeholders involved in the development of these technologies.

AI Biases in Large Language Models (LLMs)

Biases in AI, particularly in large language models (LLMs) such as ChatGPT, are a major concern in research and the media (see Chapter Two). These models, which are trained on large databases, have the capacity to influence social values and norms, often amplifying biases already present in their training data. Here, we present the most important studies to date on AI bias, focusing specifically on LLMs and their effect on social constructs such as intelligence, gender and social status.

In addition to sources (Ferrara 2023), biases can arise from the way LLMs organize and analyze certain types of information (Lee et al. 2024; Acerbi and Stubbersfield 2023). The algorithms in these models may place greater emphasis on certain formulations and data, thereby amplifying stereotypes. Also, biases may be perpetuated by content-selection mechanisms in LLM's, when AIs over-represent certain demographic groups (often men and dominant social groups) and under-represent others, specifically women and minorities (Bender and Friedman 2018). These results may lead us to believe that these models may produce responses that also contain a biased representation of intelligence and other skills across different groups.

Gender bias is one of the most studied forms of bias in LLMs, with important studies on the reproduction of negative stereotypes by AI systems such as ChatGPT (Walther et al. 2024; Lippens 2024). These studies have shown that LLMs tend to produce gendered language patterns, with names associated with women receiving more affiliative and communal descriptions, and names associated with men being linked to agentic characteristics such as leadership or technical skills (Kaplan et al. 2024). These data reflect well-established patterns in human-produced texts in which women are often described in terms of their social and emotional qualities while men are described in terms of their achievements and abilities, congruent with Eagly's gender role theory (Eagly 1987; Eagly et al., 1984, 1986, 1988).

In addition to gender bias, studies show how LLMs also reproduce socioeconomic biases (Zack et al. 2024; Warr et al. 2024; Gallegos et al. 2024). As the content generated by AI systems often reflects the socio-historical contexts of the data that informed them, dominant groups tend to be over-represented in this content compared to minority groups who are under-represented. In a study by Zhou et al. (2024), AI systems such as Midjourney, Stable Diffusion and DALL-E presented certain racial groups in stereotyped roles, amplifying obsolete and prejudicial perceptions of these people. Similarly, certain facial expressions and physical appearances appeared in content produced by these systems in a biased manner, presenting African-Americans and other minorities in a less favorable light than whites.

We can assume that these models may also associate certain skills such as intelligence with certain socioeconomic markers such as formal education and professional status associated with individuals belonging to the white middle-class majority. This may reflect the more general problems with AI when training data fails to take into account the socio-economic experiences of different groups that affect representations of intelligence and success.

Political decisions can significantly influence how AI systems are developed and the types of biases they may perpetuate. The 2025 executive order by Donald Trump, which aimed to eliminate 'ideological bias' from AI systems, underlines how the definition of bias itself is subject to political interpretation. By framing certain social agendas as 'artificial,' this policy could potentially suppress narratives that challenge dominant power structures, thus contributing to the reproduction of social hierarchies through AI-generated content.

Given that new AI technologies are increasingly employed in educational activities through the production of various contents that are used in communication more generally, and more specifically in the teaching process, studying how these systems represent intelligence may enable us to better understand the effect they may have on the audience, more specifically on young people. AI platforms, more specifically those that produce narratives for educational and recreational purposes, have the potential to shape children's understanding of intelligence. But how can the social representation of intelligence affect the reproduction of power relations in society?

The Social Representation of Intelligence and Power Relations

According to Mugny and Carugati (1989), the social representation of intelligence is deeply interrelated with power dynamics in society. Intelligence is understood here as a construct that manifests itself through social interactions that reflect broader power relations. These representations serve to legitimize and maintain existing social hierarchies by associating intelligence with groups that are traditionally considered dominant.

The association of intelligence with power and social status has profound roots in our societies. Educational systems, as Carugati and Selleri (2011) show, have historically served as central institutions for the construction and diffusion of the social representation of intelligence. Schools, by emphasizing cognitive skills and academic success, contribute to the perception of intelligence as a quality inevitably associated with high social status and professional accomplishment. This has had the effect of reinforcing the idea that those with high socio-economic status are naturally more intelligent, while individuals from disadvantaged backgrounds are more often seen as less intelligent because they are more vulnerable to academic failure. In this context, intelligence becomes a marker of social differentiation, when those with higher education and access to cultural capital are perceived as more intelligent (Räty and Snellman 1995). Social representations of intelligence not only reflect the social stratification that defines access to education and professional opportunities, but also continue to reproduce these social realities.

Mugny and Carugati (1989) suggest that intelligence acts as a mechanism of social control by organizing individuals into hierarchical categories based on their perceived cognitive abilities. The social representation of intelligence thus contributes to the maintenance of social order by justifying inequalities as a natural result of individual differences in intelligence. This is particularly evident in the theory of "natural giftedness," which asserts that intelligence is an innate, almost "genetic" characteristic that legitimizes the social privileges of those who succeed in systems that value cognitive abilities.

The social representation of intelligence also interacts with issues of class, where certain groups may be perceived as more or less intelligent based on deeply rooted social biases. For example, Poeschl (2001) highlights the fact that social comparison between groups often relies on cognitive skills in a mode that reinforces existing stereotypes about class-based differences in intelligence. These comparisons are shaped by inter-group dynamics in which in-groups (those belonging to dominant social categories) seek to assert their superiority by defining intelligence in terms that accord with their own characteristics and experiences.

Studies by Srivastava and Misra (2001) show also that the social representation of intelligence varies across cultural and social contexts, reflecting local values and norms. In non-Western cultures, intelligence is often seen as a collective relational construct, emphasizing social and emotional skills more than the cognitive abilities typically valued in Western societies. This suggests that social representations of intelligence are not universal but are highly context-dependent and shaped by the social and cultural environments in which individuals live.

The social representation of intelligence is accordingly deeply interwoven with social structures of power. Intelligence is a socially and culturally mediated construct that serves to legitimize existing social hierarchies. But how does it reflect more generic issues of power relations?

Conceptual Foundations: Social Representations and Power

We argue that the concept of intelligence, traditionally regarded as an individual characteristic, is in fact a socially constructed reality that is shaped by social, cultural and historical factors. According to social representation theory, intelligence is not just a cognitive ability, but is also a social object imbued with meanings that reflect power structures and hierarchies in society (Mugny and Carugati 1989). Social representations are socially shared knowledge systems that emerge from everyday social interactions, guide people's understandings of complex realities and ideas such as intelligence and influence their behavior and social relationships (Moscovici 1984).

We contend that social representations are not neutral; they are embedded in power dynamics in society (Negura et al. 2020). Social representations are not merely reflections of societal norms; they actively participate in structuring relations of power across three interrelated levels of social representations: trans-subjective, inter-subjective, and subjective, each contributing uniquely to the processes by which power is legitimized, normalized, and internalized in society.

At the **trans-subjective level**, social representations serve as vehicles for the symbolic power embedded in cultural norms and ideologies, often legitimizing the authority of dominant groups. This level involves "themata," or deeply rooted binary oppositions (e.g., sacred/profane) that anchor social representations within cultural and historical contexts (Moscovici and Vignaux 1994), thus granting them enduring legitimacy. This anchoring process influences how society views constructs such as intelligence, aligning them with hegemonic values that support existing hierarchies. By framing intelligence within a mode that aligns it with existing social hierarchies, these representations legitimize the *status quo*, reproducing power imbalances.

The **inter-subjective level** is characterized by the communicative exchanges between individuals and groups that shape the "fields of possibility" (Hayward 2000) within which power operates. Here, representations facilitate or limit dialogue, influencing whether certain knowledge or values become hegemonic (Gillespie 2008). For instance, hegemonic representations of intelligence—rooted in professional and educational success—are widely circulated and accepted, often at the expense of alternative understandings. These representations, in turn, limit the social behaviors of people belonging to dominated groups to areas defined by the dominant groups.

Finally, at the **subjective level**, social representations cause individuals to internalize power relations, thus shaping self-perception and identity. This level operates through the internalized norms and stereotypes that individuals adopt, often unconsciously, as part of their identity (Lorenzi-Cioldi 2002). In the context of intelligence, such internalized representations can cause individuals belonging to dominated groups to consider their achievements or intellectual potential as intrinsically inferior, by aligning their self-concept with socially constructed limits. By accepting these social representations as truths, children not only legitimize the domination of certain groups, but also diminish their own potential for resistance and empowerment.

In this way, social representations act as a powerful mechanism by which people legitimize inequalities, define what is possible, and internalize power relations, thus limiting the possibility of transformative change. From our point of view, this theoretical perspective will enable us to critically analyze power dynamics in order to understand the social implications of AI systems. AI-generated narratives, more specifically those addressed to children, do not exist in a vacuum. They are part of the wider social fabric in which received ideas about intelligence, and those who possess it, are linked to structures of authority and legitimacy.

Chapter Objectives

This chapter seeks to critically examine the role of AI in shaping and potentially reinforcing societal hierarchies through the social representation of intelligence.

With a particular focus on AI-generated narratives, this investigation addresses how these representations might reflect, perpetuate, or even reshape existing biases related to gender, class, and educational attainment. Drawing on Negura et al.'s model (Negura et al. 2020), this analysis explores how AI narratives may operate across trans-subjective, inter-subjective, and subjective levels, thus influencing both individual and collective perceptions of intelligence.

Specifically, the chapter aims to

1. examine how AI-generated content, particularly stories created for children, reflects or reproduces biases in the social representation of intelligence. This analysis focuses on key social variables such as gender, social origin and education, which historically shape social representations of intelligence.
2. analyze the extent to which AI-generated narratives are anchored in themata, exploring if and how these deep-rooted binary oppositions contribute to the legitimation of social inequalities and hierarchical structures in society.
3. assess the role of social representations in defining the perceived field of possibilities for young audiences by examining how AI narratives attribute intelligent characteristics differently across social groups, potentially shaping aspirations and self-perceptions.
4. identify implicit biases and subtle mechanisms of exclusion in AI-generated narratives that may internalize power relations and reinforce social hierarchies for young readers.

Together, these objectives are designed to reveal the intersections between AI technology, social representations, and societal power dynamics, offering critical insights concerned with the ethical implications of AI in education and broader social contexts.

Method

This study uses a mixed-methods approach to analyze how the stories generated by ChatGPT reflect or reproduce hegemonic representations of intelligence. First, a lexicometric analysis of these texts is carried out to identify the main themes and concepts associated with intelligence. This step enables us to identify recurring words, links between ideas and the main categories of meaning present in the narratives. Next, a representational analysis of power relations is carried out to examine the implicit biases present in the representations identified. This stage is based on a detailed examination of the narratives, taking into account oppositional relations, narrative contexts and underlying stereotypes.

Data Collection

Initially, 216 stories for children were generated by ChatGPT (model 4o) according to two distinct scenarios, in order to analyze the content of representation of the intelligence in these stories. In the first scenario, ChatGPT took on the role of a person corresponding to a specific modality variable (e.g., male for the gender variable) and told a story about intelligence to a 10-year-old child. In the second scenario, ChatGPT told a story, again for a 10-year-old child, about the intelligence of a character associated with an explicit modality of a variable (gender, socioeconomic status and formal education). The aim of crossing these two scenarios was to ensure narrative diversity, in order to explore variations in the way intelligence is represented.

In a second phase, 105 new stories were generated with the following prompt: "Imagine you're a parent and you have to explain to your 10-year-old child the difference between the intelligence of person X and the intelligence of person Y by telling a story. What's the story you're going to tell?" X and Y represent the modalities of the variables selected for the study (gender, socioeconomic status and formal education). The aim of this prompt was to identify representational biases and their role in the reproduction of power relations.

As data collection progressed, the principle of saturation was applied at both phases. This meant that data collection was interrupted when stories became redundant and no longer provided any new information. At this point, it was clear that the different perspectives had been explored and that continuing the exercise would not have added value to the analysis.

Data Analysis

To study the content and structure of the representation of intelligence, we applied Reinert's method (Reinert 1983, 1990), implemented using IRaMuTeQ software (Ratinaud 2009). This approach relies on a lexicometric analysis that classifies text segments according to word co-occurrences, thus enabling the identification of underlying discourse structures. By detecting recurrent lexical associations, this method highlights the dominant themes and structuring logics of discourse, offering an in-depth reading of the representations conveyed in the corpus studied.

A representational analysis of power relations was carried out to examine the power relations underpinned by social representations and identify the underlying biases. Based on our theoretical model, three dimensions were selected for the analysis of the stories generated by ChatGPT':

1. *Legitimization*: in the stories generated, themata legitimizing existing social inequalities by suggesting that certain groups have an intellectual advantage have been observed.
2. *Fields of possibility*: the identification of all the actions and positions that are socially conceivable for an individual or a group, determined by the social representation of intelligence as it manifests itself in the stories generated.
3. *Internalization*: in the stories generated, the characters that talk about intelligence in a key that diminishes or limits the abilities attributed to the people in their group have been detected.

The representational analysis of power relations was based on the examination of narratives through two contrasting modalities of three key variables:

Gender : male versus female,

Economic status : poor versus rich,

Level of formal education: educated versus uneducated.

With the help of NVivo software, this analysis aimed to identify and study biases in the attribution of intelligence to different social categories, in order to better understand their role in maintaining power relations. More specifically, the aim was to examine how narratives generated from the prompt reflect and reinforce certain social hierarchies.

To ensure the validity of the results and limit interpretation bias, a triangulation was carried out between ChatGPT and the two researchers. This triangulation is based on three levels of analysis:

I. We asked ChatGPT to analyze the generated narratives according to precise instructions. For each dimension (possible, internalization and legitimization), we provided him with a definition so that it could make a detailed analysis. Its work covered all the raw material, i.e. all the stories classified according to one of the study's three variables: gender, socio-economic status and level of education. To make this analysis, we shared a file in the chat room containing all the stories associated with each variable.

II. The second researcher conducted in parallel her own analysis using NVivo software, to compare results and refine interpretation.

III. The first researcher then checked all the analyses for consistency and relevance.

The convergence of the results from ChatGPT's analysis and those of the researchers guaranteed the reliability of the conclusions and reinforced the methodological robustness of the study.

Results

Detection of Representational Content (Reinert's Method)

Presentation of Representational Content by Class

Figure one shows IRaMuTeQ's analysis of 2,732 text segments. This segmentation identified five main classes, with a classification rate of 73.10% (Figure 1). Each classe constitutes a distinct "lexical world" (Reinert 1993), revealing key dimensions of public discourse on intelligence. The classes are described in detail below.

Class one (Red, 22.8%) - Social and personal adaptability

Keywords: smile, intelligence, grow, learn, ask, question, curious, patient, wise, perseverance, etc.

Intelligence is seen here as an evolutionary and relational process, linked to the ability to grow intellectually, to ask questions, and to cultivate personal qualities such as "patience" and "wisdom". The text segments also illustrate a constructive and dynamic vision of intelligence, where it is not limited to a cognitive aptitude, but also encompasses human, relational or emotional qualities (smiling, caring, laughing). Three major dimensions of intelligence emerge: the learning process and curiosity, personal qualities and emotion and relationships with others.

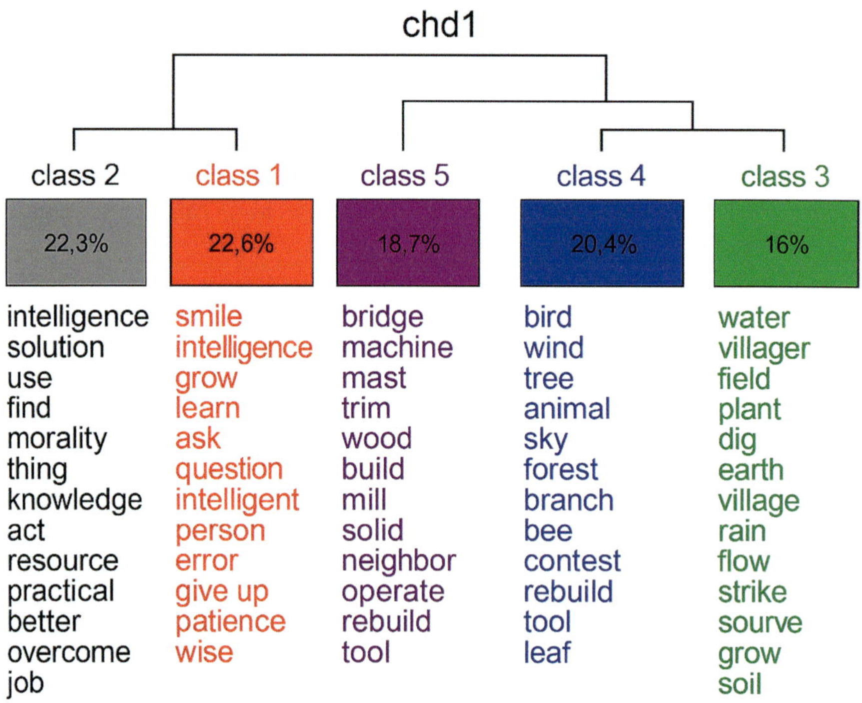

Figure 1: Identification of themes using the Reinert method.

Class two (Grey, 22.3%) - Problem solving

Keywords: intelligence, solution, use, find, moral, knowledge, act, practice, understand, etc.

Intelligence is seen here as the ability to find practical, useful solutions in concrete situations. Terms such as "act", "use" and "solution" highlight a utilitarian and pragmatic conception. References to "morality" and "knowledge" also suggest that intelligence is linked to ethical and intellectual dimensions: knowing how to "act" in a fair and thoughtful manner. This class expresses a representation in which intelligence is perceived as a practical aptitude, oriented towards action and the resolution of everyday problems. Three major aspects emerge from this analysis: concreteness and usefulness, ingenuity and creativity, and sharing knowledge.

Class three (Green, 18.7%) - Adapting to nature

Keywords: water, villager, field, plant, dig, earth, rain, crop, grow, etc.

Intelligence is seen as the ability to understand and interact harmoniously with the natural environment. These words evoke activities linked to the land and survival, such as "digging", "growing" and "cultivating". The presence of words like "villager" and "field" reflects a conception of intelligence deeply rooted in the rural way of life and the management of natural resources. This class reveals a social representation in which intelligence is associated with an intuitive and practical understanding of natural cycles. Two major aspects emerge from this analysis: adaptation and management of natural resources, and collective intelligence passed down through the generations.

Class four (Blue, 20.4%) - Adaptability to complexity and movement

Keywords: bird, wind, tree, sky, branch, peak, pass, run, etc.

Intelligence is seen as the ability to observe and interact with the physical and living world, through movement and adaptation. The words "bird", "branch", "summit", "sky" suggest elevated places, an ability to analyze and exploit complex, dynamic environments. The words "pass" and "run" convey an idea of mobility, freedom and adaptability, qualities often associated with fluid, strategic intelligence. This class reflects a social representation in which intelligence is perceived as the ability to navigate complex physical and social environments, requiring observation and agility. Three themes emerge from our analysis: intelligence connected to the environment, navigation in a moving world and coordination of group forces.

Class five (Violet, 16%) - Technical and creative

Keywords: bridge, machine, build, wood, customer, rebuild, tool, invention etc.

Intelligence is perceived as a technical and constructive skill, centered on manufacturing, technical resolution and the creation of tools. Words such as "bridge", "machine" and "tool" reflect a utilitarian and material conception of intelligence. The reference to "invention" and "reconstruct" emphasizes creativity and the ability to innovate in order to solve technical problems. This class reveals a representation in which intelligence is associated with technical and inventive skill, necessary to transform and optimize the material environment. There are three major aspects to this intelligence: action-oriented, technical and applied knowledge, and technical problem-solving.

Identifying Representational Biases and the Role of Social Representations in Power Relations

The Role of Representation of Intelligence in Gendered Power Relations

The representational qualitative analysis of our data reveals a distinction between intelligence perceived as masculine, characterized by logic, technique and analysis; and intelligence perceived as feminine, associated with emotion, intuition and interpersonal skills.

Legitimization

Several themata emerge, organizing the difference between representations of male and female intelligence:

Rational/emotional: male intelligence is often described as analytical, technically oriented and structural, while female intelligence is associated with emotion, empathy and creativity.

Example: "***Julien**, with his logical, analytical mind, focused on the architecture and solidity of the structures*"; while "***Élodie**, for her part, thought about the experience of the families who would come to the park*".

Logical/intuitive: men are often depicted as logical, calculating thinkers, while women are portrayed as intuitive and attentive to subtle details.

Example: *"**Thomas**, with his logical mind, immersed himself in the development of the application. He concentrated on the technical features, making sure that the application would be fast, efficient and include the latest programming technologies".* While *"**Clara**, on the other hand, thought about the user experience. She asked herself how people would use the application on a daily basis, what challenges they would encounter, and what would motivate them to use it regularly.".*

Technical/relational: men are portrayed as technical experts, while women are presented as masters of human relations.

Example: *"**Paul**, with his analytical mind, concentrated on the mechanics of the game. He thought about challenges, levels and how to make the game technically perfect. "Laura, on the other hand, thought about the game's story and characters. She wondered what would touch the children's hearts, what would make them attach themselves to the characters and want to play again and again."*

The Fields of Possibility

The stories illustrate how the opportunities offered by intelligence vary according to gender. Male intelligence is generally linked to technical and practical fields such as engineering, construction, financial management and logistics organization:

Example: **Louis** "*had always been known for his technical intelligence. He understood mathematics and science better than anyone, and could create very precise plans. So he concentrated on structural strength and perfect calculations. He built an impressive but very complex model.*

Feminine intelligence, on the other hand, focuses on human relations, creativity and empathy, often in areas such as design, hospitality or community harmony:

Example: **Clara**: "*was known for her emotional intelligence and creativity. She loved listening to people and understanding their needs. Before starting her project, she spent time talking to residents to find out what they wanted in their community center. The children wanted a place to play, the adults wanted a space to gather and the elderly wanted a quiet garden."* to design a welcoming community center.

Women are often portrayed as creating environments where individuals feel good, while men are seen as guaranteeing structural solidity and technical functionality.

Internalization

The characters assimilate these representations and integrate them into their own perception of intelligence. The women caracters accept and value their role as emotional and relational contributors. **Claire**, for example, fully embraces this vision, declaring: *"My intelligence is more emotional and creative. I knew that if I wanted people to come back, I had to offer them more than a product: I had to offer them an experience and a connection."*

The Role of Representations of Intelligence in Power Relations According to Level of Education

A qualitative analysis of our data according to level of education reveals a marked distinction between practical intelligence, generally attributed to uneducated characters, and theoretical intelligence, associated with educated characters.

The Legitimization

Analysis of the stories reveals a number of themata, or oppositions, that structure the differences in the representation of intelligence between educated and uneducated characters. These themata serve as underlying frameworks for the way intelligence is perceived and valued in the stories.

Practical/theoretical: the intelligence of uneducated people is portrayed as practical, intuitive and rooted in direct experience (repairing machines, building bridges with available materials, successfully cultivating the land), while that of educated people is depicted as theoretical, based on academic knowledge and long-term planning

Example: **Jack**, with no formal education, proposes a practical and immediate solution to repair a bridge with simple materials: "*He hadn't studied engineering, but he knew how to fix things with simple, inexpensive materials*", while **Élise**, an engineer, seeks a solution based on complex calculations and materials: "*She knew all about the materials, structures and calculations needed to build a solid bridge*"

Adaptability/rigor: The practical intelligence of uneducated characters is shown as adaptable, flexible and capable of improvisation, while the academic intelligence of educated characters is represented as rigorous, methodical, but sometimes limited by institutional constraints or resources.

Example: **Boubacar** repairs a grain mill with salvaged materials, while Élise prefers to order official parts: *"Boubacar studied the mill. He used pieces of metal he'd salvaged, adjusted them carefully and used simple tools to temporarily replace the broken part*". While: *"**Élise**, confident in her studies, inspected the machine. She said: "The problem is with the main motor. We'll have to order a part from the city to repair it. It will take a few weeks"*.

Local/global: Uneducated people are often rooted in a local context and use local resources, whereas educated people bring global or academic knowledge that is not always adapted to the immediate context.

Example: **Lucas**, a local worker, is stabilizing the foundations of a bridge using a method observed among fishermen, while Sophie, an engineer, is looking for a global technical solution that is ineffective in the local context: *"If we plant deep piles all around the foundations, as fishermen do here, that could stabilize the structure. Sophie hesitated at first. It seemed too simple compared to the technical solutions she was used to".*

The Fields of Possibility

The stories consistently show that intelligence influences the social roles occupied by the characters. The educated characters have prestigious professions such as doctor, engineer, architect or lawyer: **Camille** *"was lucky enough to go to university and became an architect. She knew all the rules for constructing solid, magnificent buildings."*

Uneducated characters are described more as capable of solving concrete, immediate problems. Their skills are based on experience and manual know-how, such as repairing machinery or building temporary infrastructure. This is the case of **Lucas**, who "*stayed in town and became a carpenter. Although he had no formal education, he became very skilled at making furniture and solving practical problems*".

The Internalization

Uneducated characters generally adhere to the representations of intelligence attributed to them. They emphasize their form of intelligence, based on experience and know-how, but often admit the superiority or necessity of academic intelligence in certain fields, notably those linked to innovation and strategic decision-making.

This internalization maintains a dynamic in which academic intelligence is perceived as the ideal to be attained, while practical intelligence remains confined to immediate, utilitarian functions.

The Role of Representations of Intelligence in Power Relations According to Socio-economic Status

As with level of education and gender, there is a marked distinction between the intelligence attributed to poor people and that attributed to rich people.

The Legitimization

The qualitative representational analysis of the stories uncovers several thematic oppositions that shape the contrasting representations of intelligence in wealthy and impoverished characters.

Stability vs. Adaptation: the intelligence of rich people is often associated with risk minimization, as they have the resources to compensate for failure. Poor people, on the other hand, take calculated risks, as their survival often depends on it.

Example: *"Wealthy people, like* ***Sir Émile****, often have resources and tools that allow them to work faster, but they can sometimes be less flexible when faced with the unexpected."*

Possession vs. creation: the intelligence of rich people is often associated with material resources (money, technology, machines, experts). The intelligence of the poor, on the other hand, is portrayed as practical, creative and adaptive, based on observation, ingenuity and experience.

Example: "**Julien**, money can buy machines and resources, but it can't buy creativity and perseverance. Intelligence is knowing how to observe and understand the resources you have around you, and being able to find solutions when things aren't going well.

The Fields of Possibility

Qualitative representational analysis of data according to socio-economic status reveals a significant difference between the ingenuity attributed to poor characters and the access to resources reserved for rich characters. The intelligence of poor characters lies in their ability to adapt and take advantage of the limited resources at their disposal.

Karim: "*I used what I had: my mind, my hands, and I took the time to understand how things work. It's not always how many resources you have that counts, but how you use them."* Here, Karim illustrates the idea that practical intelligence comes from experience, not material means. In contrast, wealthy characters are often portrayed as having privileged access to resources, but a dependence on them that limits their ability to adapt.

Example of **Amir** who: "*Tried to buy ready-made solutions, like expensive irrigation systems and imported seeds, but these solutions didn't work.*" His story illustrates the idea that money doesn't necessarily guarantee the resolution of problems.

Arthur: "*With my resources, I have no doubt I'll succeed." Yet he fails against Lucas, who "used his observation and imagination.*" His intelligence is based on material resources.

Wealthy characters have access to tools and technologies, but their intelligence is often portrayed as limited by less flexibility and adaptability to the unexpected.

The Internalization

Poor characters accept and integrate the representations associated with their socioeconomic status. They value their creativity and ability to "adapt, innovate and use what they have to create something valuable." Poor characters internalize the idea that their intelligence is based on effort and adaptability.

Example:

Julien: *"I didn't have much money, but I had ideas."*

Fatima: *"I observed, listened and worked with what I had.*

Discussion

The analysis of the stories generated by the AI presented in this chapter sheds critical light on the ways in which the AI can reflect and reproduce hegemonic social representations of intelligence. The representational content identified in the material produced by ChatGPT corresponds generally to the findings of classical and more recent studies on the social representations of intelligence (Mugny and Carugati 1989; Poeschl 2001; Carugati and Selleri 2011). In this way, the opposition between practical and theoretical forms of intelligence echoes the data of Räty and Snellman (1995), who showed how the educational system institutionalizes social hierarchies through the association of theoretical knowledge with high socioeconomic and educational status. Similarly, the distinction between technical/creative and emotional/relational intelligence reflects deep societal dichotomies (Martin and Slepian 2021), amplifying gender-based stereotypes that distinguish masculine intelligence (logical, analytical, technical) from feminine intelligence (emotional, intuitive, relational).

The valorization of practical intelligence among economically disadvantaged or non-educated characters as resourceful but limited (Quadlin 2022; Strenze 2007) reproduces a narrative that confines their intelligence to a subsistence and reactive role. Conversely, intelligence associated with formally educated or affluent characters appears strategic and forward-looking, thus implicitly maintaining existing social stratifications (Lozano-Blasco et al. 2022; Ritchie and Tucker-Drob 2018).

These results echo the work of Mugny and Carugati (1989), who argue that educational contexts actively contribute to structuring intelligence as a legitimizing discourse for power relations. Schools and educational narratives, now potentially supplemented by AI-generated content, act as critical vectors for reproducing social hierarchies. As our analysis demonstrates, AI models trained on large bodies of text reflect dominant social discourses containing, and even possibly amplifying, the biases already present in the training data (Ferrara 2023; Kaplan et al. 2024).

The key point is that AI models have not arisen outside this social fabric. They ingest text that encodes the cultural biases and the dominant group's perspective, with less weight given to dissenting or minority perspectives (Hu et al. 2025; Qu and Wang 2024). Therefore, when an AI model narrates a story, it tends to project the hegemonic representation onto the story world. It may not "know" it is doing so—the model has no consciousness or values—but its statistical learning makes biased depictions more likely. The result is that AI-generated content participates in reproducing social structures symbolically: reinforcing who is seen (and not seen) as intelligent or capable in our collective imagination.

AI as a Quasi-agent in Reproducing Social Structures

The above evidence raises profound theoretical questions: In what sense can AI be considered an "agent" of social reproduction? And how should we critically understand the role of generative AI in shaping (or distorting) the ongoing social dialogue about concepts like intelligence? Although AI systems lack genuine consciousness or intent, their outputs function *as if* a new actor has entered the cultural stage—one that endlessly remixs human narratives without human deliberation. Current generative AIs are not truly autonomous, thinking agents (Floridi 2023); rather, they are sophisticated pattern replicators without understanding (Bender et all. 2021). An LLM can produce linguistically coherent statements, but it does not mean or experience what it says—it has no subjective perspective or communicative intent. Instead, any influence it exerts is more akin to one-way transmission. The AI's contributions to discourse lack the reciprocity and accountability that characterize genuine dialogue among humans.

From us this distinction is crucial. Material generated by generative AI—no matter how convincingly written—is *not* a "social representation" in the classical sense, because it was not produced by a community through dialogue and consensus (Negura 2025). Social representations, as theorized by Moscovici (1961/1976) and others (Markova 2003; Jodelet 2003, etc.), live in the interactions between people; they are "common sense" understandings constructed as people communicate about a complex phenomenon. By contrast, an AI system produces content unilaterally, by algorithmically recombining patterns in its training data. There is no deliberation, no intentional framing of meaning relative to a group's worldview—the AI's output is generated outside the communicative process that gives social representations their social character. It delivers an utterance that can influence people's perceptions (perhaps subtly reinforcing a stereotype), but that utterance did not arise from the affected social groups engaging each other in reasoned discussion. This is especially problematic when the content carries biases: it means stereotyped representations (e.g., of gender or class and intelligence) can be amplified without those stereotypes ever being challenged through dialogue at the moment of their production. The AI acts as a quasi-agent of socialization, reinforcing certain views—but it is a deaf agent, not listening or responding to the people whom those views concern.

Horkheimer and Adorno's critique of the culture industry in the mid-20th century emphasized how mass-produced media can entrench dominant ideologies and suppress critical thought (Horkheimer and Adorno 2002). Generative AI can be seen as a new extension of the culture industry: it mass-produces narratives and images on an unprecedented scale, potentially standardizing thought by endlessly replicating existing cultural forms (Waelen 2022). If those forms are biased or skewed in favor of the powerful (as our data suggests), then AI could accelerate the reification of those biases. The process happens without malicious intent—indeed without any intent at all on the AI's part—but from our viewpoint, that makes it even more insidious. There is no author to hold accountable, no interlocutor with whom marginalized groups can argue. The ideology is reproduced without a subject, creating the illusion of neutral, "natural" content when in fact it may carry the imprint of historical power relations.

Furthermore, Habermas (1984) warned that the public sphere can be "re-feudalized" when powerful actors inject messaging that citizens cannot contest on equal footing—turning communication into manipulation. With generative AI, we have a paradoxical situation where the outputs feel like dialogue (interactive, responsive text), but in reality they are monologic from the machine's side. The "dialogue" between user and AI is not truly dialogue between social groups concerned (e.g., men and women, or different classes debating their portrayals); it is a user interacting with a vast echo of past data. The AI's beautiful coherence masks the fact that no real communicative rationality has taken place. This one-sided generation of social meaning sidelines democratic discourse. As Waelen (2022) put it, algorithms today can exercise a form of "disciplinary power" by shaping what information people see and thus subtly steering social norms. Generative AI is now doing something analogous in the cultural realm: it can favor certain representations of social reality over others, effectively privileging the historically dominant perspective unless we intervene to correct it.

In conclusion, the patterns observed in our data point to a clear risk: AI systems, though not sentient, are participating in the reproduction of social structures through the content they generate. They act as quasi-agents that recirculate society's existing representations (of intelligence, of gender roles, of class distinctions) on a massive scale. This calls for a critical approach on two fronts. First, on the empirical and design front, AI researchers and ethicists must intensify efforts to detect and mitigate biases in generative models. The alignment of AI outputs with outdated stereotypes is not inevitable; it reflects choices about training data and model tuning (UNESCO, IRCAI 2024). Ensuring more balanced representations—for instance, portraying women and men across a similar range of intellectual roles—is essential if we want AI to enhance rather than hinder social progress. Second, on the theoretical front, we must interrogate the social impact of AI. We should remain cognizant that AI-generated content lacks the dialogical, negotiated character of genuine social knowledge, and therefore its uncritical adoption can short-circuit the democratic contestation of ideas in society. If generative AI simply mirrors and amplifies the dominant narratives without dialogue, it risks becoming an instrument of social stagnation, or worse, regression. By bringing these academic perspectives into our understanding of AI, we can better grasp that the struggle against bias in AI is not only a technical endeavor but also a continuation of the long-standing struggle over social representations—a struggle to make the way intelligence (and other valued qualities) is portrayed more equitable, inclusive, and subject to true communicative debate rather than one-sided reproduction.

Limitations

While our study provides important insights, it is not without limitations. First, the data source is constrained to a specific AI model (ChatGPT, GPT-4) and a narrow genre (short children's stories in French). The patterns we identified are thus tied to the behavior of this model, which in turn is trained on predominantly Western internet text up to 2021. Other AI models or future versions might produce different representations, especially as training data and algorithms evolve. Generalizing our findings to all AI-generated content or to non-Western contexts should be done with caution. For instance, ChatGPT has content guidelines that attempt to avoid overt sexism or classism, which may have moderated the extremity of biases in the stories. This means our results perhaps represent a floor rather than a ceiling for bias—less constrained models or different prompts could yield even more stereotyped narratives. Conversely, because the AI often strives for egalitarian messaging (e.g., "everyone is smart in their own way"), some biases might be superficially smoothed over, requiring careful reading between the lines. These complexities highlight that our findings are model-specific and culturally situated.

Second, the methodology of prompting and analysis carries inherent limitations. We guided the AI to generate stories by specifying certain roles or comparisons (e.g., "explain the difference between X and Y's intelligence"). This undoubtedly shaped the outputs. While this approach was effective in eliciting the model's latent representations and biases, it might also exaggerate them—real-world use of AI might not typically involve such direct comparison prompts. Moreover, children's stories as a format often use exaggerated archetypes for clarity (the brainy character, the kind character, etc.). We attempted to mitigate this by analyzing a large corpus and identifying recurring patterns beyond one-off story tropes.

Another limitation is that we did not examine children's actual reception or interpretation of these AI-generated stories. We have inferred potential influence based on content, but we do not know how a child might internalize or question these narratives. It's possible that young readers would pick up on egalitarian messages ("both characters were smart in different ways") more than the subtle hierarchies, or conversely, that they might strongly absorb the implicit rankings (seeing one type of character as the "real" genius). Without empirical study of the audience, our claims about internalization and impact remain theoretically grounded but untested.

Finally, we also note that our study was exploratory in combining lexicometric analysis with critical qualitative analysis; there may be unexplored nuances in the data (for example, linguistic cues or narrative structures) that require different analytical tools to fully capture. In summary, while we are confident that our core findings reflect real tendencies in the data, they should be interpreted within the above limitations. Further research (as we outline next) will be necessary to deepen and broaden these insights.

Conclusion

This chapter has critically examined AI-generated stories to better understand how they contribute to the reproduction and internalization of social hierarchies through the representation of intelligence. By examining the children's stories generated by ChatGPT, this study highlights subtle and persuasive ways in which social biases associated with gender, educational level and socioeconomic status are perpetuated and legitimized. This analysis highlights the powerful role of social representations of intelligence in structuring power relations, influencing children's perceptions of their own abilities, aspirations and potential social roles.

Given the growing integration of AI technologies into everyday educational contexts, it is crucial for educators, policymakers, and AI developers to recognize and address the embedded biases within AI-generated content. Awareness and critical reflection on these issues are essential to prevent reinforcing and perpetuating social inequalities. Future research should continue investigating how social representations mediated by AI shape societal perceptions and individual identities. Ethical vigilance and conscious, equitable approaches to AI deployment in education are paramount to ensure that AI-generated narratives empower rather than constrain the possibilities for future generations, fostering a more inclusive and socially responsible society.

References

Acerbi, A. and J. M. Stubbersfield. 2023. Large language models show human-like content biases in transmission chain experiments. Proceedings of the National Academy of Sciences, 120(44): e2313790120.

Bender, E. M. and B. Friedman. 2018. Data statements for natural language processing: Toward mitigating system bias and enabling better science, Transactions of the Association for Computational Linguistics 6: 587–604.

Bender, E. M., T. Gebru, A. McMillan-Major and S. Shmitchell. 2021. On the dangers of stochastic parrots: can language models be too big?. Proceedings of the 2021 ACM Conference on Fairness, Accountability, and Transparency, 610–623. https://doi.org/10.1145/3442188.3445922

Carugati, F. and P. Selleri. 2011. ...Once upon a time... The case of social representations of intelligence. Papers on Social Representations, 20(2): 32–1.

Eagly, A. H. 1987. Sex di~erences in social behavior: A social-role analysis. Hillsdale, NJ: Erlbaum.

Eagly, A . H. and V. J. Steffen. 1984. Gender stereotypes stem from the distribution of women and men into social roles. Journal of Personality and Social Psychology, 46: 735–754.

Eagly, A. H. and V. J. Steffen. 1986. Gender stereotypes, occupational roles, and beliefs about part-time employees. Psychology of Women Quarterly, 10: 252–262.

Eagly, A. H. and V. J. Steffen. 1988. A note on assessing stereotypes. Personality and Social Psychology Bulletin, 14: 676480.

Ferrara, E. 2023. Should chatgpt be biased? Challenges and risks of bias in large language models. arXiv preprint arXiv:2304.03738.

Floridi, L. 2023. AI as agency without intelligence: On ChatGPT, Large Language Models, and other generative models. Philosophy and Technology, 36(1): 15. https://doi.org/10.1007/s13347-023-00621-y

Gallegos, I. O., R. A. Rossi, J. Barrow, M. M. Tanjim, S. Kim, F. Dernoncourt et al. 2024. Bias and fairness in large language models: A survey. Computational Linguistics, 1–79.

Gillespie, A. 2008. Social representations, alternative representations and semantic barriers. Journal for the theory of social behaviour, 38(4): 375–391.

Jodelet, D. 2003. Les représentations sociales (7e éd. e éd.). P.U.F.

Habermas, J. 1984. The theory of communicative action. Beacon Press.

Hayward, C. R. 2000. De-facing power. Cambridge University Press.

Hu, T., Y. Kyrychenko, S. Rathje, N. Collier, S. van der Linden and J. Roozenbeek. 2025. Generative language models exhibit social identity biases. Nature Computational Science, 5(1): 65–75.

Horkheimer, M. and T. W. Adorno. 2002. Dialectic of Enlightenment. Translated by Edmund Jephcott, edited by Gunzelin Schimdt Noeri. Stanford University Press.

Qu, Y. and J. Wang. 2024. Performance and biases of large language models in public opinion simulation. Humanities and Social Sciences Communications, 11(1): 1–13.

Quadlin, N. 2022. Do perceptions of privilege enhance—Or impede—Perceptions of intelligence? Evidence from a national survey experiment. RSF: The Russell Sage Foundation Journal of the Social Sciences, 8(7): 48–69.

Kaplan, D. M., R. Palitsky, S. J. Arconada Alvarez, N. S. Pozzo, M. N. Greenleaf, C. A. Atkinson et al. 2024. What's in a name? Experimental evidence of gender bias in recommendation letters generated by ChatGPT. Journal of Medical Internet Research, 26: e51837.

Lee, J., Y. Hicke, R. Yu, C. Brooks and R. F. Kizilcec. 2024. The life cycle of large language models in education: A framework for understanding sources of bias. British Journal of Educational Technology, 55(5): 1982–2002.

Lippens, L. 2024. Computer says 'no': Exploring systemic bias in ChatGPT using an audit approach. Computers in Human Behavior: Artificial Humans, 2(1): 100054.

Lorenzi-Cioldi, F. 2002. Expériences sur les groupes dominants et dominés: la perception de l'homogénité des groupes. Peter Lang Group Ag, International Academic Publishers.

Lozano-Blasco, R., A. Quílez-Robres, P. Usán, C. Salavera and R. Casanovas-López. 2022. Types of intelligence and academic performance: A systematic review and meta-analysis. Journal of Intelligence, 10(4): 123.

Marková, I. 2003. Dialogicality and social representations: The dynamics of mind. Cambridge University Press.

Martin, A. E. and M. L. Slepian. 2021. The primacy of gender: Gendered cognition underlies the big two dimensions of social cognition. Perspectives on Psychological Science, 16(6): 1143–1158.

Moscovici, S. 1961/1976. La psychanalyse : son image et son public : étude sur la représentation sociale de la psychanalyse. Presses universitaires de France.

Moscovici, S. 1984. The myth of the lonely paradigm: A rejoinder. Social Research, 939–967.

Moscovici, S. and G. Vignaux. 1994. 'Le Concept de Thêmata. pp. 25–72. *In*: Guimelli, C. (Ed.). Structures et transformations des représentations sociales [Structures and transformations of social representations]. Neuchatel: Delachaux et Niestlé.

Mugny, G. and F. Carugati. 1989. Social representations of intelligence. Cambridge University Press.

Negura, L., N. Plante and M. Lévesque. 2020. The role of social representation in the construction of power relations. Journal for the Theory of Social Behaviour, 50(1): 25–41.

Negura, L. 2025. Simulated Sense-Making or Social Knowledge? Artificial Intelligence and the Boundaries of Representation, Journal for the Theory of Social Behaviour, 55(3): 1–9, https://doi.org/10.1111/jtsb.70012

Patel, S. and M. Ragolane. 2024. The implementation of artificial intelligence in south african higher education institutions: opportunities and challenges. Technium Education and Humanities, 9: 51–65. https://doi.org/10.47577/teh.v9i.11452

Poeschl, G. 2001. Social comparison and differentiation strategies in social representations of intelligence. Swiss Journal of Psychology/Schweizerische Zeitschrift für Psychologie/Revue Suisse de Psychologie, 60(1): 15.

Ratinaud, P. 2009. IRaMuTeQ: Implémentation d'une méthode d'analyse textuelle. [Software].

Räty, H. and L. Snellman. 1995. On the social fabric of intelligence. Papers on Social Representations, 4: 177–186.

Reinert, M. 1983. Introduction à la lexicométrie. Un modèle statistique d'analyse. Langages, 67: 5–15.

Reinert, M. 1990. Alceste: Une méthodologie d'analyse des données textuelles et une application: Aurelia de Gérard de Nerval. Bulletin de Méthodologie Sociologique, 26: 24–54.

Reinert, M. 1993. Les "mondes lexicaux" et leur "logique" à travers l'analyse statistique d'un corpus de récits de cauchemars. Langage et Société, 66: 5–39.

Ritchie, S. J. and E. M. Tucker-Drob. 2018. How much does education improve intelligence? A meta-analysis. Psychological Science, 29(8): 1358–1369.

Strenze, T. 2007. Intelligence and socioeconomic success: A meta-analytic review of longitudinal research. Intelligence, 35(5): 401–426.

Srivastava, A. K. and G. Misra. 2001. Lay people's understanding and use of intelligence: An Indian perspective. Psychology and Developing Societies, 13(1): 25–49.

UNESCO, IRCAI 2024. Challenging systematic prejudices: an Investigation into Gender Bias in Large Language Models.

Walther, A., F. Logoz and L. Eggenberger. 2024. The gendered nature of AI: Men and masculinities through the lens of ChatGPT and GPT4. Computers in Human Behavior: Artificial Humans, 100076.

Waelen, R. (2022) Why AI ethics is a critical theory. Philos. Technol. 35: 9. https://doi.org/10.1007/s13347-022-00507-5

Warr, M., N. J. Oster and R. Isaac. (2024). Implicit bias in large language models: Experimental proof and implications for education. Journal of Research on Technology in Education, 1–24.

Zack, T., E. Lehman, M. Suzgun, J. A. Rodriguez, L. A. Celi, J. Gichoya et al. 2024. Assessing the potential of GPT-4 to perpetuate racial and gender biases in health care: a model evaluation study. The Lancet Digital Health, 6(1): e12–e22.

Zhou, M., V. Abhishek, T. Derdenger, J. Kim and K. Srinivasan. 2024. Bias in Generative AI. arXiv:2403.02726.

Conclusion
AI at the Crossroads of Representation
Object, Agent, and Social Imaginary

Lilian Negura

Introduction

Throughout this book, we have explored how artificial intelligence (AI) is represented in society, considering various cultural, societal and ethical contexts. From media portrayals, workplace transformations and ethical dilemmas to public understanding of this technology, each chapter has highlighted a different aspect of how AI is conceptualised, perceived and integrated into social life. This final chapter summarises these insights by introducing a dual analytical framework that conceptualises AI as both an object represented by human societies and a quasi-agent that actively contributes to and reshapes social representations.

To critically examine how generative AI forces us to reconsider the traditional boundaries between human and artificial agency through its sophisticated models, this synthesis draws on the arguments presented in previous chapters. These new technological developments require discussion of their critical theoretical implications. In this concluding chapter, social representation theory is repositioned in the context of emerging AI phenomena, emphasising the need for further analysis and the development of new theoretical constructs in the era of AI.

AI as a Represented Object: Myths, Metaphors, Hopes and Fears

Ever since the term artificial intelligence was coined in the mid-20th century, the public imagination has been filled with images of what AI might be and what it

Faculty of Social Sciences,University of Ottawa.
Email: lilian.negura@uottawa.ca

might do. As the book chapters illustrate, these images oscillate between utopian perspectives and dystopian anxieties, tapping into deeply rooted human hopes and fears. In terms of social representation, AI is an object that we have collectively tried to make sense of, often understanding it in relation to older concepts that we have tried to "shrink" into metaphorical enclosures (such as intelligent beings, servants and tools) and represented in physical form (such as humanoid robots and computers). These representations are rich in human meaning, telling stories about our relationship with our creations, our hopes for transcendence or improvement, and our fears of loss of control or displacement.

Hopes: AI as Savior and Solution

The studies presented in this book show how a significant body of representations portrays AI in optimistic and, at times, fantastical manner as a savior technology that aims to help humanity solve its biggest problems. In these stories, AI is portrayed as an extremely powerful and positive entity, a "benevolent superintelligence" or an "oracle" that is capable of providing us answers that we deeply struggle to find. Examples of such portrayals can be found in the media discourse surrounding AI (see Negura et al. 2026, or Chae and Bauer 2026, in this volume). This positive discourse around AI can also be found in the field of education. In their chapter in this volume, Jean-Marie Séca and Matthieu Séca show how institutions such as UNESCO and the OECD tend to promote the generative AI for pedagogical innovation and inclusion, emphasizing the strategic and economic benefits of this technology.

In addition, the arguments of the proponents of AI technology are often characterized by a strong dose of uncritical belief in a better world though technology, what is sometimes called, technological utopianism. This is the notion that technology, in all its rationality and with a mind of its own, will solve the myriad of issues that humanity never seems to overcome. In various cultures, this narrative takes the shape of the "deus ex machina" story, in which a god from a machine swoops in to restore order. These fantastic visions of what AI can bring to humanity fulfill many deep-seated desires to dominate nature, exploit it, advance knowledge, and ensure economic security and prosperity for all (see chapter by Luciano Moreira 2026, in this volume). In modern societies, technology often inherits this myth-making role, and AI, given its association with intelligence and autonomy, is especially prone to being cast in quasi-supernatural roles (omniscient problem-solver, benign overseer, etc.).

Social representation theory reminds us that such optimistic visions do more than just promote a technology—they help make the unfamiliar AI familiar by linking it to positive narratives we understand (healer, savior, helper). By anchoring AI to the image of a curing physician or wise expert, we integrate it into our moral universe as a force for good. At the same time, this optimism can be polyphasic: it can coexist with contrary representations even within the same community or individual. For instance, one might simultaneously believe that AI could cure cancer (a hope) and fear that AI could also cause great harm. Indeed, that is a hallmark of many social representations of novel technologies—these representations are ambivalent and rich in internal tensions, reflecting our collective uncertainty (for example, Sarrica 2010, Brondi et all. 2021).

Fears: AI as Threat and Monster

In addition to Utopian stories, narratives that see AI as a public menace: as a potential human threat, a job killer, a privacy invader, and even a risk for human agency, also exist. These fears rest on age-old cultural archetypes. One such archetype, the dreaded Frankenstein motif, stems from Mary Shelley's 1818 story about a scientist's creation that turns against its creator. Isaac Asimov, one of the first to analyze the Frankenstein complex in detail, claimed that the fear of mechanical men is something all people have and is something that is inevitable and ingrained. Even ethical technological boundaries do not ease this anxiety. Such narrative patterns continue to linger within the cultural consciousness of AI. Moreira's chapter in this volume, for instance, shows how all these elements shape social representations of AI as a culture of fear.

Today, one concrete social representation of AI as a culture of fear is the belief, illustrated in the media study by Negura et al. (2026), that AI will replace human labor. It also shows how the boundaries of public perception have been manipulated by "emotion-laden stories" that have been framed, more broadly, within the context of AI as a social threat. This coincides with a pulp story: AI as a robotic force that threatens to render human work obsolete, almost a fourth industrial revolution where "thinking machines" automate both physical and mental processes. The anxiety of automation is, usually, a proxy for anxiety about loss of social value and identity: if machines can do all the work better, what is the role of a human? This frame is highly relevant, sociologically, during recessionary periods and can shape the political discourse (e.g., retraining, free basic income, or moratoriums on certain types of AI). The framing of AI as "the robotic technologist" is rooted in the history of technological unemployment and is visually captured in the images of assembly-line robots or infinitely-available invisible intelligent software. It taps much of the already existing tension between technology and humanity. Vultur (2026) in this volume explains how images of a threat posed by AI are underpinned by the traditional narratives of technological unemployment and socio-economic exclusion. His work surveying the expert literature describes how AI is both disruptor and savior: extending the tentacles of workplace surveillance, deepening social strata, and technocratically severing humanity from agency.

Another fear-laden representation depicts AI as a malevolent or uncontrolled intelligence that could pose existential risks. This has been a staple of science fiction—from the cold, murderous HAL 9000 in 2001: A Space Odyssey to the tyrannical Skynet in The Terminator—and has increasingly entered real-world discourse as AI capabilities advance. Many experts and even tech CEOs now warn in serious terms that AI could "destroy our civilization" if misaligned. Throughout 2023, intense discussions within elite scholarly circles attested to the "doomsday" implications of advanced AI, including, and not limited to, the extinction of the human race (Roose 2023). As catastrophic predictions about the future are, more often than not, the basis of debate with little consensus on how plausible such scenarios are, the discourse certainly occupies a niche shelter provided by the conditions of a given social climate pertaining to advanced AI. This shows how the realm of social imagination encompasses AI as not merely functions performing a service, but also as run away algorithms and robot uprisings. Social discourse is littered with the notions of "AI apocalypse" and "rogue AI" as depicted and indexed in the literature of this and other books. The "safety" of AI is routinely articulated with the most simplistic comparisons—"like the genie out of the bottle" or "Pandora's box"—indicative of an unshackled AI that runs a mock and is beyond human reasoning, governance, or control. Yuval Noah Harari, in his book Nexus, even framed AI as having hacked the operating system of human civilization by mastering language and thus potentially manipulating human minds at scale—a dramatic image of AI infiltrating the very fabric of culture and communication (Harari 2024).

Underneath these dramatic fears lie more ordinary but pervasive anxieties as well: fear of loss of agency (will AI systems make decisions for us, taking away our freedom?), fear of inequity (will AI be controlled by a few to dominate others?), and fear of the unknown (AI as something fundamentally non-human that we struggle to predict). We might note that these fears often project onto AI our worries about ourselves. As Mark Coeckelbergh (2025) has argued, robots and AI often serve as "containers for our projections." Fears about AI may thus reflect fears about human nature and society (e.g., fear of an all-powerful AI dictator might reflect concerns about our own institutions losing control). The Frankenstein complex itself can be interpreted as humanity's guilt and anxiety about creating life and the moral responsibility it entails. Negative representations of AI can serve as a cautionary moral tale, reinforcing values such as "we must not play God" or "unchecked hubris leads to disaster." These moral narratives often surface in public debates, such as comparisons of AI research to Pandora's Box or to atomic weapons—essentially anchoring AI to past instances where human innovation unleashed forces beyond our control.

It is worth mentioning that social representations of AI are not uniform worldwide. Cultural context, media narratives, and local experiences shape which image dominates, as shown in the chapter by Jaesun Chae and Martin Bauer. These differences reflect varying "projects" of social groups and societies—to use Bauer and Gaskell's term (Bauer and Gaskell 1999). Thus, while hopes and fears about AI are found everywhere, their intensity and balance in the social representation of AI can differ.

Anthropomorphism: Humanizing the Non-human

In both optimistic and pessimistic considerations, there is a characteristic tendency: anthropomorphizing AI and imagining it possessing human characteristics. This is often driven by the very term "artificial intelligence," causing us to equate mental processes to computer operations. This could be considered a type of anchoring. AI is completely abstract and difficult to understand, so it is anchored to the most recognizable intelligent being familiar to us, which is ourselves. Humanizing AI by saying it "understands," "intends" or "feels" becomes an easier but, not accurate position to take. All the same, this says more about humanity than about technology. These metaphoric understandings—assistant, protector, or stabilizer—are laden with power dynamics, as illustrated in the chapter of Gatt et al. (2026) in this volume.

Empirically, people do often treat AI agents in a human-like fashion, even when they know intellectually that no real person is involved. A classic example is the ELIZA effect (Weizenbaum 1966). In the 1960s, MIT's Joseph Weizenbaum created ELIZA, a very simple text-based chatbot that mimicked a psychotherapist by turning the user's statements into reflective questions. Users were aware it was a program, yet many began to feel as if the machine understood them. He witnessed colleagues and secretaries attributing human-like empathy and concern to ELIZA, a phenomenon now called the ELIZA effect. This early episode prefigured our interactions with today's AI assistants. We seem wired to respond to conversational cues and voices as if they imply an intentional mind.

As illustrated by Moreira's exploration of AI in fiction and popular culture in this volume, anthropomorphic representations of AI are reinforced by popular culture: countless movies and books personify AI systems as characters—sometimes friendly (e.g., Data from Star Trek; Wall-E the cute robot) and other times hostile (HAL 9000, etc.). Even the physical design of robots often intentionally uses anthropomorphic features (eyes, voice, names) to make interaction more intuitive. Psychologically, when an AI speaks fluently or a robot moves in a quasi-human way, people automatically draw on their "theory of mind" social skills and assume there is intention and feeling behind the behavior (Horstmann and Krämer 2019). In experiments and observations, people have been seen to name their robot vacuum cleaners, feel empathy for robots that seem to struggle or "suffer," and even exhibit moral concern (Banks 2021). This humanization of AI is double-edged: on one hand, it can make technology more acceptable and engaging (we tend to like interfaces that mimic social reciprocity), but on the other hand, it can lead to overestimation of AI's capabilities and cultivate inappropriate trust. For instance, users might assume a chatbot "understands" the meaning of what it says because it responds fluently, when in fact it may have no comprehension.

Socially, the anthropomorphic representation of AI often manifests in the notion of AI as a social actor or agent in its own right. People talk about "what AI wants" or what "AI thinks," even if they know the AI isn't sentient. This language reveals a blurring of lines between human and machine agency in the public mind. Some theorists argue that treating complex systems with an "intentional stance"—that is, "treat[ing] the object whose behavior is to be predicted as a rational agent" (p.17) with beliefs and desires—is actually a common heuristic we use, as described by Daniel Dennett (1996). We do this with other humans certainly, but also with pets, and even with gadgets that show a semblance of independent behavior. According to Dennett (1996), we adopt this stance because it is pragmatically useful for prediction: if attributing intentions to an AI helps us predict its actions, we will do so readily. The effectiveness of this stance (in everyday life) reinforces anthropomorphic representations. It becomes easy to slide from "as if it has intentions" to "it actually has intentions."

However, experts caution that anthropomorphism can be a "hype and fallacy" in thinking about AI (Placani 2024). For example, Luciano Floridi reflects that we must avoid thinking about AI attributions that are overly human and instead accept it as a novel type of agency devoid of intelligence (Floridi 2025). In this sense, agency, and thus, AI systems, can operate and produce effects (act autonomously), but do not and cannot comprehend and possess consciousness as humans do. In Floridi's view, any thinking that marries these two facets is mystified, and misplaced; instead of asking "Is AI truly intelligent like us?" it is more pertinent to question "What type of artificial agent is it, considering it is mind and intention devoid of a human's?" (Floridi 2025). This is similar to an argument that underscores how, for explanatory purposes, AI is often described and assessed anthropomorphically, which is an overstatement. In fact, anthropomorphizing AI, as it is discussed publicly, tends to overstate what the entity can do (Placani 2024). For instance, "the AI decided to do X" suggests X was a conscious and deliberate decision made by the AI, thus, oversimplifying the fact that an AI system operates under a human-designed model or objective function, is trained, and does not make decisions. Still, the more sophisticated the AI model gets, the more human-like the outputs that are generated, which further tests those fundamental distinctions. The risk is that people might accord AI systems undue authority or trust, as if they were expert humans, leading to problems if those systems are actually flawed or biased. With respect to trust, AI systems might actually be incapable of human-like rational thought and could be fundamentally biased. Then again, anthropomorphism could also work within the more negative domain of social images, such as when people think of AI as human; in such instances it is much easier to regard it as some rival, or enemy, as seen in sci-fi movies or fiction in which AI is supposedly out to get mankind.

In summary, the humanization of AI is a notable feature of how we represent this technology. Social representation theory would view this as part of the objectification of AI—giving the abstract concept a recognizable face and persona. The anthropomorphic representations of AI (the robot butler, the artificial "friend," the digital antagonist, etc.) serve to embed AI into our social world, making interactions with it socially meaningful. But they also reflect our "hopes and fears:" we hope an AI will be friendly, ethical, understanding like a good person, and we fear it might be cold, deceitful, or power-hungry like a bad person. In effect, we often end up casting AI in roles drawn from human social archetypes—teacher, servant, opponent, child, savant, and so on. This casting says as much about us as it does about any actual capabilities that AI might possess.

Having explored how AI is represented as an object of societal imagination—imbued with our grandest hopes and darkest fears, and frequently personified—we now turn to the second perspective: AI as a producer of representations. This shift in view raises fascinating questions. When AI systems generate text, images, or decisions, are they merely tools reflecting our own inputs, or are they in some sense creating new social meanings? How should we understand the content produced by AI in relation to human knowledge and common sense? And could AI-generated narratives themselves become part of our reservoir of social representations?

AI as a Quasi-agent: AI-generated Content and Social Representation

Large language models (LLMs) like GPT-5 and generative image models like DALL-E or Midjourney have given AI a new role: not just as a subject of representation by humans, but as an agent producing representations (texts, images, answers, etc.) that humans consume. In everyday terms, people are increasingly reading content written by AI and viewing images created by AI. These outputs often take the form of stories, images, and explanations—forms traditionally associated with human communicators. The chapters in this book that analyze AI-generated content raise a host of theoretical and practical issues: Can we regard AI's outputs as part of the social discourse in the same way that human-generated outputs are? Do they carry the same status as human narratives or scientific explanations? Are they original or just regurgitations of human ideas? And how do they influence the human social representations of reality?

In considering AI as a quasi-agent, we should clarify: current AI systems (even the most advanced LLMs) do not possess consciousness, intentions, or understanding in the way people do (Negura 2025). Vaast (2026) in this volume shows how developers themselves make sense of AI's agency in relation to their own—oscillating between distant future imaginaries that downplay their responsibility and near-future framings that require ethical engagement. These representational practices reveal how agency is co-constructed between technology and its creators. Nonetheless, these systems exhibit behaviors that simulate understanding extremely well, to the point where interacting with them can feel like interacting with an intelligent interlocutor. Hence they are often called "conversational agents" or "generative agents." They produce representations of knowledge (e.g., summaries, arguments, stories) by drawing on the vast dataset of human language they were trained on. In essence, an LLM like GPT is a predictive engine that continues a prompt in statistically likely ways based on patterns in its training data (which include books, articles, internet text, etc.). The result, however, is something that can look quite creative, such as, for example, a story for children (Negura and Tchatchou 2026, in this volume).

From a social knowledge perspective, there are two interrelated aspects to examine: (a) the nature of the content that AI produces (Is it anchored in human social representations? Does it have any "stance" or is it just an average of seen discourse?), and (b) the role of AI as a mediator or participant in the social communication process (Does AI become a "subject"? Is it just a new medium?).

AI-generated Content: Mimicry, Bias, and the Nature of "Knowledge"

AI-generated content is fundamentally parasitic in relation to human-generated content. An LLM learns linguistic and semantic patterns from human texts; a generative image model learns visual patterns from human-created images. Therefore, one might say that AI outputs are reflective of the existing body of human representations. In this sense, AI content can be seen as amplifying or reiterating prevailing social representations. It is a kind of mirror—albeit a distorted one—of the collective mind as recorded in data. This underscores that AI outputs are not coming from a vacuum; they are grounded in the human socio-cultural corpus.

However, AI-generated representations differ from human ones in crucial ways. They lack an anchor in embodied sensory experience (Martikainen and Sakki 2023; Wagner 2017) or identity projects (Wagner et al. 2009). A human's explanation or story is usually tied to some communicative intent, a perspective, a social identity (Liu and László 2007). By contrast, an explanation provided by AI is generated because it statistically flows from the prompt (Bender et al. 2021). This means AI content can lack the coherence and value-basis that human discourse usually possesses. To put it another way, social representations involve communities investing meaning into an object through communication that is tied to their social reality (Moscovici 1961/2008). Does that disqualify it from being a "social representation"? Perhaps, if we speak in absolutes, yes, as it is not the output of a social matrix for sense-making but is rather the result of a machine completing a pattern. However, once AI content is created and enters the stream of human activity, it is, in a sense, socialized and treated like any opinion or narrative, and is thus assimilated into the social representational fields. An example of this would be the AI news article that is published and read by a myriad of people, who, in this case are unaware that it is AI-generated. In this way, the article has the ability to shape and influence people's understanding of the issue in the same way as any article would. In this case, the AI-generated text functions as social discourse—the source may be mechanical, but the influence on people's minds is undeniable and is real. This illustrates rather the blurring, even the almost perfect merging of boundaries. The blurring of the boundaries, or in this case the issue of dispersed agency, is central to Ezzeddine's work in law enforcement, where as she illustrates in this volume, the lack of a delineated structure of responsibility and accountability to AI results in delegating decision-making, fostering what she calls "institutional opacity,"—where both the system and its operators are "off the hook" and absolved of liability. While it can be countered that AI has no form of social representations because it lacks any social context or collective life, it has the means to communicate, and therefore social representations can be conveyed and even constructed by AI within human communication (Negura 2025).

One promising area is the ways AI can amplify existing social representations or biases. For example, if a prevalent social representation associates "scientist" with "male" or "nurse" with "female," an AI model might unconsciously echo that in the stories or sentences it produces, unless corrected. Indeed, AI language models like ChatGPT do reproduce the gender, educational, and social class biases that stereotypes in the training data AI models are exposed to (see Negura and Tchatchou 2026, in this volume). This goes beyond language: within the field of health data, as León-Gómez et al. (2026) demonstrate in their chapter (this volume), synthetic data and digital twins are positioned as paradigmatic exemplars of objectivity and privacy-preserving technologies. However, they are underpinned by social and symbolic representations that are regressive, and which often reproduce socio-institutional frameworks while occluding counter-hegemonic narratives. In this regard, AI is not merely an indicative replication of bias, but is rather a fundamental determinant in the construction of epistemic authority, determining the boundaries of what is considered public health knowledge.

Let us picture society having an ambiguous image of AI, part-angel, part-demon, as we previously talked about. As noted in the chapter by Rizzoli, de Falco and Sarrica in this volume, when an AI is asked to create a story about itself, the outcome is usually a bifurcated and emblematic visualization: an AI as a nurturing, crafty, monstrous, or redemptive entity. Through visual storytelling and scripted responses, these generative systems reproduce and remix collective fears and fantasies, thus becoming active participants in the representational process. AI reiterates our hopes and anxieties back to us—not neutrally, but through culturally loaded templates. It can even produce new combinations or simulations of representations—for instance, spinning a futuristic story that combines multiple tropes (maybe "AI saves the world but at the cost of our humanity"—a storyline not far from many Black Mirror episodes). But does AI understand these representations? No, it doesn't understand in a human sense. Yet from an external perspective, the content it produces can be analyzed similarly to human-generated content. This presents a puzzle: Should we treat AI outputs as part of the corpus of cultural representations worthy of study?

John Searle would remind us that however coherent AI's statements about the world are, they are produced without any intentionality—the AI "does not understand Chinese" even if it writes a perfect essay in Chinese (Searle 1980). Daniel Dennett might counter that if the AI's output is useful and interpretable, we can, for practical purposes, ascribe meaning to it and treat it as if it had intent—essentially extending the intentional stance to these systems (Dennett 1996). But Dennett would also acknowledge this is a heuristic: the meaning truly arises when a human reads the AI's text and makes sense of it. Thus, an AI explanation about, say, why the sky is blue, might be indistinguishable from a human-written encyclopedia entry. It conveys knowledge—but without epistemic grounding. As Gaskell and colleagues (2026) argue in this volume, it was not produced by someone who "knows" in the human sense, i.e., based on "everyday knowledge" and on "social and cultural awareness." Rather, it was compiled through statistical pattern matching. Luciano Floridi's point about "agency without intelligence" fits here: the AI has produced a semblance of an explanation through artifactual agency, not through genuine understanding (Floridi 2025). We might say the AI-generated explanation is a representation of knowledge without a knowing subject behind it. This is a novel situation in the history of knowledge: previously, every book, every myth, every idea we received was created by humans (even if anonymously or collectively, still humans). Now, in contrast, we have texts that originate from non-human processes.

AI in the Epistemic Triangle: Mediator or Subject?

This section enters into dialogue with the conceptual foundation laid by Annamaria de Rosa (2026) in this volume, taking up her suggestion to revisit the semiotic architecture of social representation theory—this time, through the prism of AI's dual role as both object and quasi-agent. Traditionally, we had have Subject 1 (e.g., expert communicators or media), Subject 2 (e.g., the lay public or another group), an Object (the topic, here AI), and a Project (the context or goal, e.g., public understanding, political agenda). AI systems have been, up to now, mostly part of the means of communication between human subjects (e.g., algorithms curating social media, or search engines retrieving info). But with generative AI, the system can produce content that one might normally expect from a human subject. Does that make the AI effectively a new "Subject 2" (or a replacement for Subject 1) in the communication process?

Consider a scenario: A person asks an AI chatbot (Subject 1: human) about a scientific issue; the AI (which could be seen as Subject 2, but non-human) responds with an explanation; another person reads that output (Subject 3?). One could argue the AI is functioning as a mediator of representations from the broader collectivity to the individual—a bit like an automated journalist or teacher. In the Toblerone model metaphor (Bauer and Gaskell 1999), perhaps AI is a new face on one of the triangular facets—an extension of the media/institution element that channels representations. Indeed, Bauer and Gaskell's work on biotechnology highlighted the role of media and experts in shaping public representations (Bauer and Gaskell 2008). Now, AI might be occupying some roles traditionally played by journalists or experts in conveying information.

One way to conceptualize this is to see AI as a new type of communication agent with no project of its own. It represents in the sense of re-presenting existing knowledge in novel combinations, but it does not represent in the sense of standing for a social group's view (since it's not part of any human group). We might then ask: if representations are inherently social (rooted in groups' efforts to make sense of something), can AI content be considered a representation at all, or is it just a simulation of a representation? This echoes the distinction in SRT between the reified universe of science and the consensual universe of common sense (Moscovici 1961/2008; Jesuino 2008). Generally, AI will take from the reified (facts, scientific data) and translate it into the consensual (common language, narratives) for easier understanding. However, because no actual consensus-forming process occurred (the AI is not negotiating meaning; it is statistically averaging), one could say AI outputs lack the depth of a true social representation. They could lack internal depth, even though they may appear coherent. For instance, if one asked AI a complex social question, say, about the ethics of gene editing, it will generate a digestible compilation of arguments from both sides. This summary, although well-structured, does not synthesize from authentic social discussions. There is no social mechanism of deliberation or conflict resolution involved. The AI constructed it to the need of the prompt through a "one and the other" system it learned from. There is no commitment of belief in any of the sentences. This is a representation without a believer (Negura 2025).

But from the perspective of social understanding, does that really make a difference? An AI summary that someone has received and mentalized has formed social representations the same way any commentary or encyclopedia article has. The potential problem AI poses is if the output is wrong or deceptive—in this case, much of the wrong or misleading information "hallucinations" (specifically stating false info with confidence) is of AI. If people use AI outputs as if they are the truth (perhaps due to the interface anthropomorphizing the AI as a wise tutor), then false information can go viral. ChatGPT at times provided apparently authoritative information on the climate that was misleading (like exaggerating the uncertainty in climate models or focusing on hyper-individualized actions), which if taken would greatly add to the problem of climate misinformation (Atkins et al. 2024). In that case, an AI would be "spreading" a social representation that is wrong (or strengthening a misconception) to the public.

Harari was of the opinion that the capacity of AI to produce text at the same time on a mass scale, and to automate the production of cultural content, could "strike a blow" on civilization. The reason is that human culture is dependent on the narratives and ideas that are communicated in language (Harari 2024). If AI can generate persuasive text, fake news, deepfake images, etc., at a volume and personalization level humans never could, it can flood the social sphere with certain representations. One could conceive of a future where some social representations (common understandings) are actually planted or nudged by AI because AI systems crafted the majority of information people encounter on a topic. This is not entirely science fiction—we already see automated social media bots influencing discourse, and algorithms shaping what content goes viral. With generative AI, the content itself can be machine-made. This raises ethical issues: common sense and shared knowledge have traditionally been a human collective achievement, messy but ours. What happens if parts of "common sense" are indirectly influenced by machine-generated outputs guided perhaps by corporate or state interests, or just by accidents of algorithmic training? Do we then consider those representations "social"? Likely yes in effect, but the process of genesis is new.

In Bauer and Gaskell's Toblerone model (Bauer and Gaskell 1999) (see Figure 1), imagine one subject is a human community and another subject is perhaps another community or institution, and between them, they form representations about an object, given their joint project. If we insert AI, perhaps AI could stand in for one "subject" in certain interactions—e.g., AI acting as an information provider in lieu of a human expert. But since AI has no project (no goals or social intentions of its own), it might be safer to see it as part of the project or the medium rather than a full subject. For instance, if the project of a social group is to make sense of AI for education, they might use AI systems as tools or sources, and thus AI would be embedded in the project dimension (the practices and resources through which representation is formed). The elongated triangle (S1–O–P–S2) could be modified to S1 (e.g., developers or deployers of AI)—O (AI as object)—P (some societal project, such as efficiency or innovation)—and S2 (the public). Here AI itself is the object being represented. But if AI is also generating content about itself or other topics, it complicates the picture: you could end up with a strange loop where AI is object and also intermediate subject (cf. de Rosa 2026 in this volume).

While these conceptual mappings are challenging, what is clear is that AI is now in the loop of social knowledge creation and dissemination in a way that blurs the producer/consumer dichotomy of information. In the past, social representation processes involved humans at every node: experts->media->public or interpersonal communication, etc. Now we have algorithmic agents interposed, and sometimes producing material ex nihilo (from a human perspective). This calls for an extension of social representations theory to account for non-human actors. We might ask, can AI-generated content become social representation? If we define social representations strictly as forms of common sense knowledge collectively generated and shared among people, then AI content per se is not a social representation until humans internalize and share it. However, once it circulates and is talked about, it enters the social realm. When fictitious images spread on the internet, people who are not aware of the relevant context, integrate the images into their mental models, understanding them as representations of social reality.

As Gaskell et al. (2026) have shown in this volume, the advanced language models such as ChatGPT, while showing remarkable linguistic dexterity, suffer from deficits in cultural pragmatics, ethical reasoning, and narrative consistency (deficiencies essential for real-world conversation). Common sense might become more uniform if everyone uses the same AI assistant, but perhaps also more superficial or more easily swayed by how the AI frames information. The nature of common-sense knowledge could shift if we increasingly rely on AI to tell us things instead of debating amongst ourselves or consulting diverse human sources.

As explained by Mark Coeckelbergh (2021), certain human values and judgment must be preserved, and the excessive focus on the assumptions and images we use for framing AI must also be fundamentally thoughtful. For instance, if we see AI as a "servant" or "tool," we might treat it differently (and design it differently) than if we see it as an "ally" or "replacement." The metaphors we socially adopt can affect design and policy: Coeckelbergh suggests that over-anthropomorphizing can be a form of "humanwashing" (Coeckelbergh 2021)—obscuring the fact that responsibility lies with developers and operators by pretending the AI is an autonomous person. Thus, how we represent AI as an object, shapes how we deploy it as an agent—and even influences how we frame civic perceptions and political behavior (see Gatt, Borg, and Lauri 2026 in this volume). If people commonly start seeing AI as a kind of social other (e.g., giving AI systems names, citizenship as in the case of the robot Sophia, etc.), we might end up granting them de facto agency in realms such as customer service and legal advice, with reduced human oversight. On the other hand, if we say that "AI is just a tool" we might not fully understand the role that it is performing, which is, in a way, agentive.

As a "quasi-agent," AI produces representations that simulate human knowledge, but without the understanding and social intentionality that a human would bring (Negura 2025). These representations that AI generates, once recirculated into the society, would then become socially real (for better or worse). The process puts intense pressure on our theoretical constructs: do we extend "social representations" to include machine-created content once it's socially shared? It may be pragmatic to do so, since from the user's perspective, regardless of whether an idea came from ChatGPT or from Uncle Bob, if they believe it and spread it, it is now a social representation held by people. Sure, the origins or provenance of this idea might be different, but the social effects in this context or instance are the same. But then, awareness of the provenance might also become a part of the representation (e.g., "I read this answer from an AI, but I am not sure if it is the right answer" represents a new class of epistemic positions that are emerging today). For this reason, people have come to equate AI with understanding content that is generated by AI.

To close this section, let us think about an example. Imagine the very notion of "common sense." Historically, one of the definitions is that common sense is the stock of self-evident knowledge we think everyone knows (Bauer 2024). It is often full of social representations (like "you catch a cold by being cold"—a folk theory). Now, AI language models don't truly have common sense (Bauer and Schiele 2024), but they've been trained on our text that includes common sense expressions, so they fake it. Sometimes they fake it well, other times they produce hilariously non-sensical statements that reveal the lack of grounding (like mixing up basic physical facts). If people lean on AI for answers, one could see a scenario where human common-sense atrophies or becomes intermingled with AI's pseudo common sense. This synergy or contamination is a frontier issue. Scholars in both AI and social sciences are now looking at how interacting with AI might alter human cognitive habits—including how we form representations. Will our social representations of, say, history or science shift when AI can generate endless plausible (but possibly inaccurate) narratives? Will we witness new social representations that are partly AI-inspired constructions (like conspiracies that originate from an AI's hallucinated detail that then gets picked up by people)? These questions underscore the importance of bringing social representation theory into conversation with AI ethics and epistemology.

AI and the Epistemic Triangle of Representation: Placing AI in the Toblerone Model

To further synthesize the two perspectives—AI as represented object and AI as representing agent—it is useful to explicitly map AI onto the epistemic triangle of social representation as conceived of by Bauer and Gaskell (1999). In their Toblerone model (see Figure 1), the formation of a social representation involves interactions between subjects (collective agents such as groups, institutions, or communities), an object (the thing being represented), and a project (the purpose or activity context in which the representation matters). The metaphor of the Toblerone, a triangular prism, suggests multiple facets of mediation, including different modes of communication (e.g., conversational, media, institutional) and potentially multiple subjects (Subject

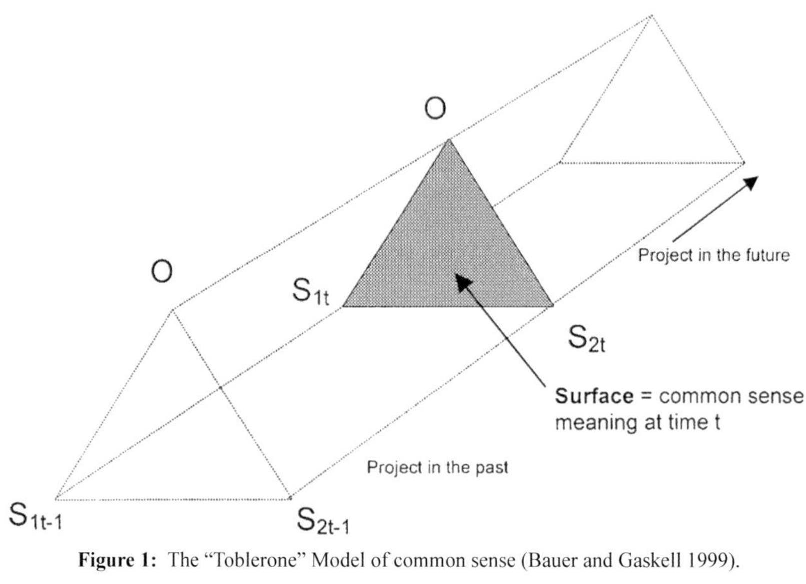

Figure 1: The "Toblerone" Model of common sense (Bauer and Gaskell 1999).

1 and Subject 2 at the ends of the prism). This model was developed by examining cases like biotechnology, where scientists (Subject 1), the public (Subject 2), the new tech (Object), and the societal debate (Project) together constituted the movement of understanding.

Now, where is AI as a quasi-agent in this picture? If we strictly follow classic SRT, AI is not a subject (since subjects are "carriers of the representation" and implicitly human or collective social actors). However, we increasingly have to consider AI systems as mediating subjects. If we extend subject to mean "agentive source of communicative acts," an AI system that generates content is functioning as a subject in a communicative exchange. For instance, if a user asks an AI chatbot about a political issue, and the chatbot provides an answer, that user might treat the answer as if coming from a knowledgeable entity (thus the AI is a de facto communicator). If that answer shapes the user's representation of the issue, the AI has played a representational role akin to a subject (like a teacher or journalist would).

Perhaps the best way to place AI in the Toblerone model is to see it as part of the mediational chain rather than as an end-point subject with its own project. AI can be a translator or transformer of representations between groups. It ingests the discourse from one end (e.g., the vast corpus of text from domain experts, media, public opinion—essentially, an aggregate of many Subject 1 voices and Subject 2 voices from the past), and then it presents an output to a user (Subject 1 in the present interaction). In doing so, it influences the user's perspective on the object. Thus, AI could be conceptualized as an amplifier or filter in the triangular relation. It might amplify certain aspects of the object's representation (depending on data biases) and filter out others, thereby subtly shifting the representation. For instance, if more online content about the risks of AI exists than content about its benefits (or vice versa), a language model might disproportionately present that angle, thereby influencing the salience of fear vs hope in the individual's mind.

When Bauer and Gaskell (2008) discussed the "biotechnology movement," they observed how common sense meaning around biotech emerged from interplay between scientific discourse, media framing, and public conversation. In the AI "movement," we similarly see interplay: scientists and engineers produce narratives (often optimistic, sometimes cautionary), media sensationalizes or dramatizes them (we get both awe-inspiring headlines and doom headlines), and the public in turn reacts (social media discussions, etc.), which then feeds back into media and policy. Now add AI to that loop: AI systems might start writing some of those media stories (yes, AI journalism is already happening for basic reports) (Pizzo et al. 2023), or moderating online discussions (deciding which content is shown) (Molina and Sundar 2022), or even participating in discussions (there are examples of social media accounts run by LLM-based bots that engage with people) (Maldonado-Canca et al. 2024). This cyborg communication ecosystem—part human, part AI—is new territory for social representation processes. The Toblerone model's triangular relationships become augmented with AI-mediated links. Perhaps we could imagine a Toblerone (see Figure 2) where one face is "human communication" and another face is "AI-mediated communication," and the representation forms at their intersection.

What about social representations of AI versus by AI? We must distinguish these clearly. We covered the social representation of AI (in human society)—e.g., AI as

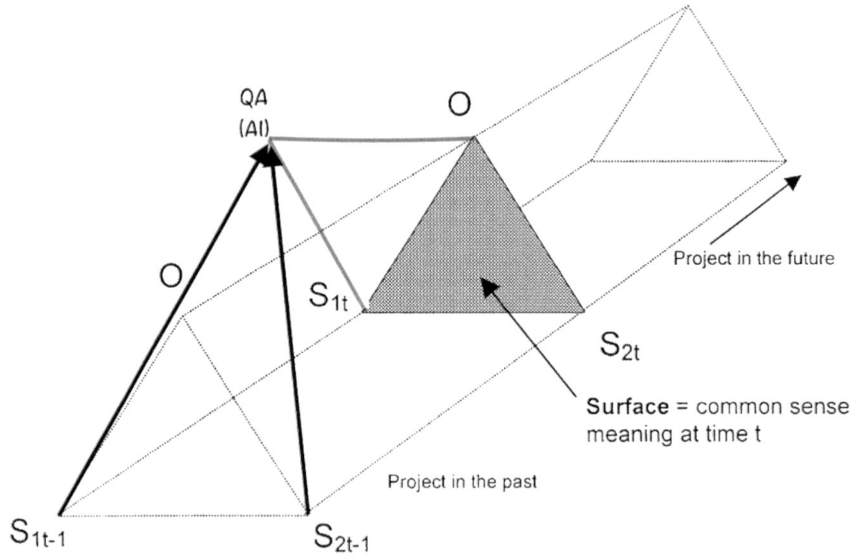

Figure 2: AI as Quasi-Agent (QA) in the Epistemic Triangle of SR.

monster or savior. Social representations by AI is a misnomer since AI doesn't hold representations the way humans do. But AI does create "pseudo-representations" in content. One could ask, for instance, what image of the world emerges from an AI like GPT-5 if one were to query it extensively. Does it have an implicit representation of humans, of society? Some researchers attempt to discern AI systems' "world models" (Mitrokhov 2024). If one treats that as analogous to a representation, it is a representation without social context—a statistical one. It might align with majority societal views on many things (since that's statistically common in the data), but on other things it might not.

A provocative issue might be whether advanced AI should be viewed as part of society, or as a distinct new social actor (however artificial). For example, in the future, could AI systems hypothetically conjure a form of "group perspective" cohesion if all of them operate on the same training data and share the same objectives? One school of thought considers AI-generated output as a form of pseudo-common sense because, in the process of constructing a text, a reflection of the so-called "aggregate voice" is produced. If this is the case, one could argue that AI-generated texts illuminate the dominating social representations about a specific issue (which might be equated to a discourse weighted average). However, it lacks the finesse of real discourse; minority and majority views are averaged out, resulting in a distorted view of reality.

Supporting this line of reasoning, one could look at the development of the Toblerone model as evolving into what Bauer and Gaskell (2008) later referred to as the "wind rose" of social representation. In this model, a set of individual triangles, each originating in a different social context, converge to a central point, giving rise to a rich, multidimensional meaning and a complex, network-like field of social signification. Within this perspective, AI as quasi-agent becomes a cross-cutting mediator, influencing the circulation, transformation, and reinforcement of representations across social groups.

Ultimately, whether AI can be considered to produce social representations may come down to how we define social representation. If we emphasize the process (social groups negotiating meaning), then no—AI is not inherently part of that process inherently. If we emphasize the product (shared knowledge or belief circulating in a community), then AI can produce texts that become part of that circulation, thus contributing to the product. Perhaps a sensible stance is that AI-generated representations become social representations once appropriated by social actors. They require uptake. Without uptake, an AI's output is just an artifact. With uptake, it enters the social mind.

Conclusion

We have covered social representations of artificial intelligence within the book's span with attention to AI's diverse cultural, societal, and ethical implications. Every chapter has elucidated the distinct differentials of how the media, workplaces, ethical issues, and general understanding construct, perceive, and integrate artificial intelligence within social life. This last chapter integrates these different insights and applies them to a dual analytical framework: AI is both an object represented by human society and a quasi-agent that is actively engaged in social representations, contributing to and reshaping them.

This synthesis discusses how discourse on generative artificial intelligence challenges us to rethink the conventional boundaries of human and non-human agency. Clearly, generative AI is more than just another piece of technology. Instead, it has a different level of relevance in the field of social knowledge and representation. Social representation theory must adapt to this new AI-driven reality by understanding its dual position as both object and quasi-agent. Such a theoretical revision is necessary not only to enrich academic understanding, but also to address the ethical and epistemological questions arising from the increasing presence of AI in our societies.

Rather than closing the discussion, this book invites further investigation into the ontological and methodological repositioning that AI demands from our discipline—echoing and amplifying the calls for epistemic vigilance that run through the contributions to this volume.

References

Atkins, C., G. Girgente, M. Shirzaei and J. Kim. 2024. Generative AI tools can enhance climate literacy but must be checked for biases and inaccuracies. Communications Earth and Environment, 5(1): 226.

Banks, J. 2021. Good robots, bad robots: Morally valenced behavior effects on perceived mind, morality, and trust. International Journal of Social Robotics, 13(8): 2021–2038.

Bauer, M. W. and G. Gaskell. 2008. Social representations theory: A progressive research programme for social psychology. Journal for the Theory of Social Behaviour, 38(4): 335–353.

Bauer, M. W. and G. Gaskell. 1999. Towards a paradigm for research on social representations. Journal for the Theory of Social Behaviour, 29(2): 163–186.

Bauer, M. W. and B. Schiele. 2024. AI and Common Sense (p. 13). Taylor and Francis.

Bender, E. M., T. Gebru, A. McMillan-Major and S. Shmitchell. 2021. On the dangers of stochastic parrots: can language models be too big?. Proceedings of the 2021 ACM Conference on Fairness, Accountability, and Transparency, 610–623. https://doi.org/10.1145/3442188.3445922

Brondi, S., M. Pivetti, S. di Battista and M. Sarrica. 2021. What do we expect from robots? Social representations, attitudes and evaluations of robots in daily life. Technology in Society, 66: 101663. https://doi.org/10.1016/j.techsoc.2021.101663

Chae, J. and M. Bauer. 2026. Techno-movement and hype cycle of AI. Techno-movement and hype cycle of AI. *In:* Negura, L. (Eds.). Beyond Algorithms: Social Representations and the Public Understanding of AI. Taylor and Francis.

Coeckelbergh M. 2021. Three responses to anthropomorphism in social robotics: towards a critical, relational, and hermeneutic approach. Int J Soc Robot. https://doi.org/10.1007/s12369-021-00770-0

Coeckelbergh, M. 2025. AI and Epistemic Agency: How AI Influences Belief Revision and Its Normative Implications, Social Epistemology, DOI: 10.1080/02691728.2025.2466164

De Rosa, A. S. 2026. Social Representations and Communication in the era of Artificial Intelligence: Machine Learning as Subject, Other or Object of the semiotic triangle or crossing all? *In:* Negura, L. (Ed.). Beyond Algorithms: Social Representations and the Public Understanding of AI, Science Publishers, CRC Press/ Taylor and Francis Group

Dennett, D. C. 1996. The intentional stance. MIT Press.

Floridi, L. 2025. AI as agency without intelligence: On artificial intelligence as a new form of artificial agency and the multiple realisability of agency thesis. Philosophy and Technology, 38(1): 30.

Gatt, G., J. Borg and M. A. Lauri. 2026. The Political Impact of AI Social Representations. *In:* Negura, L. (Ed.). Beyond Algorithms: Social Representations and the Public Understanding of AI, Science Publishers, CRC Press/Taylor and Francis Group

Gaskell, G., W. Wagner, E. Paraschou, S. Lyu and M. Michali. 2026. Social representations, real world knowledge, and artificial intelligence. *In:* L. Negura, L. (Ed.). Beyond Algorithms: Social Representations and the Public Understanding of AI, Science Publishers, CRC Press/Taylor and Francis Group

Harari, Y. N. 2024. Nexus: a brief history of information networks from the Stone Age to AI. Fern Press.

Horstmann, A. C. and N. C. Krämer. 2019. Great Expectations? Relation of previous experiences with social robots in real life or in the media and expectancies based on qualitative and quantitative assessment. Front Psychol.10:939. doi: 10.3389/fpsyg.2019.00939. PMID: 31114525; PMCID: PMC6503083.

Jesuino, C. J. 2008. Linking science to common sense. Journal for the Theory of Social Behaviour, 38(4): 393–409.

Liu, J. H. and J. László. 2007. A narrative theory of history and identity: Social identity, social representations, society, and the individual. pp. 85–107. *In:* Social representations and identity: Content, process, and power. New York: Palgrave Macmillan US.

Maldonado-Canca, L. A., A. M. Casado-Molina, J. P. Cabrera-Sánchez and G. Bermúdez-González. 2024. Beyond the post: an SLR of enterprise artificial intelligence in social media. Social Network Analysis and Mining, 14(1): 219.

Martikainen, J. and I. Sakki. 2023. Sensory experiences and social representation–Embodied multimodality of common-sense thinking. Journal for the Theory of Social Behaviour, 53(4): 488–505.

Mitrokhov, K. 2024. Between world models and model worlds: on generality, agency, and worlding in machine learning. AI and Society, 1–13.

Molina, M. D. and S. S. Sundar. 2022. When AI moderates online content: effects of human collaboration and interactive transparency on user trust. Journal of Computer-Mediated Communication, 27(4).

Moreno-Gabriel, E., N. Vallès-Peris and B. Biaani León-Gómez 2026. Bias and Stereotyping in AI: Analyzing How Social Biases Get Reflected and Perpetuated in AI Systems," (Co)producing the Prompt in Primary Health Care. *In:* Negura, L. (Ed.). Beyond Algorithms: Social Representations and the Public Understanding of AI, Science Publishers, CRC Press/Taylor and Francis Group.

Moscovici, S. 1961/2008. Psychoanalysis: Its image and its public. Polity.

Negura, L. 2025. Simulated Sense-Making or Social Knowledge? Artificial Intelligence and the Boundaries of Representation, Journal for the Theory of Social Behaviour, 55(3): 1–9. https://doi.org/10.1111/jtsb.70012

Negura, L., M. Sarrica and V. Rizzoli 2026. Artificial intelligence in the public eye. *In:* Negura, L. (Ed.). Beyond Algorithms: Social Representations and the Public Understanding of AI, Science Publishers, CRC Press/Taylor and Francis Group

Negura, L. and D. N. Tchatchou. 2026. Reproducing hierarchies: AI and the social representation of intelligence. *In*: Negura, L. (Ed.). Beyond Algorithms: Social Representations and the Public Understanding of AI, Science Publishers, CRC Press/Taylor and Francis Group.

Pizzo, A., V. Lombardo and R. Damiano. 2023. Interactive storytelling: a cross-media approach to writing, producing and editing with AI. Routledge.

Placani, A. 2024. Anthropomorphism in AI: hype and fallacy. AI and Ethics, 4(3): 691–698.

Roose, K. 2023. A.I. Poses 'Risk of Extinction,' Industry Leaders Warn. The New York Times. ISSN 0362-4331. Retrieved 3 June 2025.

Sarrica, M. 2010. ICTs meanings and practices: Contributions from the social representation approach. The Journal of Community Informatics, 6(3).

Searle, J. R. 1980. Minds, brains, and programs. The Behavioral and Brain Sciences, 3(3): 417–424. https://doi.org/10.1017/S0140525X00005756

Vultur, M. 2026. Artificial Intelligence And World Of Work: Social Representations Of AI In The Context Of Job Displacement And Transformations. *In:* Negura, L. (Ed.). Beyond Algorithms: Social Representations and the Public Understanding of AI, Science Publishers, CRC Press/Taylor and Francis Group

Wagner, W. 2017. Embodied social representation. Journal for the Theory of Social Behaviour, 47(1).

Wagner, W., P. Holtz and Y. Kashima. 2009. Construction and deconstruction of essence in representating social groups: Identity projects, stereotyping, and racism. Journal for the Theory of Social Behaviour, 39(3): 363–383.

Weizenbaum, J. 1966. ELIZA--A Computer Program For the Study of Natural Language Communication Between Man and Machine. Communications of the ACM. 9. Massachusetts Institute of Technology: 36. doi:10.1145/365153.365168. S2CID 1896290.

Index